MOBILE MESSAGING TECHNOLOGIES AND SERVICES

MOBILE MESSAGING
TECHNOLOGIES AND SERVICES
SMS, EMS and MMS

Gwenaël Le Bodic
Alcatel, France

JOHN WILEY & SONS, LTD

This publication is designed to provide accurate and authoritative information in regard to the subject matter covered. It is sold on the understanding that the Publisher is not engaged in rendering professional services. If professional advice or other expert assistance is required, the services of a competent professional should be sought.

Other Wiley Editorial Offices

John Wiley & Sons Inc.,
111 River Street, Hoboken, NJ 07030, USA

Jossey-Bass, 989 Market Street, San Francisco,
CA 94103–1741, USA

Wiley-VCH Verlag GmbH,
Boschstr. 12, D–69469 Weinheim, Germany

John Wiley & Sons Australia Ltd, 33 Park Road,
Milton, Queensland 4064, Australia

John Wiley & Sons (Asia) Pte Ltd, 2 Clementi Loop 02–01,
Jin Xing Distripark, Singapore 129809

John Wiley & Sons Canada Ltd, 22 Worcester Road,
Etobicoke, Ontario, Canada M9W 1L1

Library of Congress Cataloguing in Publication Data

Le Bodic, Gwenaël.
    Mobile messaging technologies and services: SMS, EMS, and MMS/Gwenaël Le Bodic.
       p.cm.
    Includes bibliographical references and index.
    ISBN 0-470-84876-6
     1. Personal communication service systems. 2. Radio paging. 3. Multimedia systems.
    4. Cellular telephone systems. I. Title

    TK5103.485. L423 2002
    621.3845–dc21
                                                            2002033100

British Library Cataloguing in Publication Data

A catalogue record for this book is available from the British Library

ISBN 0470 84876 6

Typeset in 10/12pt Times by Deerpark Publishing Services Ltd, Shannon, Ireland.
Printed and bound in Great Britain by Biddles Ltd, Guildford and King's Lynn
This book is printed on acid-free paper responsibly manufactured from sustainable forestry in which at least two trees are planted for each one used for paper production.

To Marie-Amélie and Louise

Contents

Foreword

The advent of Cellular Mobile Communications will no doubt go down in history as one of the most significant 'inventions' that took place in the 20th century and has had a vast impact on how we conduct our business and social lives.

The main attraction for GSM in the late 1980s was that it would provide mobility for speech communications. Some spare unused capacity in the signalling channel was identified and so in the background, the GSM standards body invented a simple 'two way pager like' messaging service called the Short Message Service (SMS). SMS had problems of latency because of its 'store and forward' characteristics and also a severe bandwidth problem of 160 characters. Disparaging comments emerged such as 'Why would you want to send a text message to somebody rather than talk to them?'

Little did the proponents of such 'messages of gloom' or the inventors of SMS realize the impact that SMS would have on GSM customers and network operators' revenue streams within a decade.

Today, hundreds of millions of short messages are sent every day and SMS is the biggest revenue generator for most network operators other than speech.

The growth of SMS traffic has been slowing over the past two years and has been showing signs of reaching a plateau within the next year or two. This may be attributed to the fact that SMS is beginning to look like an old technology because, with the advent of 3GPP, the focus has shifted to a far more sophisticated messaging service called the Multimedia Messaging Service (MMS).

Network operators recognized the need to encourage further growth in SMS whilst awaiting MMS and one mechanism to address this requirement, and also provide customers with a new SMS experience, was the invention of the Enhanced Messaging Service (EMS) by the 3GPP standards body during the past 12 months. This has not de-focussed the attention on MMS and will hopefully provide SMS users with a new experience and an introduction to MMS.

The invention of MMS has brought new challenges for its inventors. For the first time, the Internet and the mobile telecommunications environments must be merged. This will require a change in the cultural thinking of the traditional users of these two very different worlds. For example, the Internet does not seem to know any bounds of bandwidth or storage capacity and the general perception is that the use of the Internet is 'free'. By contrast, the mobile telecommunications environment is limited in terms of bandwidth, storage capacity and the cost for using it is certainly not 'free' and is unlikely to be.

Despite such differences, MMS has been identified as being the most important service for 3GPP. It has no option but to succeed in order to satisfy the public expectations for mobile access to the Internet and for the communication of various message media types already in widespread use on the Internet.

The 21st century has only just begun. What greater testimony to the inventors and developers of MMS could there be if by the end of the 21st century, 3GPP MMS was heralded as being one of the most significant inventions of that age.

Ian Harris, C.Eng. FIEE

Ian Harris is the Chairman of the 3GPP standards group responsible for 'Messaging'. He was with Vodafone for 18 years from their very beginning and was their design authority for a number of Vodafone's Value Added Services – in particular SMS. He has worked in GSM standards since that work began in 1988 and is one of the few 'original designers' of SMS. Since January 2002 he has been a consultant with Teleca Ltd heading up their work in messaging mobile telecommunications.

Preface

Communications in general, and messaging in particular, have always been of key importance for the organization of human societies. First messaging systems can be traced back to early civilizations such as the American Indians. Amerindians have been known to communicate using smoke signals. With this method, the exchange of complex messages could be carried out by agreeing on a common set of smoke signals between communicating tribes. Later, the exchange of hand-written letters allowed a more reliable messaging service where message carriers were foot or horse couriers or even 'carrier pigeons'. In the modern world, messaging has benefited from advances in high technology. With the development of mobile communications networks, mobile messaging has become a very popular and reliable way for communicating with almost anybody, at anytime and from almost anywhere. All message services in the telecommunication world are based on a *store-and-forward* paradigm where messages are kept temporarily if users are not able to immediately retrieve them.

The Short Message Service (SMS) has proved to be a very popular messaging service, supported by most GSM, TDMA and CDMA mobile networks. An entire chapter of this book describes the Short Message Service introduced by the European Telecommunications Standards Institute (ETSI) for GSM and GPRS networks. The 3rd Generation Partnership Project (3GPP) is now the organization responsible for maintaining SMS technical specifications. In its simplest form, the Short Message Service allows users to exchange short messages composed of a limited amount of text. In more advanced SMS extensions, short messages can be concatenated in order to increase the amount of data that can be exchanged between mobile users. The first SMS message is believed to have been sent in December 1992 from a personal computer to a European mobile phone network. In 2001, an estimated 102.9 billion SMS messages were exchanged worldwide. Gartner Dataquest, one of the industry's major research agencies, expects the number of SMS messages to grow to 146 billion in 2002 and to peak at around 168 billion in 2003 before declining.

The Enhanced Messaging Service (EMS), an application-level extension of SMS, supersedes limited SMS features by allowing elements such as pictures, animations, text formatting instructions and melodies to be inserted in short or concatenated messages.

With SMS, the most common usage scenario is the exchange of a short text message between two mobile users. This *person-to-person scenario* is also applicable in the EMS case. Furthermore, the ability to create content-rich EMS messages introduces new business opportunities. For instance, EMS allows content providers to generate revenue by pushing compelling content to selected mobile devices. In this book, this usage scenario is referred to as the *machine-to-person scenario*.

In the machine-to-person scenario, the messaging service is usually perceived in two different ways by the user. On the one hand, the content provider can generate messages composed of elements such as text, pictures, animations and melodies (weather forecasts, news updates, etc.). This scenario is a direct extension of the person-to-person scenario where the user reads the message and then replies, forwards or deletes it. On the other hand, specific messaging services also enable the customization of mobile devices according to user requirements. This is known as the *download service* where the mobile handset receives a download message containing elements such as melodies or animations. In this situation, the user does not read the message in the normal way. Instead, elements extracted from the download message are stored in the mobile device and can be used as ring tones or switch-on/off[1] animations, for instance.

Recently, several standardization organizations initiated the work on the development of the Multimedia Messaging Service (MMS). The Multimedia Messaging Service defines a framework for the realization of services enabling the exchange of multimedia messages. The Multimedia Messaging Service encompasses the identification and definition of a large number of high-level multimedia features which so far have only been provided by fixed messaging systems such as electronic mail. MMS features include the exchange of multi-media messages choreographed as 'slideshows' (similar to Microsoft Powerpoint presentations). A slideshow is constructed as a series of slides, each slide being composed of text, audio, images and/or video organized over a predefined graphical layout. The deployment and operation of MMS requires significant network resources in terms of equipment and airtime. Consequently, advanced network technologies, such as GPRS and UMTS, are desirable for MMS.

The development of SMS and EMS was largely driven by the availability of underlying technologies. Unlike SMS and EMS, high-level service requirements for MMS have been identified first and appropriate technical realizations have then been developed accordingly. For MMS, the 3GPP has concentrated on identification of high-level service requirements and architectural aspects. Additionally, the 3GPP has provided technical realizations for several interfaces allowing communications between elements of the MMS architecture. To complement the 3GPP work, the WAP Forum has defined technical realizations for MMS on the basis of WAP and Internet transport technologies. MMS technical specifications have reached a fairly mature stage and MMS commercial solutions are appearing on the market. Business analysts have already identified the high potential of MMS and this has led to significant investments in the service from network operators, device manufacturers and service providers.

Chapter 1 introduces the basic concepts of mobile technologies and services. It includes an overview of the three generations of mobile network technologies along with a description of supported services. Chapter 2 demystifies the working procedures of selected standardization development organizations. For this purpose, Chapter 2 presents the standardization processes of relevant organizations such as the 3GPP, the WAP Forum, the IETF and the W3C. How to identify and retrieve the necessary technical specifications produced by these organizations is of particular interest. Chapter 3 provides an in-depth description of the Short Message Service (SMS). Reference materials are provided for any engineer willing to develop SMS-based applications. Chapters 4 and 5 describe two application-level extensions of SMS: the basic Enhanced Messaging Service (EMS) and the extended EMS. Chapter 6 provides a comprehensive description of the Multimedia Messaging Service (MMS). Finally, Chapter 7 provides an introduction to other messaging services and technologies relevant to the world of mobile communications.

---

[1] Switch-on and switch-off animations are short animations displayed when the mobile device is switched on or off.

The author would like to gratefully acknowledge the time and effort of many people who reviewed the content of this book. The book has benefited from constructive comments from experts involved in many fields of the vast world of mobile communications. In particular, the author is grateful to Olivier Barault, Philippe Bellordre, Luis Carroll, François Courau, Dave Chen, Cyril Fenard, Peter Freitag, Arthur Gidlow, Ian Harris, Pieter Keijzer, Hervé Languille, Josef Laumen, Marie-Amélie Le Bodic, Jérome Marcon, Jean-Luc Ricoeur, Ngoc Tanh Ly, Frédéric Villain and Paul Vincent.

The team at John Wiley & Sons, Ltd involved in the production of this book, provided excellent support and guidance. Particularly, the author is thankful to Mark Hammond, Sarah Hinton and Zoë Pinnock for explaining the book production process and for their continuous support during the entire process.

In addition, the author is thankful to Alcatel Business Systems and Bijitec for providing the illustrations for this book.

Gwenaël Le Bodic, PhD

About the author

Gwenaël Le Bodic is a messaging expert for Alcatel's Mobile Phone Division. One of his activities for Alcatel is participating in and contributing to the development of messaging technologies and services in the scope of the 3GPP standardization process. A certified engineer in computer sciences, Gwenaël Le Bodic obtained a PhD in mobile communications from the University of Strathclyde, Glasgow. For two years, he has been a researcher for the Mobile Virtual Centre of Excellence in the United Kingdom. Gwenaël Le Bodic is the author of many research publications in the field of mobile communications. He can be contacted at gwenael@lebodic.net.

Notational Conventions

This book uses the augmented Backus-Naur Form (BNF) as described in [RFC-2616] for certain constructs.

In Chapters 3 and 6, many tables define the composition of sets of information elements. Each information element is associated with a status (mandatory, optional or conditional) which is represented in the corresponding table with the following graphical notation:

- ●   The plain dot means that the associated information element is mandatory in the set.
- ○   The empty dot means that the associated information element is optional in the set.
- C   The C means that the associated information element is present in the set only if a specified condition is fulfilled.

In addition, in Chapter 6, several tables summarize which operations have been defined over several interfaces. For each operation the following graphical notation is used:

- ■   The plain square means that the associated operation has been defined. No square means that the associated operation has not been defined in the corresponding technical specification.

The book illustrates technologies and services defined by standardization organizations and other parties. When relevant, a pointer to the original work is provided in the text of the book in the following form: [3GPP-23.040]. The list of all pointers, along with detailed corresponding references, is provided at the end of the book. Similarly, a list of all acronyms and abbreviations used in this book is provided. Hexadecimal values are often used for the description of values that can be assigned to various parameters. In this book, a hexadecimal value is always prefixed with '0x'. For instance, the value '0x1A' represents the hexadecimal value '1A'.

1

Basic Concepts

This chapter outlines the basic concepts of mobile communications systems and presents the required background information necessary for a clear understanding of this book.

First, an overview of the evolution of mobile communications systems is provided. This encompasses the introduction of first generation analogue systems supporting only voice communications to the forthcoming deployment of third generation systems supporting voice and multimedia services.

The Global System for Mobile, commonly known as GSM, has been a major breakthrough in the domain of mobile communications. Elements composing a typical GSM network are presented. Another important milestone is the introduction of the General Packet Radio Service (GPRS) allowing the support of packet-based communications in evolved GSM networks. The architecture of a GPRS network is presented. Next to be deployed are Universal Mobile Telecommunications Systems (UMTS). These systems will support advanced multimedia services requiring high data rates. UMTS services and supporting technologies are also introduced. Finally, the Wireless Application Protocol (WAP) is described. WAP is an enabling technology for developing services such as browsing and multimedia messaging.

The last section provides pointers to books and reference articles for anybody wishing to further explore the topics outlined in this chapter.

1.1 Generations of Mobile Communications Networks

In France, in 1956, a very basic mobile telephony network was implemented with vacuum electronic tubes and electron-mechanical logic circuitry. These devices used for wireless communications had to be carried in car boots. In these early days of mobile telephony, service access was far from being ubiquitous and was reserved for a very limited portion of the population. Since the introduction of this experimental network, mobile communications technologies benefited from major breakthroughs commonly categorized in three generations. In the 1980s, *first generation mobile systems* arrived in Nordic countries. These first generation systems were characterized by analogue wireless communications and limited support for user mobility. Digital communications technology was introduced with second generation (2G) mobile systems in the 1990s. Second generation systems are characterized by the provision of better quality voice services available to the mass market. Second generation systems benefited from the cellular concept in which scarce radio resources are used simultaneously by several mobile users without interference. The best known 2G system is the Global System for Mobile (GSM) with 646 million subscribers worldwide (January 2002).

Other major 2G systems include cdmaOne (based on CDMA technology), with users in the Americas (55 million subscribers) and Asia (40 million subscribers) and Japanese Personal Data Cellular (PDC) with the iMode technology for mobile Internet (30 million subscribers). In the near future, third generation (3G) mobile systems will be deployed worldwide. With 3G systems, various wireless technologies such as satellite and terrestrial communications will converge with Internet technologies. Third generation services will encompass a wide range of multimedia and cost-effective services with support for worldwide user mobility. The migration to 3G systems will be facilitated by the introduction of intermediary evolved 2G systems, also known as 2.5G systems.

1.2 Telecommunications Context: Standard and Regulation

In the telecommunications environment, standard development organizations (SDOs) provide the necessary framework for the development of standards. These standards are technical documents[1] defining or identifying the technologies enabling the realization of telecommunication network technologies and services. The prime objective of SDOs is to develop and maintain widely accepted standards allowing the introduction of attractive services over interoperable networks. The actors that are involved in the standardization process are network operators, manufacturers and third party organizations such as content providers, equipment testers and regulatory authorities. One of the main objectives of tele-communications regulation authorities is to ensure that the telecommunications environment is organized in a sufficiently competitive environment and that the quality of service offered to subscribers is satisfactory.

In the early days of mobile communications, various regional SDOs developed specifica-tions for network technologies and services independently. This led to the development of heterogeneous networks where interoperability was seldom ensured. The lack of interoper-ability of first generation mobile systems prevented the expansion of a global international mobile network that would have certainly greatly improved user experience. With second and third generations systems, major SDOs decided to gather their efforts in order to ensure that mobile communications networks will appropriately interoperate in various regions of the world. In 1998, such an effort was initiated by several SDOs including ARIB (Japan), ETSI (Europe), TTA (Korea), TTC (Japan) and T1P1 (USA). The initiative was named the Third Generation Partnership Project (3GPP). The 3GPP standardization process is presented in Chapter 2.

1.3 Global System for Mobile

Before the introduction of the Global System for Mobile (GSM), mobile networks imple-mented in different countries were usually incompatible. This incompatibility made imprac-ticable the roaming of mobile users across international borders. In order to get around this system incompatibility, the Conférence Européenne des Postes et Télécommunications (CEPT) created the Groupe Spécial Mobile[2] committee in 1982. The main task of the committee was to standardize a pan-European cellular public communication network in

---

[1] Technical documents are also known as technical specifications, reports or recommendations.
[2] The name Groupe Spécial Mobile was later translated to Special Mobile Group (SMG).

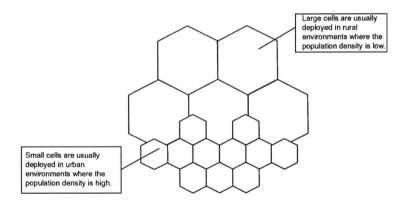

Figure 1.1   Cellular concept

the 900 MHz radio band. In 1989, the European Telecommunications Standards Institute (ETSI) took over responsibility for the maintenance and evolution of GSM specifications. In 2000, this responsibility was transferred to the 3GPP. The initiative was so successful that networks compliant with the GSM standard have now been developed worldwide. Variations of the GSM specification have been standardized for the 1800 MHz and 1900 MHz bands and are known as DCS 1800 and PCS 1900, respectively. In January 2002, there were more than 470 GSM operators offering services in 172 countries.

A GSM network is characterized by digital voice communications and support of low-rate data services. The GSM air interface is based on Time Division Multiple Access (TDMA). With TDMA, a radio band is shared by multiple subscribers by allocating one or more timeslots on given radio carriers to each subscriber. The length of a GSM timeslot is 0.5769 ms. With GSM, the transfer of data can be carried out over circuit-switched connections. For these data communications, bit rates up to 14.4 kbit/s can be achieved on single-slot connections. The single-slot configuration is called Circuit Switched Data (CSD). Higher bit rates up to 57.6 kbit/s can be attained by allocating more than one slot for a data connection. This multi-slot configuration is called High Speed CSD (HSCSD).

One of the most popular GSM services is the Short Message Service (SMS). This service allows SMS subscribers to exchange short text messages. An in-depth description of this service in provided in Chapter 3. Application-level extensions of SMS are presented in Chapters 4 and 5.

1.3.1 Cellular Concept

Radio bands available for wireless communications in mobile networks represent very scarce resources. In order to efficiently use these resources, GSM networks are based on the cellular concept. With this concept, the same radio resources (characterized by a frequency band and a timeslot) can be utilized simultaneously by several subscribers without interference if they are separated by a minimum distance. The minimum distance between two subscribers depends on the way radio waves propagate in the environment where the two subscribers are located. In a GSM network, the smaller the cells, the higher the frequency reuse factor, as shown in Figure 1.1.

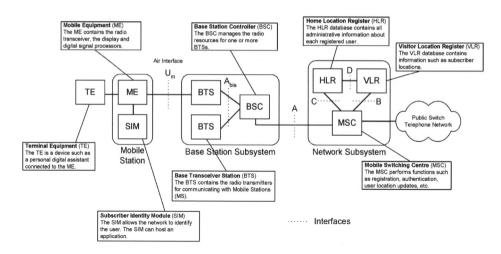

Figure 1.2   GSM architecture

In a GSM network, a fixed base station transceiver manages the radio communications for all mobile stations located in a cell. Each geometrical cell in Figure 1.1 represents the radio coverage of one single base station.

1.3.2 GSM Architecture

The main elements of the GSM architecture [3GPP-23.002] are shown in Figure 1.2.

The GSM network is composed of three subsystems: the base station subsystem (BSS), the network subsystem (NSS) and the operation subsystem (OSS). The OSS implements functions that allow the administration of the mobile network. Elements of the OSS are not represented in the GSM architecture shown in Figure 1.2 and are not discussed further in this book. Elements of the BSS and NSS are described in the following sections.

1.3.3 Mobile Station

The Mobile Station (MS) is a device that transmits and receives radio signals within a cell site. A mobile station can be a basic mobile handset, as shown in Figure 1.3, or a more complex Personal Digital Assistant (PDA). Mobile handset capabilities include voice communications, messaging features and contact directory management. In addition to these basic capabilities, a PDA is usually shipped with an Internet microbrowser and an advanced Personal Information Manager (PIM) for managing contacts and calendaring/scheduling entries. When the user is moving (i.e. while driving), network control of MS connections is switched over from cell site to cell site to support MS mobility. This process is called handover.

The mobile station is composed of the Mobile Equipment (ME) and the Subscriber Identity Module (SIM). The unique International Mobile Equipment Identity (IMEI) stored in the ME identifies the device when attached to the mobile network.

The SIM is usually provided by the network operator to the subscriber in the form of a

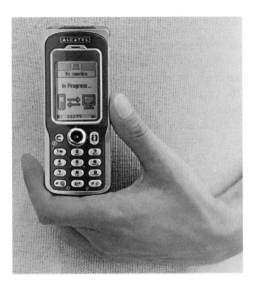

Figure 1.3 Mobile station handset (reproduced by permission of Alcatel Business Systems)

smart card. The microchip is often taken out of the smart card and directly inserted into a dedicated slot in the mobile equipment. A SIM microchip is shown in Figure 1.4.

The most recent mobile stations can be connected to an external device such as a PDA or a personal computer. Such an external device is named a Terminal Equipment (TE) in the GSM architecture.

A short message is typically stored in the mobile station. Most handsets have SIM storage capacities. High-end products sometimes complement the SIM storage capacity with additional storage in the mobile equipment itself (e.g. flash memory). It is now common to find handsets shipped with a PIM. The PIM is usually implemented as an ME internal feature and enables elements such as calendar entries, memos, phonebook entries and of course messages to be stored in the ME. These elements are managed, by the subscriber, with a suitable graphical interface. These PIM elements remain in the PIM even when the SIM is removed from the mobile handset. Alternatively, simple elements such as short messages and phonebook entries can be directly stored in the SIM. A SIM can contain from 10 short messages to 50 short messages on high-end solutions. Storing elements in the SIM allows messages to be retrieved from any handset simply by inserting the SIM in the desired handset. The benefit of storing messages in the ME is that the ME storage capacity is often significantly larger than the SIM storage capacity.

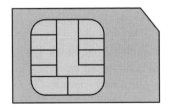

Figure 1.4 SIM microchip

1.3.4 Base Transceiver Station

The Base Transceiver Station (BTS) implements the air communications interface with all active MSs located under its coverage area (cell site). This includes signal modulation/ demodulation, signal equalizing and error coding. Several BTSs are connected to a single Base Station Controller (BSC). In the United Kingdom, the number of GSM BTSs is estimated around several thousand. Cell radii range from 10 to 200 m for the smallest cells to several kilometres for the largest cells. A BTS is typically capable of handling 20–40 simultaneous communications.

1.3.5 Base Station Controller

The BSC supplies a set of functions for managing connections of BTSs under its control. Functions enable operations such as handover, cell site configuration, management of radio resources and tuning of BTS radio frequency power levels. In addition, the BSC realizes a first concentration of circuits towards the MSC. In a typical GSM network, the BSC controls over 70 BTSs.

1.3.6 Mobile Switching Centre and Visitor Location Register

The Mobile Switching Centre (MSC) performs the communications switching functions of the system and is responsible for call set-up, release and routing. It also provides functions for service billing and for interfacing other networks.

The Visitor Location Register (VLR) contains dynamic information about users who are attached to the mobile network including the user's geographical location. The VLR is usually integrated to the MSC. Through the MSC, the mobile network communicates with other networks such as the Public Switched Telephone Network (PSTN), Integrated Services Digital Network (ISDN), Circuit Switched Public Data Network (CSPDN) and Packet Switched Public Data Network (PSPDN).

1.3.7 Home Location Register

The Home Location Register (HLR) is a network element containing subscription details for each subscriber. A HLR is typically capable of managing information for hundreds of thousands of subscribers.

In a GSM network, signalling is based on the Signalling System Number 7 (SS7) protocol. The use of SS7 is complemented by the use of the Mobile Application Part (MAP) protocol for mobile specific signalling. In particular, MAP is used for the exchange of location and subscriber information between the HLR and other network elements such as the MSC.

For each subscriber, the HLR maintains the mapping between the International Mobile Subscriber Identity (IMSI) and the Mobile Station ISDN Number (MSISDN). For security reasons, the IMSI is seldom transmitted over the air interface and is only known within a given GSM network. The IMSI is constructed according to [ITU-E.212] format. Unlike the IMSI, the MSISDN identifies a subscriber outside the GSM network. The MSISDN is constructed according to [ITU-E.164] format (e.g. +33 612345678).

1.4 General Packet Radio Service

In its simplest form, GSM manages voice and data communications over circuit-switched connections. The General Packet Radio Service (GPRS) is an extension of GSM which allows subscribers to send and receive data over packet-switched connections. The use of GPRS is particularly appropriate for applications with the following characteristics:

- Bursty transmission (for which the time between successive transmissions greatly exceeds the average transfer delay).
- Frequent transmission of small volumes of data.
- Infrequent transmission of large volumes of data.

These applications do not usually need to communicate permanently. Consequently, the continuous reservation of resources for realizing a circuit-switched connection does not represent an efficient way to exploit scarce radio resources. The basic concept behind the GPRS packet-based transmission lies in its ability to allow selected applications to share radio resources by allocating radio resources for transmission only when applications have data to transmit. Once the data have been transmitted by an application, radio resources are released for use by other applications. Scarce radio resources are used more efficiently with this mechanism. GPRS allows more radio resources to be allocated to a packet-based connection than to a circuit-switched connection in GSM. Consequently, a packet-based connection usually achieves higher bit rates (up to 171.2 kb/s). In addition, GPRS can offer 'always on' connections (sending or receiving data at any time) and enables separate allocations for uplink and downlink channels.

1.4.1 GPRS Architecture

The main elements composing the GPRS architecture [3GPP-23.060] are shown in Figure 1.5.

A GPRS mobile station operates in one of the three following modes [3GPP-22.060]:

- Class A: the mobile station supports simultaneous use of GSM and GPRS services (attachment, activation, monitoring, transmission, etc.). A class A mobile station may establish or receive calls on the two services simultaneously.
- Class B: the mobile station is attached to both GSM and GPRS services. However, the mobile station can only operate in one of the two services at a time.
- Class C: the mobile station is attached to either the GSM service or the GPRS service but is not attached to both services at the same time. Prior to establishing or receiving a call on one of the two services, the mobile station has to be explicitly attached to the desired service.

Before a mobile station can access GPRS services, it must execute a GPRS attachment procedure to indicate its presence to the network. The mobile station must then activate a Packet Data Protocol (PDP) context with the network in order to be able to transmit or receive data.

The GPRS air interface is identical to that of the GSM network (same radio modulation, frequency bands and frame structure). GPRS is based on an evolved GSM base station subsystem. However, the GPRS core network is based on a GSM network subsystem in

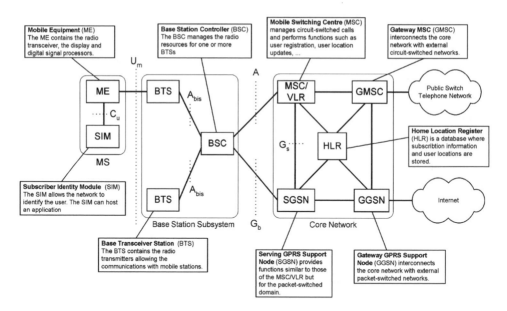

Figure 1.5   GPRS architecture

which two additional network elements have been integrated: serving and gateway GPRS support nodes. In addition, Enhanced Data Rate for Global Evolution (EDGE) improves GPRS performances by introducing an enhanced modulation scheme.

1.4.2 Serving GPRS Support Node

The Serving GPRS Support Node (SGSN) is connected to one or more base station subsystems. It operates as a router for data packets for all mobile stations present in a given geographical area. It also keeps track of the location of mobile stations and performs security functions and access control.

1.4.3 Gateway GPRS Support Node

The Gateway GPRS Support Node (GGSN) ensures interactions between the GPRS core network and external packet-switched networks such as the Internet. For this purpose, it encapsulates data packets received from external networks and routes them toward the SGSN.

1.5 Universal Mobile Telecommunications System

Since 1990, focus has been given to the standardization of third generation mobile systems. The International Telecommunication Union (ITU) has initiated the work on a set of standards named the International Mobile Telecommunications 2000 (IMT-2000) for the definition of technologies and services for 3G systems. In this family of IMT-2000[3] standards, the Universal Mobile Telecommunications System (UMTS) encompasses the definition of new radio access techni-

---

[3] IMT-2000 was formerly known as Future Public Land Mobile Telecommunications System (FPLMTS).

ques along with a new service architecture. The UMTS aims at providing services such as web browsing, messaging, mobile commerce, videoconferencing and other services to be developed according to emerging subscribers' needs with the following objectives:

- High transmission rates encompassing circuit and packet-switched connections.
- High spectral efficiency and overall cost improvement.
- Definition of common radio interfaces for multiple environments.
- Portability of services in various environments (indoor, outdoor, suburban, urban, rural, pedestrian, vehicular, satellite, etc). This service portability is also known as the Virtual Home Environment concept [3GPP-22.121][3GPP-23.127]. The provisioning of services in heterogeneous environments is enabled with an Open Service Architecture. (OSA) [3GPP-22.127][3GPP-23.127].

The UMTS extends 2G voice and data capabilities to multimedia capabilities with access to higher bandwidth targeting 384 kb/s for full area coverage and 2 Mb/s for local area coverage. The UMTS will become the basis for new mobile telecommunications networks with highly personalized and user-friendly services. The UMTS should provide a convergence of communications technologies such as satellite, cellular radio, cordless and wireless LANs. The network operator NTT DoCoMo introduced 3G services to the Japanese market in October 2001. Elsewhere, the commercial introduction of UMTS networks and services for the mass market is expected in 2003 or 2004.

1.5.1 3G Services

Second generation networks provide voice and limited data services. In addition to these 2G services, 3G systems offer multimedia services adapted to the capabilities of multimedia devices and network conditions with a possibility to provide some content specifically formatted according to the subscriber location. The UMTS Forum in [UFRep9][UFRep13] classifies 3G services into six groups as illustrated in Figure 1.6.

- Mobile Internet access: a mobile access to the Internet with service quality close to the one offered by fixed Internet Service Providers. This includes full Web access, file transfer, electronic mail and streaming video and audio.
- Mobile Intranet/Extranet access: a secure framework for accessing corporate Local Area Networks (LANs) and Virtual Private Networks (VPNs).
- Customized infotainment: a device-independent access to personalized content from mobile portals.
- Multimedia messaging service: a means of exchanging messages containing multimedia contents including text, images, video and audio elements. The multimedia messaging service can be considered as an evolution of SMS where truly multimedia messages can be exchanged between subscribers. An in-depth description of the multimedia messaging service is provided in Chapter 6.
- Location-based services: location-aware services such as vehicle tracking, local advertisements, etc.
- Rich voice and simple voice: real-time, two-way voice communications. This includes Voice over IP (VoIP), voice-activated network access and Internet-initiated voice calls.

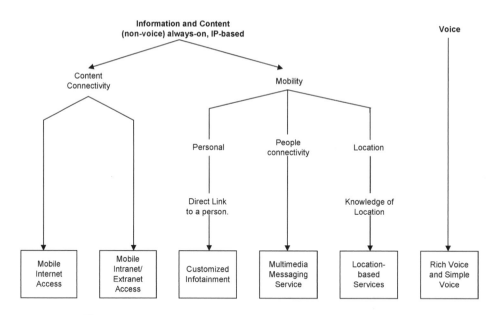

Figure 1.6   3G service categories. Source: UMTS Forum [UF-Rep-9]

Mobile videophone and multimedia real-time communications should also be available on high-end multimedia devices.

In the scope of the 3GPP standardization process, the UMTS specification work was divided into two distinct phases. The first phase UMTS, named UMTS release 99 (also known as release 3), is a direct evolution from 2G and 2.5G networks (GSM and GPRS networks). The second phase UMTS, also known as UMTS release 4/5, is a complete revolution introducing new concepts and features.

1.5.2 First Phase UMTS

The UMTS architecture [3GPP-23.101] has to meet the requirements of various UMTS services. These requirements range from real-time voice traffic and bursty data access to mixed multimedia traffic. UMTS is intended to offer a true global service availability. To meet this objective, the UMTS architecture includes terrestrial segments complemented by satellite constellations where necessary.

1.5.3 UMTS Architecture

The first phase UMTS architecture is based on evolved GSM and GPRS core networks and a specifically tailored Universal Terrestrial Radio Access Network (UTRAN). Two duplexing methods defining how the received signal is separated from the transmitted signal have been defined:

- Universal Terrestrial Radio Access/Time Division Duplex (UTRA/TDD). This method achieves bi-directional transmission by allowing the use of different time slots over the same radio carrier for the transmission of sent and received signals.

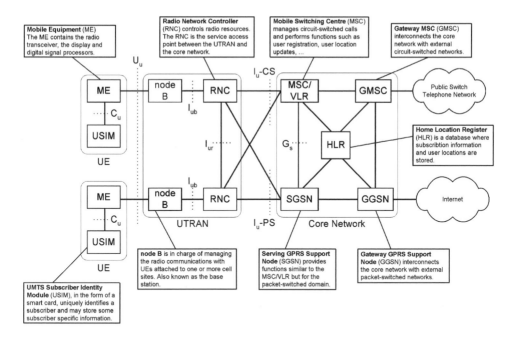

Figure 1.7 UMTS architecture

- Universal Terrestrial Radio Access/Frequency Division Duplex (UTRA/FDD). This method achieves bi-directional transmission by allowing sent and received signals to be transmitted over two separate and symmetrical radio bands for the two links.

The name Wideband CDMA (WCDMA) is also used to identify the two UTRA operating modes (TDD and FDD). Elements composing the first phase UMTS architecture are shown in Figure 1.7.

Elements of the UMTS architecture are grouped into three sub-systems: the User Equipment (UE), the access network, also called UTRA Network (UTRAN), and the switching and routing infrastructure, also known as the Core Network (CN). Elements of the UMTS architecture support both circuit-switched connections and packet-switched connections.

1.5.4 User Equipment

The UE, usually provided to the subscriber in the form of a handset, is itself composed of Mobile Equipment (ME) and an UMTS Subscriber Identity Module (USIM). The ME contains the radio transceiver, the display and digital signal processors. The USIM is a 3G application on an UMTS IC card (UICC) which holds the subscriber identity, authentication algorithms and other subscriber-related information. The ME and USIM are interconnected via the Cu electrical interface whereas the UE is connected to the UTRAN via the Uu radio interface. An UE always supports at least one of the operating modes of UTRA: TDD or FDD. In order to allow a smooth transition to the UMTS, it is expected that UEs will initially be capable of communicating with legacy systems such as GSM and GPRS. UMTS UEs supporting legacy systems are called multi-mode UEs. The 3GPP classifies multi-mode UEs into four categories [3GPP-21.910]:

- Type 1: type 1 user equipment operates in one single mode at a time (GSM or UTRA). It cannot operate in more than one mode at a time. While operating in a given mode, the user equipment does not scan for or monitor any other mode and switching from one mode to another is done manually by the subscriber.
- Type 2: while operating in one mode, type 2 user equipment can scan for and monitor another mode of operation. The user equipment reports to the subscriber on the status of another mode by using the current mode of operation. Type 2 user equipment does not support simultaneous reception or transmission through different modes. The switching from one mode to another is performed automatically.
- Type 3: type 3 user equipment differs from type 2 user equipment by the fact that the type 3 UE can receive in more than one mode at a time. However, a type 3 UE cannot emit simultaneously in more than one mode. Switching from one mode to another is performed automatically.
- Type 4: type 4 user equipment can receive and transmit simultaneously in more than one mode. Switching from one mode to another is performed automatically.

1.5.5 UTRA Network

The UTRAN is composed of nodes B and Radio Network Controllers (RNCs). The node B is responsible for the transmission of information in one or more cells, to and from UEs. It also participates partly in the system resource management. The node B interconnects with the RNC via the Iub interface. The RNC controls resources in the system and interfaces the core network.

1.5.6 UMTS Core Network

The first phase UMTS core network is based on an evolved GSM network sub-system (circuit-switched domain) and a GPRS core network (packet-switched domain). Consequently, the UMTS core network is composed of the HLR, the MSC/VLR and the GMSC (to manage circuit-switched connections) and the SGSN and GGSN (to manage packet-based connections).

1.5.7 Second Phase UMTS

The initial UMTS architecture presented in this chapter is based on evolved GSM and GPRS core networks (providing support for circuit-switched and packet-switched domains, respectively). The objective of this initial architecture is to allow mobile network operators to rapidly roll out UMTS networks on the basis of existing GSM and GPRS networks. From this first phase UMTS architecture, the next phase is to evolve to an architecture with a core network based on an enhanced packet-switched domain only. The objective is to allow a better convergence with the Internet by using IP-based protocols whenever possible. At the end of 1999, the 3GPP started the work on the specification of an 'all-IP' architecture. In this architecture, the MSC function is split into a control plane part (MSC server) and a user plane part (media gateway). The core network of the second phase UMTS is built on the IP Multimedia Subsystem, abbreviated IMS [3GPP-22.228][3GPP-23.228]. IMS introduces the capability to support IP-based multimedia services such as Voice over IP (VoIP) and Multimedia over IP (MMoIP) . In IMS, call control is managed with the Session Initiation Protocol (SIP), published by the IETF in [RFC-3261], and all network elements are based on IPv6.

1.6 Wireless Application Protocol

The Wireless Application Protocol (WAP) is the result of the collaborative work of many wireless industry players. This work has been carried out in the scope of the WAP Forum. The forum, launched in 1997 by Phone.Com (now Openwave), Motorola, Nokia and Ericsson, produces technical specifications enabling the support of applications over existing and forthcoming wireless platforms (GSM, GPRS, UMTS, etc.). For this purpose, the WAP Forum identifies and defines a set of protocols and content formats. The WAP Forum standardization process is presented in Chapter 2.

1.6.1 Technology Overview

The WAP technology is an enabler for building applications over various wireless platforms. The objective of the WAP Forum is to provide a framework for the development of applications with a focus on the following aspects:

- Interoperability: applications developed by various parties and hosted on devices produced by different manufacturers interoperate in a satisfactory manner.
- Scalability: mobile network operators are able to scale services to subscribers' needs.
- Efficiency: the framework offers a quality of service suited to the capabilities of underlying wireless networks.
- Reliability: the framework represents a stable platform for deploying services.
- Security: the framework ensures that user data can be safely transmitted over a serving mobile network which may not always be the home network. This includes the protection of services and devices and the confidentiality of subscriber data.

In line with these considerations, the WAP technology provides an application model which is close to the World Wide Web model (also known as the web model). In the web model, content is represented using standard description formats. Additionally, applications, known as web browsers, retrieve the available content using standard transport protocols. The web model includes the following key elements:

- Standard naming model: some content available over the web is uniquely identified by a Uniform Resource Identifier (URI) .
- Content type: objects available on the web are typed. This allows web browsers to correctly identify the type to which a specific content belongs.
- Standard content format: web browsers support a number of standard content formats such as HyperText Markup Language (HTML).
- Standard protocols: web browsers also support a number of standard protocols for accessing content on the web. This includes the widely accepted HyperText Transfer Protocol (HTTP).

The wheel has not been reinvented with WAP and the WAP model borrows a lot from the successful web model. However, the web model, as it is, does not efficiently cope with characteristics of mobile networks and devices. To cope with these characteristics, the WAP model leverages the web model by adding the following improvements:

- The push technology allows content to be pushed directly from the server to the mobile device without any prior explicit request from the user.
- The adaptation of content to the capability of WAP devices is allowed with a mechanism known as the User Agent Profile (UAProf) .
- The support of advanced telephony features by applications, such as the handling of calls (establishment and release of calls, placing a call on hold or redirecting the call to another user, etc.).
- The External Functionality Interface (EFI) allows 'plug-in' modules to be added to browsers and applications hosted in WAP devices in order to increase their overall capabilities.
- The persistent storage allows users to organize, access, store and retrieve content from/to remote locations.
- The multimedia messaging service, which is extensively described in Chapter 6, is a significant enhancement of the WAP model over the web model. This service allows the exchange of multimedia messages between subscribers.

The WAP model uses the standard naming model and content types defined in the web model. In addition, the WAP model includes:

- Standard content formats: browsers in the WAP environment, known as microbrowsers, support a number of standard content formats/languages including Wireless Markup Language (WML) and Extensible HTML (XHTML). WML and XHTML are both applications of Extended Markup Language (XML). See Box 1.1 for a description of markup languages for WAP-enabled devices.
- Standard protocols: microbrowsers communicate with protocols, which have been optimized for mobile networks, including HTTP from the web model and the Wireless Session Protocol (WSP).

Box 1.1 Markup Languages for WAP-enabled devices

The Hypertext Markup Language (HTML) is the content format commonly used in the World Wide Web. HTML enables a visual presentation of information (text, images, hyperlinks, etc.) on large screens of desktop computers. Extensible Markup Language (XML) is another markup language which is generic enough to represent the basis for the definition of many other dedicated languages. Several markup languages supported by WAP-enabled devices are derived from XML. This is the case for WML and XHTML. The Wireless Markup Language (WML) has been optimized for rendering information on mobile devices with limited rendering capabilities. The Extensible Hypertext Markup Language (XHTML) is an XML reformulation of HTML. Both WML and XHTML are extensible since the formats allow the addition of new markup tags to meet changing needs.

Table 1.1 WAP Forum specification suites

WAP Forum specification suite	Delivery date	Description
WAP 1.0	April 1998	Basic WAP framework. Almost no available commercial solutions
WAP 1.1	June 1999	First commercial solutions with support for: – Wireless Application Environment (WAE) – Wireless Markup Language (WML) – WML Script
WAP 1.2	Nov. 1999	Additional features: – Push – User Agent Profile (UAProf) – Wireless Telephony Application (WTA) – Wireless Identity Module (WIM)
WAP 1.2.1	June 2000	Bug fixes
WAP 2.0	July 2001	Convergence with Internet technologies. Additional features: – support of MMS over WAP 1.x (3GPP release 99) – HTTP, TCP, persistent storage – XHTML, SyncML, provisioning, – etc.

First WAP technical specifications were made public in 1998 and have since evolved to allow the development of more advanced services. The major milestones for WAP technology are reflected in the availability of what the WAP Forum calls specifications suites. Each specification suite contains a set of WAP technical specifications providing a specific level of features as shown in Table 1.1.

With WAP specification suites 1.x, the WAP device communicates with a web server via a WAP gateway. Communications between the WAP device and the WAP gateway is performed over WSP. In addition, WAP specification suite 2.x, allows a better convergence of wireless and Internet technologies by promoting the use of standard protocols from the web model.

1.6.2 WAP Architecture

Figure 1.8 shows the components of a generic WAP architecture. The WAP device can communicate with remote servers directly or via a number of intermediary proxies. These proxies may belong to the mobile network operator or alternatively to service providers. The primary function of proxies is to optimize the transport of content from servers to WAP devices.

Supporting servers, as defined by the WAP Forum, include Public Key Infrastructure (PKI) portals, content adaptation servers and provisioning servers.

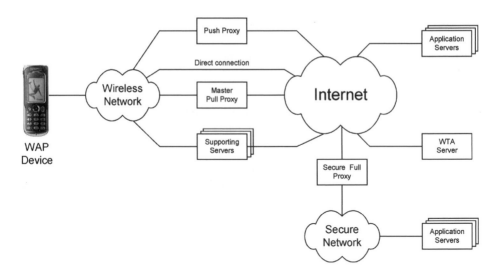

Figure 1.8   Generic WAP architecture

1.6.3 Push Technology

In a typical client/server model, a client retrieves the selected information from a server by explicitly requesting the download of information from the server. This retrieval method is also known as the pull technology since the client pulls some data from a server. Internet browsing is an example of models based on pull technology. In contrast, another technology has been introduced in the WAP model and is known as push technology. With push technology, a server is able to push some data to the WAP device with no prior explicit request from the client. In other words, the pull of information is always initiated by the client whereas the push of information is always initiated by the server.

The push framework, defined by the WAP Forum in [WAP-250], is shown in Figure 1.9. In the push framework, a push transaction is initiated by a Push Initiator (PI). The Push Initiator, usually a web server, transmits the content to be pushed along with delivery instructions (formatted with XML) to a Push Proxy Gateway (PPG). The PPG then delivers the push

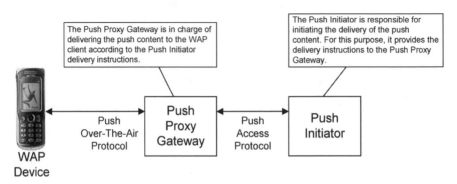

Figure 1.9   The push framework

content to the WAP device according to the delivery instructions. The Push Initiator interacts with the Push Proxy Gateway using the Push Access Protocol (PAP). On the other side, the Push Proxy Gateway uses the Push Over-The-Air (OTA) Protocol (based on WSP or HTTP) to deliver the push content to the WAP device.

The Push Proxy Gateway may implement network-access-control policies indicating whether or not push initiators are allowed to push content to WAP devices. The Push Proxy Gateway can send back a notification to the Push Initiator to indicate the status of a push request (delivered, cancelled, expired, etc.).

In addition to application-specific push contents (MMS, provisioning, and WTA), three generic types of contents can be pushed in the WAP environment: Service Indication (SI), Service Loading (SL) and Cache Operation (CO). Push SI provides the ability to push content to subscribers to notify them about electronic mail messages awaiting retrieval, news headlines, commercial offers, etc. In its simplest form, a push SI contains a short message along with a URI. Upon receipt of the push SI, the message is presented to the subscriber who is given the possibility of starting the service (retrieve the content) to which the URI refers. The subscriber may decide to start the service immediately or to postpone it. In contrast to push SI, push SL provides the ability to push some content to the WAP device without subscriber explicit request. A push SL contains a URI that refers to the push content. Upon receipt of the push SL, the push content is automatically fetched by the WAP device and is presented to the subscriber. Push CO provides a means for invalidating objects stored in the WAP device's cache memory.

1.6.4 User Agent Profile

The User Agent Profile (UAProf) specification was first published in the WAP 1.2 specification suite and improved in WAP 2.0. The objective of this specification is to define a method for describing the capabilities of clients and the preferences of subscribers. This description, known as a user agent profile, is mainly used for adapting available content to the rendering capabilities of WAP devices. For this purpose, the user agent profile is formatted using a Resource Description Framework (RDF) schema in accordance with Composite Capability/ Preference Profiles (CC/PP). The CC/PP specification defines a high level framework for exchanging and describing capability and preference information using RDF. Both RDF and CC/PP specifications have been published by the W3C. The UAProf, as defined by the WAP Forum in [WAP-174], allows the exchange of user agent profiles, also known as Capability and Preference Information (CPI), between the WAP device, intermediate network points and the origin server. These intermediate network points and origin servers can use the CPI to tailor the content of WSP/HTTP responses to the capabilities of receiving WAP devices.

The WAP Forum's UAProf specification defines a set of components that WAP-enabled devices can convey within the CPI. Each component is itself composed of a set of attributes or properties. Alternatively, a component can contain a URI pointing to a document describing the capabilities of the client. Such a document is stored on a server known as a profile repository (usually managed by device manufacturers or by software companies developing WAP microbrowsers). The UAProf is composed of the following components:

- Hardware platform: this component gathers a set of properties indicating the hardware capabilities of a device (screen size, etc.).

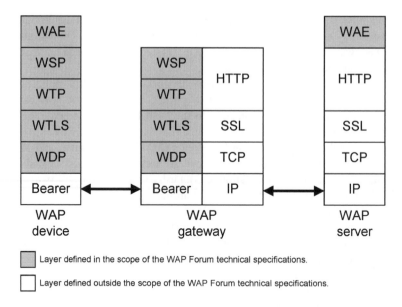

Figure 1.10   WAP 1.x legacy configuration with WAP gateway

- Software platform: this component groups a set of properties indicating the software capabilities of a device (operating system, supported image formats, etc.).
- Browser User Agent: this component gathers properties characterizing the Internet browser capabilities.
- Network characteristics: this component informs on network and environment characteristics such as the bearer capacity.
- WAP characteristics: this component informs on the device WAP capabilities. This includes information on the configuration of the WML browser, etc.
- Push characteristics: this component informs on the device's push capabilities. This includes the set of supported MIME types, the maximum message size that can be handled and whether or not the device can buffer push messages.

For a configuration involving a WAP device and a gateway communicating with WSP, RDF descriptions can be encoded in binary with the WAP binary XML (WBXML). In this context, the CPI is provided by the WAP device as part of the WSP session establishment request. The WAP device can also update its CPI at any time during an active WSP session. Note that the WAP gateway may also override the a CPI provided by a device.

1.6.5 Possible Configurations of WAP Technology

The WAP framework has been designed to offer service scalability. To fulfil the requirements of various services in heterogeneous mobile networks, it is possible to allow different configurations to coexist in the WAP environment. This section presents the three most common configurations of the WAP environment.

1.6.5.1 WAP 1.x Legacy Configuration

Figure 1.10 shows the protocol stack of the configuration defined in the WAP specification suite 1.x.[4] This configuration is also supported by the WAP specification suite 2.0. In this configuration, the WAP device communicates with a remote server via an intermediary WAP gateway. The primary function of the WAP gateway is to optimize the transport of content between the remote server and the WAP device. For this purpose, the content delivered by the remote server is converted into a compact binary form by the WAP gateway prior to the transfer over the wireless link. The WAP gateway converts the content transported between the datagram-based protocols (WSP, WTP, WTLS and WDP) and the connection-oriented protocols commonly used on the Internet (HTTP, SSL and TCP).

The Wireless Application Environment (WAE) is a general-purpose application environment where operators and service providers can build applications for a wide variety of wireless platforms.

The Wireless Session Protocol (WSP) provides features also available in HTTP (requests and corresponding responses). Additionally, WSP supports long-lived sessions and the possibility to suspend and resume previously established sessions. WSP requests and corresponding responses are encoded in binary for transport efficiency.

The Wireless Transaction Protocol (WTP) is a light-weight transaction-oriented protocol. WTP improves the reliability over underlying datagram services by ensuring the acknowledgement and retransmission of datagrams. WTP has no explicit connection setup or connection release. Being a message-oriented protocol, WTP is appropriate for implementing mobile services such as browsing.

The Wireless Transport Layer Security (WTLS) provides privacy, data integrity and authentication between applications communicating with the WAP technology. This includes the support of a secure transport service. WTLS provides operations for the establishment and the release of secure connections.

The Wireless Data Protocol (WDP) is a general datagram service based on underlying low-level bearers. WDP offers a level of service equivalent to the one offered by the Internet User Datagram Protocol (UDP).

At the bearer-level, the connection may be a circuit-switched connection (as found in GSM networks) or a packet-switched connection (as found in GPRS and UMTS networks). Alternatively, the transport of data at the bearer-level may be performed with the Short Message Service or the Cell Broadcast Service.

1.6.5.2 WAP HTTP Proxy with Profiled TCP and HTTP

Figure 1.11 shows a configuration where the WAP device communicates with web servers via an intermediary WAP proxy. The primary role of the proxy is to optimize the transport of content between the fixed Internet and the mobile network. With this configuration, Internet protocols are preferred against legacy WAP protocols. This is motivated by the need to support IP-based protocols in an end-to-end fashion, from the web server back to the WAP device. The protocol stack shown in this configuration, defined in the WAP specification suite 2.0, is shown in Figure 1.11.

---

[4] The WAP 1.x protocol stack is sometimes known as the 'legacy protocol stack'.

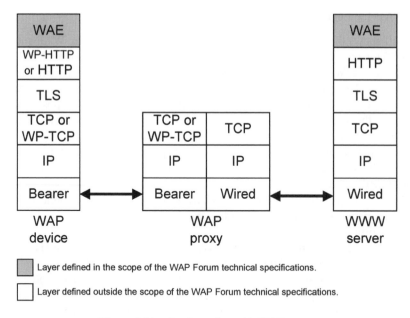

Figure 1.11   Configuration with WAP proxy

The Wireless Profiled HTTP (WP-HTTP) is a HTTP profile specifically designed for coping with the limitations of wireless environments. This profile is fully interoperable with HTTP/1.1. In addition, WP-HTTP supports message compression.

The Transport Layer Security (TLS) ensures the secure interoperability between WAP devices involved in the exchange of confidential information.

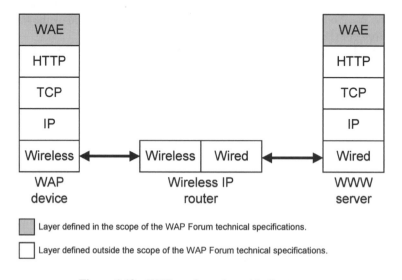

Figure 1.12   WAP configuration with direct access

The Wireless Profiled TCP (WP-TCP) offers a connection-oriented service. It is adapted to the limitations of wireless environments but remains interoperable with existing TCP implementations.

1.6.5.3 Direct Access

Figure 1.12 shows a configuration where the WAP device is directly connected to the web server (via a wireless router which provides a bearer-level connection). The protocol stack shown in this configuration is defined in the WAP specification suite 2.0.

A WAP device, compliant with the WAP 2.0 specification suite, may support all configurations by supporting WAP 1.x and WAP 2.0 protocol stacks.

Box 1.2 The Success of WAP

To most subscribers' eyes, WAP is not a very attractive service. However, there is probably some confusion on identifying what WAP really means. Over the last few years, much hype went on in promoting the appealing service to be offered by WAP-enabled devices: high-speed access to the Internet. However, WAP is not only a service for browsing the Internet and certainly not the Internet as perceived by users of desktop computers. With early versions of WAP, only limited amount of data could be transferred to small-screen mobile devices and this prevented the provision of a service equivalent to that offered to Internet users in the fixed environment. WAP was never intended to become the enabler for accessing the Internet comparable to the way it is accessed from a desktop computer today. Furthermore, WAP is more than an enabler for the browsing service, it is about a whole framework including transport technologies and execution environments enabling the realization of applications adapted to the mobile communications domain. Accessing data over the Internet is one of these applications but other applications, such as immediate and multimedia messaging services, are currently being deployed for the WAP environment. Even if WAP is now poorly perceived by subscribers, it is likely that, in the near future, advanced configurations of the WAP framework will enable the provision of successful applications to mobile subscribers and give WAP 'another chance'.

Further Reading

M. Mouly and M.B. Pautet, The GSM System for Mobile Communications, Palaiseau, France, 1992.

F. Muratore, UMTS, Mobile Communications for the Future, John Wiley & Sons, Chichester, 2001.

L. Bos and S. Leroy, Toward an all-IP-based UMTS architecture, IEEE Network Magazine, January/ February, 2001.

J. De Vriendt, P. Lainé, C. Lerouge and X. Xu, Mobile network evolution: a revolution on the move, IEEE Communications Magazine, April, 2002.

2

Standardization

In the mobile communications market, several messaging services are already available (SMS and basic EMS) and other services will emerge in the near future (extended EMS and MMS). For service providers, it becomes crucial to identify and deploy the right service at the most appropriate time in the market. In order to achieve this, it is important to build a minimum understanding of the standardization process of various bodies since it directly impacts on the commercial availability of related solutions.

This chapter presents the major milestones in the standardization of messaging technologies and services. Additionally, the working procedures of the following standardization development organizations are explained:

- *Third Generation Partnership Project* (3GPP): the 3GPP is not a standardization development organization in its own right but rather a joint project between several regional standardization bodies from Europe, North America, Korea, Japan and China. The prime objective of the 3GPP is to develop UMTS technical specifications. This encompasses the development of widely accepted technologies and service capabilities.
- *WAP Forum*: the Wireless Application Protocol (WAP) Forum is a joint project for the definition of WAP technical specifications. This encompasses the definition of a framework for the development of applications to be executed in various wireless platforms.
- *Internet Engineering Task Force* (IETF): the IETF is a large community of academic and industrial contributors that defines the protocols in use on the Internet.
- *World Wide Web Consortium* (W3C): the W3C is a standardization body that concentrates on the development of protocols and formats to be used in the World Wide Web. Well known formats and protocols published by the W3C are the Hypertext Transfer Protocol (HTTP) and the eXtensible Modelling Language (XML).

2.1 Messaging Road Map

The road map of messaging technologies and services is becoming more and more complex. This complexity is mainly due to the fact that services rely on technologies developed by a large number of standardization development organizations. This makes it difficult for service providers, operators and manufacturers to gather the necessary technical specifications that constitute the basis for software or hardware developments. This chapter facilitates the manipulation of such technical specifications by explaining the specification development process for relevant standardization bodies.

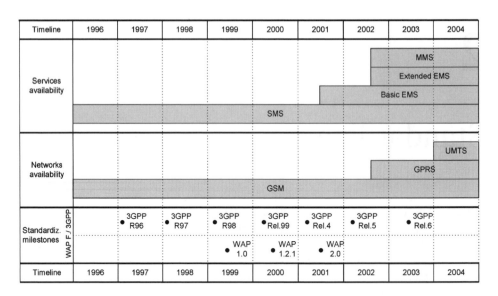

Timeline	1996	1997	1998	1999	2000	2001	2002	2003	2004
Services availability								MMS	
								Extended EMS	
						Basic EMS			
				SMS					
Networks availability									UMTS
							GPRS		
				GSM					
Standardiz. milestones	WAP F. / 3GPP	3GPP R96	3GPP R97	3GPP R98	3GPP Rel.99	3GPP Rel.4	3GPP Rel.5	3GPP Rel.6	
				WAP 1.0	WAP 1.2.1	WAP 2.0			
Timeline	1996	1997	1998	1999	2000	2001	2002	2003	2004

Figure 2.1   Relationships between availability of services and technologies

Figure 2.1 shows the relationships between the introduction of network technologies (GSM, GPRS and UMTS) and the commercial availability of messaging services. The figure also shows the major milestones for two standardization development organizations: the Third Generation Partnership Project (3GPP) and the WAP Forum.

Several messaging technologies have been developed to meet specific market requirements. One of the first messaging systems to have been introduced in mobile networks is the Short Message Service (SMS). In its simplest form, SMS allows subscribers to exchange short messages containing text only. SMS was first introduced as a basic service of GSM and has been the subject of many extensions. Initial standardization work on SMS was carried out within the scope of the European Telecommunications Standards Institute (ETSI) standardization process until the transfer of responsibility to the 3GPP. Standardization work for SMS is now carried out in the scope of the 3GPP standardization process. One of the most significant evolutions of SMS is an application-level extension called the Enhanced Messaging Service (EMS). EMS allows subscribers to exchange long messages containing text, melodies, pictures, animations and various other objects. Two versions of EMS are available and are covered in this book under *basic EMS* and *extended EMS*. Since 1998, standardization bodies have concentrated on the development of a new messaging service called the *Multimedia Messaging Service* (MMS). MMS enables subscribers to exchange multimedia messages. The standardization of MMS has reached a mature stage and initial solutions are appearing on the market.

2.2 Third Generation Partnership Project

As introduced in the previous chapter, ETSI has elaborated on the GSM standards during a period of almost 18 years. Within the scope of the ETSI standardization process, the work was carried out by the Special Mobile Group (SMG) technical committee. In 2000, the committee agreed to transfer the responsibility of the development and maintenance of the GSM standards to the Third Generation Partnership Project (3GPP). The 3GPP was set in 1989 by six standard development organizations (including ETSI) with the objective of collaborating on the development of interoperable mobile systems. These six organizations represent telecommunications companies from five different parts of the world:

- European Telecommunications Standards Institute (ETSI) for Europe
- Committee T1 for the United States
- Association of Radio Industries and Businesses (ARIB) for Japan
- Telecommunications Technology Committee (TTC) for Japan
- Telecommunications Technology Association (TTA) for Korea
- China Wireless Telecommunication Standard (CWTS) for China.

Each individual member of one of the five partners can contribute to the development of 3GPP specifications. In order to define timely services and technologies, individual members are helped by several market representative partners. At the time of writing this book, 3GPP market representatives are the UMTS Forum, the Global Mobile Suppliers Association (GSA), the GSM Association (GSMA), the Universal Wireless Communications Consortium (UWCC), the IPv6 Forum, the 3G.IP focus group and the Mobile Wireless Internet Forum (MWIF). The responsibility of these market representative partners consists of identifying the requirements for 3G services. In this process, the six partner organizations (ETSI, Committee T1, ARIB, TTC, TTA and CWTS) take the role of publishers of 3GPP specifications.

It has to be noted that parts of the 3GPP work, such as SMS and EMS, are also applicable to 2G and 2.5G systems.

2.2.1 3GPP Structure

The 3GPP standardization process strictly defines how partners should coordinate the standardization work and how individual members should participate in the development of specifications. There is a clear separation between the coordination work of 3GPP partners and the development of specifications by individual members. This separation enables a very efficient and robust standardization process. In order to achieve it, the 3GPP structure is split into the Project Coordination Group (PCG) and five Technical Specifications Groups (TSGs). The PCG is responsible for managing and supervising the overall work carried out within the scope of the 3GPP whereas TSGs create and maintain 3GPP specifications. Decisions in PCG are always made by consensus whereas decisions in TSGs may be made by vote if consensus cannot be reached. In each TSG, several working groups (WGs) create and manage specifications for a set of related technical topics (for instance CN WG5 deals with the set of technical topics related to the Open Service Architecture). If the set of technical topics is too broad, then a WG may be further split into Sub Working Groups (SWGs). This is the case for T WG2 (or also T2 for short) which deals with mobile terminal services and capabilities. T2 is split into three SWGs: T2 SWG1 deals with the mobile execution environment (MExE), T2 SWG2 deals with user equipment capabilities and interfaces and T2 SWG3 deals with messaging aspects. Figure 2.2 shows the list of 3GPP TSGs and corresponding WGs.

Activities of sub-working group T2 SWG3 encompass the development of messaging services and technologies including SMS, EMS, Cell Broadcast Service and MMS.

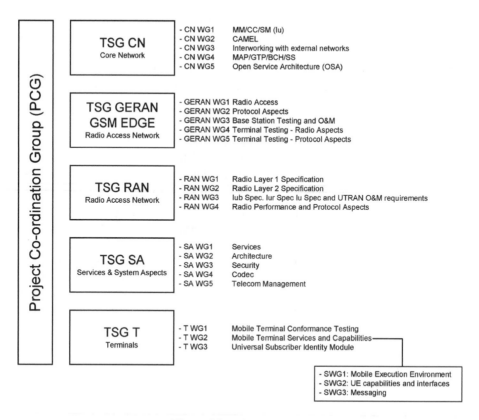

Figure 2.2 3GPP structure

2.2.2 3GPP Specifications: Release, Phase and Stage

Documents produced by the 3GPP are known as specifications. Specifications are either *Technical Specifications* (TS) or *Technical Reports* (TR). Technical specifications define a GSM/UMTS standard and are published independently by the six partners (ETSI, Committee T1, ARIB, TTC, TTA and CWTS). Technical reports are working documents that can later become technical specifications. Technical reports are usually not published by standardization organizations involved in the 3GPP.

In order to fulfil ever-changing market requirements, 3GPP specifications are regularly extended with new features. To ensure that market players have access to a stable platform for implementation and meanwhile allowing the addition of new features, the development of 3GPP specifications is based on a concept of parallel *releases*. In this process, specifications are regularly frozen. Only essential corrections are permitted for a frozen specification. New work can still be carried out but will be incorporated in the next release of the same specification. An engineer implementing a commercial solution based on one or more 3GPP standards should, as much as possible, base the work on frozen specifications. An unfrozen specification is subject to change and should never be considered as a stable platform on which to build a commercial solution. In 3GPP, technical specifications are typically frozen once a year. Consequently, releases used to be named according to the expected specification freezing date (release 98, release 99, etc.). In 1999, the 3GPP decided that releases produced after 1999 would no longer be named according to a year but according to a unique sequential number (release 5 followed release 4 which itself followed release 99).

Each 3GPP technical specification is usually categorized into one of three possible stages. The concept of characterizing telecommunications services into three stages was first introduced by the ITU in [ITU-I.130]. A *stage 1 specification* provides a service description from a service-user's perspective. A *stage 2 specification* describes a logical analysis of the problem to be solved, a functional architecture and associated information flows. A *stage 3 specification* describes a concrete implementation of the protocols between physical elements onto the elements of the stage 2 functional architecture. A stage 3 implementation is also known as a *technical realization*. Note that several technical realizations may derive from a common stage 2 specification.

2.2.3 3GPP Specifications: Numbering Scheme

Each 3GPP technical document (report or specification) is uniquely identified by a reference as shown in Figure 2.3. The reference starts with the prefix '3GPP' and is followed by two letters identifying the document type ('TS' for a specification and 'TR' for a report). After the document type, follows a specification number which can take one of the following forms: *aa.bbb* or *aa.bb*. In the specification number, *aa* indicates the document intended use as shown in Table 2.1. In the document reference, the document number is followed by a version number in the format *Vx.y.z*. In this format, *x* represents the release, *y* represents the technical version and *z* represents the editorial version. Table 2.2 shows how the document version is formatted according to its associated release. The freezing date for each release is also indicated.

A 3GPP document is also given a title in addition to the reference. For instance, the following document contains the definition of MMS stage 1:

Reference   3GPP TS 23.140 V5.2.0
Title   Multimedia Messaging Service, Stage 1

Lists of available 3GPP specifications are provided in the documents listed in Table 2.3. 3GPP members can download 3GPP specifications from the 3GPP website at http://www.3gpp.org.

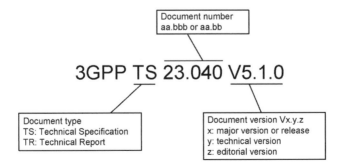

Figure 2.3   3GPP specification type, number and version

Table 2.1   3GPP specifications/numbering scheme

Range for GSM up to and including release 99	Range for GSM release 4 onwards	Range for UMTS release 1999 onwards	Type of use
01.bb	41.bbb	21.bbb	Requirement specifications
02.bb	42.bbb	22.bbb	Service aspects
03.bb	43.bbb	23.bbb	Technical realizations
04.bb	44.bbb	24.bbb	Signalling protocols
05.bb	45.bbb	25.bbb	Radio access aspects
06.bb	46.bbb	26.bbb	Codecs
07.bb	47.bbb	27.bbb	Data
08.bb	48.bbb	28.bbb	Signalling protocols
09.bb	49.bbb	29.bbb	Core network signalling protocols
10.bb	50.bbb	30.bbb	Programme management
11.bb	51.bbb	31.bbb	SIM/USIM
12.bb	52.bbb	32.bbb	Charging and OAM&P
13.bb			Regulatory test specifications
		33.bbb	Security aspects
		34.bbb	Test specifications
		35.bbb	Algorithms

Table 2.2   3GPP specifications/releases

GSM/edge release	3G release	Abbreviated name	Specification number format	Specification version format	Freeze date
Phase 2 + release 6	Release 6	Rel-6	aaa.bb (3G)	6.x.y (3G)	June 2003
Phase 2 + release 5	Release 5	Rel-5	aa.bb (GSM)	5.x.y (GSM)	March 2002
Phase 2 + release 4	Release 4	Rel-4	aaa.bb (3G)	4.x.y (3G)	March 2001
			aa.bb (GSM)	9.x.y (GSM)	
Phase 2 + release 99	Release 99	R99	aaa.bb (3G)	3.x.y (3G)	March 2000
			aa.bb (GSM)	8.x.y (GSM)	
Phase 2 + release 98		R98	aa.bb	7.x.y	Early 1999
Phase 2 + release 97		R97	aa.bb	6.x.y	Early 1998
Phase 2 + release 96		R96	aa.bb	5.x.y	Early 1997
Phase 2		PH2	aa.bb	4.x.y	1995
Phase 1		PH1	aa.bb	3.x.y	1992

Table 2.3   List of GSM/UMTS specifications produced by the 3GPP

Release	List of GSM specifications	List of UMTS specifications
Release 99	–	[3GPP-21.101]
Release 4	[3GPP-41.102]	[3GPP-21.102]
Release 5	[3GPP-41.103]	[3GPP-21.103]

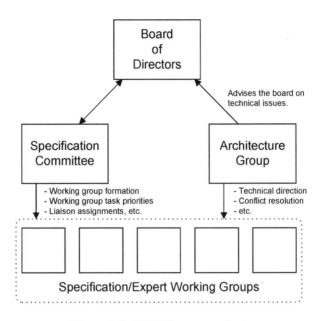

Figure 2.4 WAP Forum organization

2.3 WAP Forum Specifications

The WAP Forum concentrates on the definition of a generic platform for the development of applications for various wireless technologies. The WAP Forum is organized into functional areas as shown in Figure 2.4.

The *board of directors* is responsible for creating working groups, approving the board charter and approving specifications for publication. The *specification committee* is appointed by the board and is in charge of project management activities. It also supports the technical activities carried out by the architecture group and other specification/expert working groups. The specification committee also manages procedural and review issues, the document process and the creation of new working groups. The *architecture group* is responsible for the overall technical architecture of the WAP Forum technology. The *specification working groups* are chartered to define detailed architectures and draft technical specifications whereas *expert working groups* are chartered to investigate new areas of technology and address industry and market viewpoints.

The WAP Forum manages four types of technical documents:

- *Specification*: a specification contains technical or procedural information. At any given time, a specification is associated with a stage such as proposal, draft, etc. This stage indicates the level of maturity of the specification content.
- *Change Request* (CR): an unofficial proposal to change a specification. A change request is proposed by one or more individuals for discussion between WAP Forum members.
- *Specification Change Document* (SCD): an SCD is the draft of a proposed modification of a specification. An SCD can only be produced by the specification working group responsible for the corresponding specification. An SCD applies to a specific version of a specification.

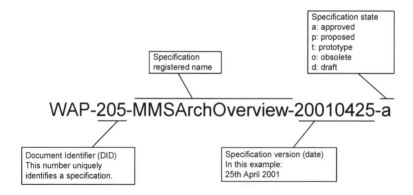

Figure 2.5   WAP Forum specification naming convention

- *Specification Implementation Note* (SIN): a SIN is an approved modification of a previously published specification. SINs are used to fix bugs or revise an existing approved specification. A SIN applies to a specific version of a specification.

A WAP Forum document is identified by a Document IDentifier (DID). A specification keeps its associated DID for its entire lifespan (all revisions of the specification and the approved specification).

WAP Forum specifications are named according to the convention outlined in Figure 2.5. A WAP Forum specification lifecycle consists of five different stages:

- *1st stage/proposal:* a proposal is a specification at a very early stage which is proposed by an organization for review by WAP Forum members. A proposal is not part of any WAP Forum specification suite.
- *2nd stage/draft specification:* a draft specification is a document which is under consideration for inclusion in a WAP Forum specification suite.
- *3rd stage/prototype specification:* a prototype specification has reached a mature stage within the WAP Forum but still needs to be publicly reviewed. Prototype specifications are not part of any WAP Forum specification suite.
- *4th stage/proposed specification:* a proposed specification is under active validation by WAP Forum members. Proposed specifications are not part of any WAP Forum specification suite.
- *5th stage/approved specification:* an approved specification has been validated by WAP Forum members. An approved specification can be part of the WAP Forum specification suite. An approved specification is considered as mature and viable for the development of WAP-based solutions.

Proposals, draft specifications, prototype specifications are not considered as complete and stable specifications. Only approved specifications should be considered as a basis for the development of WAP-based solutions. WAP Forum members can download WAP Forum specifications from the WAP Forum website at http://www.wapforum.org.

Recently, the Open Mobile Alliance (OMA) has been established by the consolidation of the WAP Forum and the Open Mobile Architecture Initiative. The mission of OMA consists of removing interoperability barriers by defining a framework of open standards for the

development of mobile services. OMA objectives and activities are explained at http://
www.openmobilealliance.org.

2.4 Internet Engineering Task Force

The Internet Engineering Task Force (IETF) is another organization which produces
technical specifications related to Internet protocols and procedures in the scope of the
Internet Standards Process. It is an international forum open to any interested party. In the
IETF, the technical work is carried out by *working groups*, each one focusing on specific
technical topics (routing, transport, security, etc.). Furthermore, working groups are
grouped into areas managed by *area directors*. Area directors are members of the Internet
Engineering Steering Group (IESG). In the IETF, the Internet Architecture Board (IAB)
provides an architectural oversight of the Internet. Both the IESG and the IAB are
chartered by the Internet Society (ISOC). In addition, the Internet Assigned Numbers
Authority (IANA) has the responsibility for assigning unique numbers for Internet protocols (applications ports, content types, etc.).

2.4.1 Internet Standards-related Publications

IETF communications are primarily performed by the publication of a series of documents
known as *Request For Comments* (RFCs). First RFCs on Internet networking were produced
in 1969 as part of the original ARPA wide-area networking (ARPANET) project. RFCs are
numbered in chronological order of creation. An RFC documenting an *Internet standard* is
given an additional reference in the form STDxxx and becomes part of the STD subseries of
the RFC series. RFCs are classified into five high level categories:

1. Standard track (including proposed standards, draft standards and Internet standards);
2. Best current practices;
3. Informational RFCs;
4. Experimental RFCs; and
5. Historic RFCs.

For instance, the following specification is published as RFC822. The specification is an
Internet standard also known as STD0011:

```
RFC822    Standard for the format of ARPA Internet text messages.
          D.Crocker. Aug-13-1982 / Status: STANDARD / STD0011
```

During the development of a specification, draft versions of the document are often made
available for informal review and comments. These temporary documents are known as
Internet drafts. Internet drafts are working documents subject to changes without notice.
Consequently, Internet drafts should not be considered as stable specifications on which to
base developments.

2.4.2 Internet Standard Specifications

Specifications subject to the Internet Standards Process fall into two categories: technical
specifications and applicability statements. A *technical specification* is a description of a

protocol, service, procedure, convention or format. On the other hand, an *applicability statement* indicates how one or more technical specifications may be applied to support a particular Internet capability.

Specifications that are to become Internet standards evolve on a scale of maturity levels known as the *standard track*. The standard track is composed of three maturity levels:

- *Proposed standard:* this is the specification at the entry level. The IESG is responsible for registering an existing specification as a proposed standard. A proposed standard is a specification that has reached a stable stage and has already been the subject of public review. However, implementers should consider proposed standards as immature specifications.
- *Draft standard:* a specification can be moved to the draft standard of the standard track if at least two implementations based on the specification exist and interoperate well. Since draft standards are considered as almost final specifications, implementers can provide services developed on the basis of draft standards.
- *Internet standard:* an Internet standard (also referred to as a standard) is a specification that is the basis of many significant implementations. An Internet standard has reached a high level of technical maturity.

RFCs can be downloaded from the IETF website at http://www.ietf.org.

2.5 World Wide Web Consortium

The World Wide Web Consortium (W3C) is a standardization body (creation in 1994) involved in the development of widely accepted protocols and formats for the World Wide Web. Technical specifications published by the W3C are known as *recommendations*. The W3C collaborates closely with the IETF. W3C activities are organized into groups: working groups (for technical developments), interest groups (for more general work) and coordination groups (for communications among related groups). W3C groups are organized into five domains:

- The *Architecture domain* includes activities related to the development of technologies which represent the basis of the World Wide Web architecture.
- The *Document formats domain* covers all activities related to the definition of formats and languages.
- The *Interaction domain* includes activities related to the improvements of user interactions with the World Wide Web. This includes the authoring of content for the World Wide Web.
- The *Technology and society domain* covers activities related to the resolution of social and legal issues along with handling public policy concerns.
- The *Web accessibility initiative* aims at promoting a high degree of usability for disabled people. Work is carried out in five primary areas: technology, guidelines, tools, education and outreach, and research and development.

To date, significant W3C contributions include the architecture of the initial World Wide Web (based on HTML, URIs and HTTP), XML, XHTML, SVG and SMIL.

The W3C *recommendation track* is the process followed by the W3C to initiate discussions on proposed technologies and to ultimately publish recommendations. The recommendation

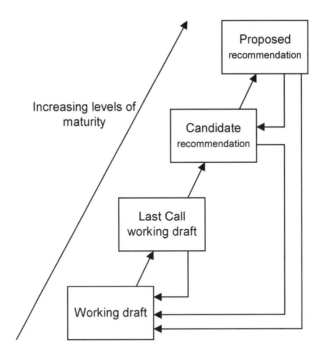

Figure 2.6   W3C recommendation track

track defines four technical specification status corresponding to four increasing levels of maturity. These status are depicted in Figure 2.6.

A *working draft* is the initial status for a technical specification in the W3C recommendation track. A working draft is a work item which is being or will be discussed within the relevant W3C working group. However, the status 'working draft' for a technical specification does not imply that there is a consensus between W3C members on the acceptability of the proposed technology. A *last call working draft* is a special working draft that is regarded by the relevant working group as fulfilling the requirements of its charter. A technical specification has the status 'last call working draft' when the relevant working group seeks technical review from other W3C groups, W3C members and the public. A *candidate recommendation* is a technical specification that is believed to fulfil the relevant working group's charter, and has been published in order to gather implementation experience and feedback. Finally, a *proposed recommendation* is a technical specification which has reached the highest level of maturity in the W3C recommendation track. A proposed recommendation obviously fulfils the requirements of the relevant working group's charter but has also benefited from sufficient implementation experience and has been carefully reviewed. A proposed W3C recommendation can be considered as a stable technical specification on which to base the development of commercial solutions.

W3C technical specifications can be retrieved from http://www.w3c.org.

Further Reading

J. Huber, D. Weiler and H. Brand, UMTS, the mobile multimedia vision for IMT-2000: a focus on
    standardization, IEEE Communications Magazine, September, 2000.
P. Loshin, Essential Email Standards, John Wiley & Sons, Chichester, 1999.
3GPP-TR-21.900, 3GPP working methods.
WAP-181-TAWP-20001213-a, WAP Work Processes, December, 2000.
RFC 2026, The Internet Standards Process – Revision 3, October, 1996.

3

Short Message Service

The Short Message Service (SMS) is a basic service allowing the exchange of short text messages between subscribers. The first short text message is believed to have been transferred in 1992 over signalling channels of a European GSM network. Since this successful trial, SMS usage has been the subject of tremendous growth. In 2001, an estimated 102.9 billion SMS messages were exchanged worldwide. Gartner Dataquest, one of the industry's main research agencies, expects the number of SMS messages to grow to 146 billion in 2002 and to peak at around 168 billion in 2003 before declining.

This chapter, dedicated to SMS, first introduces common use cases for SMS such as consumer, corporate and operator applications. Components of a typical SMS-enabled GSM architecture are presented along with basic SMS features. The four-layer transport protocol stack of SMS (application, transfer, relay and link) is presented and the transfer layer of this stack is described in detail. The transfer layer is the component which needs to be mastered by implementers for the development of SMS-based applications. An insight is also given into techniques available for exchanging messages between servers and applications running in the SIM. Interworking between SMS and Email is presented and the manipulation of messages via AT commands is illustrated.

Note that the content of this chapter also represents the basis for Chapters 4 and 5. These chapters describe two different flavours of the Enhanced Messaging Service (EMS). EMS is an application-level extension of SMS.

3.1 Service Description

Developed as part of the GSM Phase 1 ETSI technical specifications, the Short Message Service (SMS) allows mobile stations and other network-connected devices to exchange short text messages. Work on the standardization of SMS was initiated by ETSI and is now being carried out in the scope of the 3GPP. Since its initial introduction in GSM networks, SMS has been ported to other network technologies such as GPRS and CDMA. The Short Message Service allows users to exchange messages containing a short amount of text. These messages can be sent from GSM/UMTS mobile devices but also from a wide range of other devices such as Internet hosts, telex and facsimile. The SMS is a very mature technology supported by 100% of GSM handsets and by most GSM networks worldwide.

Box 3.1 The success of SMS

SMS was commercially introduced in the telecommunications market in 1992 but it was only in the late 1990s that the service became widely accepted by the mass market. In 2001, an estimated 102.9 billion SMS messages were exchanged world-wide. Gartner Dataquest, one of the industry's major research agencies, expects the number of SMS messages to grow to 146 billion in 2002 and to peak at around 168 billion in 2003 before declining. At the time of service introduction, operators did not really expect that SMS would become such a successful messaging service.

3.2 SMS Use Cases

SMS was intended to be a means of exchanging limited amounts of information between two mobile subscribers. This limited capability has become a building block for the development of more compelling services ranging from the download of ringtones to professional applications such as remote monitoring or fleet tracking. SMS use cases described in this chapter are representative of typical applications based on SMS, including consumer applications, corporate applications and operator applications.

3.2.1 Consumer Applications Based on SMS

In this category are grouped services such as person-to-person messaging, information services, download services or chat applications. Consumers have access to these services for customizing their handsets, receiving information from remote servers or simply for exchanging information between friends.

Person-to-person Messaging
This is the original use case for which SMS has been designed. This use case relates to the exchange of a short text message between two mobile subscribers. The subscriber that originates the message first composes the message using the man–machine interface of the mobile device. Usually, the message text is entered via the handset keyboard and the message text is echoed on the handset display. After composition, the subscriber enters the address of the message recipient and sends it to the serving network. The message is then transported over one or more mobile networks before reaching the recipient mobile network. If the recipient handset is able to handle the message immediately, then the message is transferred to the recipient mobile handset and the recipient subscriber is notified that a new message has arrived. Otherwise the message is kept temporarily by the network until the recipient handset becomes available. It is not easy to input text with a small handset keypad. In order to cope with this limitation, handsets are usually shipped with a *predictive text input mechanism*. These mechanisms anticipate which word the user is trying to enter by analysing the most recent characters or symbols which have been typed in and checking those against entries of a static or dynamic dictionary. Entries from the dictionary which closely match the partially entered word are then presented to the subscriber who can select one of them if appropriate. These mechanisms significantly reduce the number of keys that need to be pressed to input

Figure 3.1   Mobile device with external keyboard. Reproduced by permission of Alcatel Business Systems

text for the composition of a message. Predictive text input mechanisms are available for various different languages. With SMS, the two most well known predictive text input algorithms are T9 from Tegic and ZI from ZI corporation. It has to be noted that advanced handsets are sometimes equipped with a built-in QWERTY keyboard. Alternatively, a small external keyboard can sometimes be connected to a handset as shown in the Figure 3.1. Because of the difficulty of entering text with mobile handsets, (young) subscribers have 'invented' what is now commonly referred to as the 'SMS-speak'. SMS-speak consists of using abbreviated terms that are quick to compose on mobile handsets. For instance, 'R U OK' means 'Are you OK?' and 'C U L8ER' means 'See you later'! This is sometimes complemented by the use of text-based pictograms or 'smileys'. For instance, :-) represents a happy face and :-(represents a sad face.

Information Services
This is probably one of the most common use cases in the machine-to-person scenario. With information services, weather updates and financial reports can be prepared by value-added service providers and pushed to mobile handsets with SMS. For these services to be activated,

it is usually necessary for the user to first subscribe manually to the service prior to receiving associated reports and updates.

Voice Message and Fax Notifications
This use case is widely supported in GSM mobile networks. This use case relates to the reception of messages containing notifications for voice messages and fax waiting in a remote message inbox.

Internet Email Alerts
With email alerts via SMS, subscribers are notified that one or more Email messages are waiting to be retrieved. Such an alert usually contains the address of the message originator along with the message subject and the first few words from the Email message body.

Download Services
It has become popular for mobile subscribers to customize their mobile handset. This can be done by associating ringtones to groups of persons in the phone contact directory. With such a configuration, an incoming phone call from an identified person triggers the selected ringtone. It is also common to change switch on and off animations or to change the general look-and-feel (also known as 'skin') of the handset's graphical interface. All these objects used to customize the mobile handset can be downloaded as part of one or more short messages.

Chat Applications
During a chat session, several users can exchange messages in an interactive fashion. All messages exchanged during a session are kept in chronological order in a chat history. In the chat history, messages sent from a recipient are differentiated from messages sent from other users. Several existing mobile chat applications are based on SMS for the transport of messages.

Smart Messaging
Smart Messaging is a proprietary service developed by Nokia. This service enables the exchange of various objects via SMS. This includes the transfer of Internet configuration parameters, business cards for PIM updates, etc. An interesting feature of this service is called Picture Messaging which allows the association of one bitmap picture to the text of a message.

3.2.2 Corporate Applications Based on SMS

In this category are grouped services tailored for the need of professionals. This includes vehicle positioning and remote monitoring of machines.

Vehicle Positioning
The Global Positioning System (GPS) is a technology for determining global positions on Earth. The device determines its location by analysing signals broadcast from a group of

satellites. The location information is usually expressed in terms of longitude, latitude and sometimes altitude. A GPS receiver coupled to a handset, built-in or as an accessory, can provide the location of a person or equipment. This location information can be formatted in a short message and sent to a remote server via SMS. The server interprets locations received from several handsets and displays them on associated geographical maps. Such an application can help logisticians to keep track of a fleet of trucks or policemen to track down stolen vehicles.

Remote Monitoring

Messages can transport information about the state of remote devices. For instance, system administrators can be notified by a short message that a server is running low of resources or that a fault has been detected on a remote computer.

3.2.3 Operator Applications Based on SMS

Operators have used SMS as a building block for enabling the realization of several services including the ones listed below.

SIM Lock

Operators sometimes require handsets to be locked and usable with only one specific SIM. After a minimal subscription period, the user may request the operator to deactivate the lock in order to be able to use the mobile handset with another SIM (from the same operator or from another operator). If the operator agrees on the lock being deactivated, then the operator sends a short message containing a code allowing the device to be unlocked.

SIM Updates

With SMS, operators can remotely update parameters stored in the SIM. This is performed by sending one or more messages with new parameters to a mobile device. In the past, operators have used this method for updating voice-mail access numbers, customer service profiles (determining which network services are accessible to the subscriber), operator name for display in idle mode on the device screen and address book entries.

Message Waiting Indicator

Operators have used SMS as a simple way to update message waiting indicators on the receiving handset. With this mechanism, a short message contains the type of indicator (voice mail, etc.) to be updated along with the number of waiting messages.

WAP Push

The SMS can be used as a bearer for realizing the WAP push. With this configuration, a WSP protocol data unit or the URI of the content to be retrieved is encoded in a short message and sent to the receiving device. Upon reception of such a message, the WAP microbrowser intercepts the message, interprets the pushed content and presents the content to the subscriber.

3.2.4 Value Chain of SMS-based Applications

In the person-to-person scenario, SMS involves two persons, the message originator and the message recipient. In addition, one or more network operators may be involved in transporting the message. In the machine-to-person scenario, the business model may become more complex. Indeed, the business model involves a person, the message recipient, one or more network operators to transport the message and various intermediaries such as service providers, portal providers, SMS resellers, etc.

To cope with the high demand for transferring short messages in the mobile-to-person scenario, a business model has been put in place by operators to provide *wholesale SMS* to service providers. This allows service providers to purchase SMS resources in bulk at a wholesale price from operators and to resell short messages with customized content to subscribers.

3.3 Architecture of the GSM Short Message Service

The realization of SMS implies the inclusion of several additional elements in the network architecture (GSM, GPRS or UMTS). Figure 3.2 shows the architecture of an SMS-enabled GSM network. The two additional network elements in the architecture are the SMS centre and the Email gateway. In addition, an element called the short message entity, usually in the form of a software application in a mobile device, is necessary for the handling of messages (sending, reception, storage, etc.). The short message entity is not shown in Figure 3.2. The SMS centre, Email gateway and the short message entity are presented in the following sections.

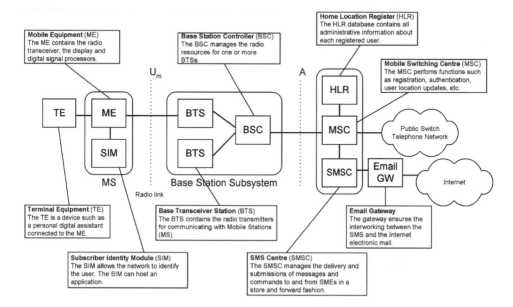

Figure 3.2   SMS-enabled GSM network architecture

3.3.1 Short Message Entity

Elements that can send or receive short messages are named Short Message Entities (SME). An SME can be a software application in a mobile handset but it can also be a facsimile device, telex equipment, remote Internet server, etc. An SME can be a server that interconnects to the SMS centre directly or via a gateway. Such an SME is also known as an External SME (ESME). Typically, an ESME represents a WAP proxy/server, an Email gateway or a voice mail server.

For the exchange of a short message, the SME which generates and sends the short message is known as the *originator SME* whereas the SME which receives the short message is knows as the *recipient SME*.

3.3.2 Service Centre

The service centre (SC) or SMS centre (SMSC) plays a key role in the SMS architecture. The main functions of the SMSC are the relaying of short messages between SMEs and the store-and-forwarding of short messages (storage of messages if the recipient SME is not available). The SMSC may be integrated as part of the mobile network (e.g. integrated to the MSC) or as an independent network entity. The SMSC may also be located outside the network and be managed by a third party organization. Practically, it is very common for network operators to acquire one or more SMSCs since SMS is now considered as a very basic service to be provided by any mobile network. In theory, one single SMSC could manage SMS for several mobile network operators. However, this latter scenario is seldom encountered in real life and an SMSC is often dedicated to the management of SMS operations in one single mobile network.

Mobile network operators usually have mutual commercial agreements to allow the exchange of messages between networks. This means that a message sent from an SME attached to a network A can be delivered to another SME attached to a mobile network B. This ability for users to exchange messages even if they are not subscribers to the same network and sometimes located in different countries is undoubtedly one of the key features that makes SMS so successful.

3.3.3 Email Gateway

The Email gateway enables an Email-to-SMS interoperability by interconnecting the SMSC with the Internet. With the Email gateway, messages can be sent from a SME to an Internet host, and vice versa. The role of the Email gateway is to convert message formats (from SMS to Email and vice versa) and to relay messages between SMS and Internet domains.

3.4 Short Message Basic Features

SMS encompasses a number of basic features. This includes message submission, message delivery, handling of status reports, requests for command execution, reply path, etc. These features are presented in the following sections.

3.4.1 Message Submission and Delivery

The two most basic features of SMS are the sending and the receipt of a short message.

Message Sending

Mobile-originated messages are messages which are submitted from an MS to an SMSC. These messages are addressed to other SMEs such as other mobile users or Internet hosts. An originator SME may specify a message validity period after which the message is no longer valid. A message which is no longer valid may be deleted by an SMSC during the message transfer. With the first GSM networks, not all handsets supported message submission. Presently, almost all handsets support message submission. This feature is also known as the Short Message-Mobile Originated (SM-MO).

Message Delivery

Mobile-terminated messages are messages delivered by the SMSC to the MS. Nearly all GSM handsets support message reception. This feature is also known as Short Message-Mobile Terminated (SM-MT).

Mobile-originated and mobile-terminated short messages can be delivered or submitted even while a voice call or data connection is under progress. Messages can be sent or received over GSM signalling channels, but also over GPRS channels. In GSM, messages are sent over SDCCH or SACCH whereas, in GPRS, short messages are sent over PDTCH channels. The choice of the bearer for transporting a message is usually made according to a predefined network policy.

3.4.2 Status Reports

It is possible for an originator SME to request that a status report be generated upon delivery of the short message to the recipient SME. The status report indicates to the originator whether or not the short message has been successfully delivered to the recipient SME.

3.4.3 Reply Path

The reply path can be set by the originator SME (or the serving SMSC) to indicate that the serving SMSC is able and willing to directly handle a reply from the recipient SME in response to the original message. In this situation, the recipient SME usually submits the reply message directly to the SMSC that serviced the submission of the original message.

This feature is sometimes used by operators to allow the message recipient to provide a reply message 'free of charge' for the message recipient.

Additionally, for networks supporting several SMSCs, operators sometimes use this feature to get reply messages to be returned to a particular SMSC. For example, an operator could have several SMSCs but only one connected to the Email gateway. In this configuration, if a message is originated from the Internet domain, then the operator uses the reply path to indicate that any message reply associated with this Email-originated message should be submitted to the SMSC connected to the Email gateway. In this situation, the SMSC can

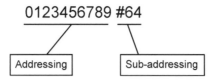

0123456789 #64

| Addressing | Sub-addressing |

Figure 3.3   Sub-addressing/format

appropriately request the Email gateway to convert reply messages into Email messages and to deliver them to the Email recipient(s).

3.4.4 Addressing Modes

With SMS, several modes are possible for addressing message recipients. The most common addressing mode consists of using the Mobile Station ISDN Number (MSISDN) in the [ITU-E.164] format (e.g. +33 612345678). However, other less commonly used addressing modes are available such as Email addressing, as defined by the IETF in [RFC-2822], or operator specific numbering schemes (short codes for instance).

An optional addressing feature of SMS consists of conveying sub-addressing information as part of a message. The sub-addressing information is appended at the end of the recipient address by the originator SME before message sending. When this scheme is applied, the SMSC extracts the sub-addressing information from the recipient address specified for the submitted message and appends it to the originator address for the message to be delivered. This optional feature can be used to maintain session identifications for the exchange of messages or for identifying a specific service code to which a message relates. The sub-addressing information is separated from the normal addressing information by the '#' delimiter (this character is part of the sub-addressing information) as shown in Figure 3.3.

The sub-addressing information is a combination of digits and '#' and '*' characters. For instance, a mobile handset may request a weather forecast update by sending a message to an application server. The identification of the requested service (e.g. service 64) can be specified as part of the sub-addressing information as shown in Figure 3.4.

A message originator has the possibility of indicating an alternate reply address as part of a submitted message. Such an alternate reply address should be used by the recipient SME if the recipient wishes to reply to the message.

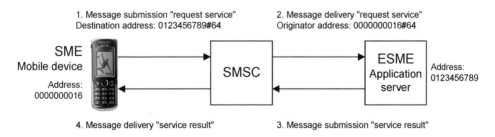

Figure 3.4   Sub-addressing/example

Sub-addressing and alternate reply address features are recent evolutions of the SMS standard and still have to be supported by SMEs and SMSCs.

Note that a limitation of SMS, and to some extent EMS, resides in its inability to support group sending, at the transport level. A message can only be addressed to one single recipient. Group sending can be emulated by the originator SME by sending the same message a number of times to the network, one time per message recipient. However, this method is not very efficient and proves to be costly for the message originator. Alternatively, operators sometimes cope with this limitation by offering a distribution list service. With this service, subscribers are able to manage (create/modify/delete) personal distribution lists. Each distribution list is identified by a short code. Once a distribution list has been configured, the subscriber can send a message to all recipients of a distribution list by sending the message just once to the corresponding short code. Upon reception of the message, the network generates and delivers separate copies of the message to all recipients. In this situation, the message is only transferred once by the originator SME to the network. This leads to a more efficient use of network resources. Unfortunately, the message originator is still charged, by the network operator, for the number of copies generated by the network, as if they had been submitted directly from the originator SME.

3.4.5 Validity Period

A message originator has the possibility of indicating a validity period for a message. This validity period defines the deadline after which the message content is to be discarded. If a message has not been delivered to the message recipient before the expiry date, then the network usually discards the message without further attempts to deliver it to the recipient. For instance, a subscriber may send a message with the following content 'please phone me in the coming hour to get your answer'. Additionally, the subscriber may wisely indicate that the message validity period is limited to one hour. In the situation where the message recipient does not turn on his/her mobile device in the hour following the message sending, then the network can decide to discard the message. Consequently, in this example, the message will never be delivered unless the recipient successfully retrieves the message in the hour following the message sending.

Note that mobile operators often assign a default validity period for messages transiting in their network (e.g. 2 days).

3.5 Technical Specification Synopsis

SMS is defined in a number of 3GPP technical specifications as shown in Table 3.1.

3.6 Short Message Layers and Protocols

The SMS protocol stack is composed of four layers: the application layer, the transfer layer, the relay layer and the link layer. This book outlines each layer and provides an in-depth description of the transfer layer. SMS-based applications are directly based on the transfer layer. Consequently, any engineer willing to develop applications, for which SMS is a building block, needs to master the transfer layer.

Table 3.1 SMS technical specification synopsis

TS Reference	Title
3GPP TS 22.205	Services and service capabilities. Note: this specification identifies the high-level requirements for SMS
3GPP TS 23.011	Technical realization of supplementary services. Note: this specification presents the use of radio resources for the transfer of short messages between the MS and the MSC or the SGSN
3GPP TS 23.038	Alphabets and language-specific information
3GPP TR 23.039	Interface protocols for the connection of SMSCs to SMEs
3GPP TS 23.040	Technical realization of the short message service
3GPP TS 23.042	Compression algorithm for text messaging services
3GPP TS 24.011	Point-to-point Short Message Service (SMS) support on mobile radio interface
3GPP TS 27.005	Use of Data Terminal Equipment – DTE-DCE interface for SMS and CBS
3GPP TS 43.041	Example protocol stacks for interconnecting SCs and MSCs

The *application layer* is implemented in SMEs in the form of software applications that send, receive and interpret the content of messages (e.g. message editor, games, etc.). The application layer is also known as SM-AL for Short-Message-Application Layer.

At the *transfer layer*, the message is considered as a sequence of octets containing information such as message length, message originator or recipient, date of reception, etc. The transfer layer is also known as the SM-TL for Short-Message-Transfer-Layer.

The *relay layer* allows the transport of a message between various network elements. A network element may temporarily store a message if the next element to which the message is to be forwarded is not available. At the relay layer, the MSC handles two functions in addition to its usual switching capabilities. The first function called *SMS gateway MSC* (SMS-GMSC) consists of receiving a message from an SMSC and interrogating the HLR to obtain routing information and further to deliver the message to the recipient network. The second function called *SMS InterWorking MSC* (SMS-IWMSC) consists of receiving a message from a mobile network and submitting it to the serving SMSC. The relay layer is also known as the SM-RL for Short-Message-Relay-Layer.

The *link layer* allows the transmission of the message at the physical level. For this purpose, the message is protected to cope with low-level channel errors. The link layer is also known as the SM-LL for Short-Message-Link-Layer.

The stack of transport protocol layers for SMS is shown in Figure 3.5. For transport purposes, an application maps the message content and associated delivery instructions onto a Transfer Protocol Data Unit (TPDU) at the transfer layer. A TPDU is composed of various parameters indicating the type of the message, specifying whether or not a status report is requested, containing the text part of the message, etc. Each parameter is prefixed by the abbreviation `TP` for Transfer Protocol such as `TP-Message-Type-Indicator` (abbreviated `TP-MTI`), `TP-Status-Report-Indication` (abbreviated `TP-SRI`), `TP-User-Data` (abbreviated `TP-UD`), etc.

At the transfer layer, the exchange of a message from the originator SME to the recipient SME consists of two to three steps. The three steps are shown in Figure 3.6. After creation by the originator SME, the message is submitted to the SMSC (step 1). The SMSC may verify

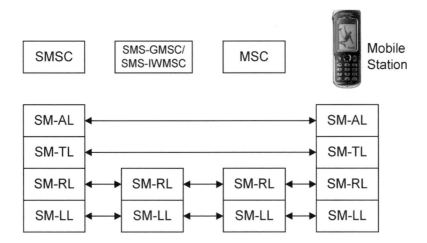

Figure 3.5 SMS protocol stack

with other network elements that the message originator is allowed to send messages (e.g. sufficient pre-paid credit, subscriber belongs to the network, etc.). The SMSC delivers the message to the recipient SME (step 2). If the recipient SME is not available for the message delivery, then the SMSC stores the message temporarily until the recipient SME becomes available or until the message validity period expires. Upon delivery of the message or upon message deletion by the network, a status report might be transferred back to the originator SME (step 3), only if this report had been requested by the originator SME during message submission.

3.6.1 SMS Interoperability Between Mobile Networks

With GSM/GPRS technologies, mobile operators can easily support the exchange of messages between distinct networks. For this purpose, operators have commercial agreements. Each mobile network counts the number of messages being sent from another network. After a given period of time, these counts are compared and there is a commercial settlement between operators. In the simplest configuration, MAP signalling transactions for SMS are allowed between the two mobile networks. In this configuration, the SMSC of the originator SME queries the HLR of the relevant destination network to obtain the necessary routing information and forwards the message directly towards the message recipient. In this case, the SMSC of the message recipient is not involved in the message delivery. At the transfer layer, the different steps involved in the exchange of a message, in this configuration, are basically those presented in Figure 3.6. This configuration is supported by the four GSM mobile network operators in the United Kingdom.

Unfortunately, things are more complicated for ensuring interoperability between networks based on different technologies (e.g. between GSM/GPRS and CDMA) or when signalling interconnections cannot be supported between distinct networks. This is the case, for instance, for many North American networks[1] where a wide variety of technologies are

---

[1] Due to the difficulty of ensuring SMS interoperability in these configurations, operators sometimes decide not to offer the ability for their subscribers to exchange messages with subscribers belonging to another network.

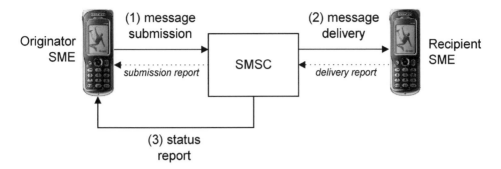

Figure 3.6   Message transfer between two SMEs

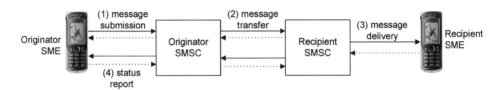

Figure 3.7   Message transfer

available. For these configurations, the exchange of messages between distinct networks can still be offered by interconnecting the two mobile networks with a gateway or interconnecting the two SMSCs with a proprietary exchange protocol. This latter configuration, considered from the transfer layer, is depicted in Figure 3.7. In this configuration, the exchange of a message between two subscribers consists of three to four steps. After creation by the message originator, the originator SME submits the message to the originator SMSC (step 1). The originator SMSC forwards the message towards the recipient SMSC (step 2) and the recipient SMSC delivers the message to the recipient SME (step 3). If a status report was requested by the message originator, then the recipient SMSC generates a status report and transfers it back to the originator SME (step 4).

3.6.2 Message Structure and Terminology

In this chapter, the following terminology is used for referring to messages.

A *message* refers to the subscriber's perception of the message composed of text and/or elements such as pictures, melodies, etc. For transport purposes and due to limitations at the transfer layer, an application may need to segment the message into several pieces called message segments. A one-segment message is also known as a *short message*.

A *message segment* is an element manipulated by an application. A message segment has a limited payload size. In order to convey a large amount of data, several message segments can be combined into a *concatenated message*.[2] The message concatenation is handled at the application layer. In order to be transported, the message segment needs to be mapped onto a TPDU at the transfer layer as shown in Figure 3.8.

---

[2] In several documents dealing with SMS, a concatenated message is also known as a long message.

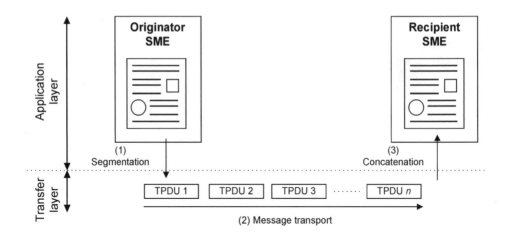

Figure 3.8   Message structure

In the scope of the download service, a message containing one or more objects being downloaded is called a *download message*.

In this book, the word *octet* refers to a group of eight bits (also known as a byte) whereas a *septet* refers to a group of seven bits.

3.6.3 SME-SMSC Transactions/Submit, Deliver, Report and Command

At the transfer layer, six types of transactions can occur between an SME and the SMSC. A TPDU type corresponds to each one of the transaction types. Figure 3.9 shows the possible transactions between an SME and the serving SMSC.

- *SMS-SUBMIT:* this transaction corresponds to the submission of a message segment from the SME to the SMSC. Upon submission of the message segment, the SMSC acknowledges the submission with the *SMS-SUBMIT-REPORT* transaction.
- *SMS-DELIVER:* this transaction corresponds to the delivery of a message segment from the SMSC to the SME. Upon delivery of the message segment, the SME acknowledges the delivery with the *SMS-DELIVER-REPORT* transaction.

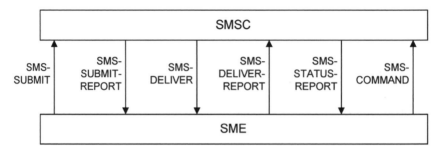

Figure 3.9   Transaction types

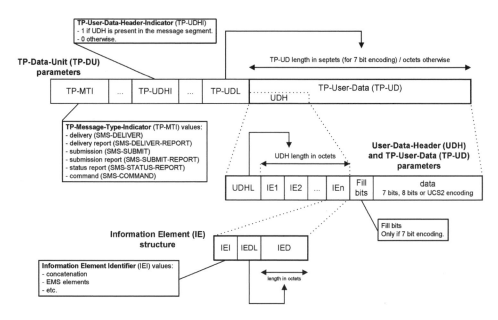

Figure 3.10   Structure of the TP-Data-Unit

- *SMS-STATUS-REPORT:* this transaction corresponds to the transfer of a status report from an SMSC back to an SME.
- *SMS-COMMAND:* this transaction corresponds to the request from an SME, usually an external SME, for the execution of a specific command by the SMSC.

3.7 Structure of a Message Segment

A message segment is associated with a number of parameters. These parameters indicate the message type, class, coding group, etc. In addition, parameters also contain the message content which is provided by the subscriber, the content provider or content which is automatically generated by a machine.

3.7.1 Transport Protocol Data Unit

As introduced in this chapter, a TPDU type has been assigned to each transaction that can occur between an SME and an SMSC at the transfer layer. Depending to its type, a TPDU is composed of a varying number of parameters organized according to a predefined TPDU layout. The partial representation of a TPDU is shown in Figure 3.10.

High-level parameters in the TPDU (the ones for which the name starts with `TP-`) inform on the transaction type (`TP-Message-Type-Indicator`), the presence of binary elements in the message such as concatenation instructions (`TP-User-Data-Header-Indicator`), etc.

One of the important parameters is the user data parameter[3] (`TP-User-Data`). If present,

---

[3] The user data parameter is not present in the Command transaction type.

this parameter contains the text part of a message segment and may also contain binary elements such as concatenation instructions, pictures, melodies, etc. To cope with the complexity of this parameter, the parameter is divided into two sub-parts. The first sub-part, known as the User-Data-Header (UDH) contains binary elements whereas the remaining sub-part contains the message text. The UDH is itself structured as a sequence of sub-parameters. The first sub-parameter, the User-Data-Header-Length (UDHL), indicates the length of the UDH in octets. A set of information elements immediately follows the UDHL.

Whatever the coding of the TP-User-Data, the User-Data-Header is always 8-bit encoded. Consequently, if the TP-User-Data is 7-bit aligned, then fill bits may be inserted between the UDH and the remaining part of the TP-User-Data. With this method, a 7-bit data part always starts on a 7-bit data boundary of the TP-User-Data as shown in Figure 3.10. This allows older handsets that do not support the User-Data-Header concept to still be able to present the 7-bit text part of the message to the subscriber.

An in-depth presentation of the TP-User-Data and the User-Data-Header are provided in Section 3.15.

3.7.2 Message Types

The transaction type associated with a message is indicated in one of the message parameters. The set of possible transaction types that can occur between an SMSC and an originator or receiving SME are the message submission (SMS-SUBMIT) and the corresponding submission report (SMS-SUBMIT-REPORT), the message delivery and the corresponding message delivery report (SMS-DELIVERY-REPORT), the transfer of a status report (SMS-STATUS-REPORT) and the submission of a command (SMS-COMMAND). A dedicated value, assigned to the TP-Message-Type-Indicator (abbreviated TP-MTI) parameter, corresponds to each transaction type.

3.7.3 Text Coding Schemes

The text part of a message can be encoded according to several text alphabets. The two text coding schemes that can be used in SMS are the GSM 7-bit default alphabet defined in [3GPP-23.038] and the Universal Character Set (UCS2) defined in [ISO-10646]. The two text alphabets are presented in Appendix C. The amount of text that can be included in a message segment needs to fit into 140 octets. Since, the two text coding schemes utilize one septet and two octets, respectively, to encode a character/symbol, the amount of text that can be included in a message segment is as shown in Table 3.2.

The value assigned to the TP-Data-Coding-Scheme parameter indicates which coding scheme has been used for encoding the message content. In its simplest form, the GSM 7-bit default alphabet is composed of 128 characters plus 9 additional characters (extension table) including the Euro sign. The Universal Character Set with 2-octet symbols (USC2) is used for encoding complex sets of non-Latin characters such as Chinese and Arabic.

3.7.4 Text Compression

In theory, the text part of a message may be compressed [3GPP-23.042]. However, none of

Table 3.2 Relation between coding scheme and text size

Coding scheme	Text length per message segment	TP-DCS	
		Bit 3	Bit 2
GSM alphabet, 7 bits	160 characters	0	0
8-bit data	140 octets	0	1
USC2, 16 bits	70 complex characters	1	0

the handsets currently available on the market support text compression. It is therefore not advised to generate messages with compressed text unless one can ensure that text decompression is supported by the recipient SME.

A message with compressed text cannot be displayed properly by an MS that does not support text decompression.

3.7.5 Message Classes

In addition to its type, a message belongs to a class. The TP-Data-Coding-Scheme (TP-DCS) parameter of the TPDU indicates the class to which the message belongs. Four classes have been defined and indicate how a message should be handled by the receiving SME (Table 3.3).

It has to be noted that in most cases, a message does not belong to any of the four classes. In this situation the message is known as a *no-class message* and is usually handled as a class 1 or 2 message by the receiving SME.

Table 3.3 Message classes

Class	Description	TP-DCS value (bits 7...4 = 00xx or bits 7...4 = 1111)	
		Bit 1	Bit 0
Class 0	*Immediate display message:* messages belonging to the class 0 are immediately presented on the recipient device display	0	0
Class 1	*Mobile equipment specific message:* if possible, messages belonging to the class 1 are stored in the ME; otherwise, class 1 messages may be stored in the SIM	0	1
Class 2	*SIM specific message:* messages belonging to class 2 are stored in the SIM	1	0
Class 3	*Terminal equipment specific message:* messages belonging to class 3 are transferred to the terminal equipment (PDA, personal computer, etc.) which is connected to the ME	1	1

Table 3.4 Coding groups

Group name	Description	TP-DCS value bit 7...4
Message marked for automatic deletion	Upon reading, a message marked for automatic deletion is deleted by the recipient SME. The message can be of any class. A configuration setting may allow the user to disable the automatic deletion	0 1 x x
Message waiting indication group: *discard message*	Messages belonging to this group are used for informing the subscriber of the status of messages waiting for retrieval. The message is processed independently from SME storage availability and is usually discarded after processing	1 1 0 0
Message waiting indication group: *store message*	Messages belonging to this group are used for informing the subscriber of the status of messages waiting retrieval. If possible, the ME updates the message waiting indicator status on the (U)SIM. Otherwise, status are stored in the ME	1 1 0 1

3.7.6 Coding Groups

A message may belong to one of three coding groups. The coding group indicates what the receiving SME should do with the message once it has been read or interpreted. The TP-Data-Coding-Scheme (TP-DCS) indicates the coding group to which the message belongs. The list of available coding groups is provided in Table 3.4.

3.7.7 Protocol Identifiers

Previous sections have shown that a short message has a type, may belong to a class and to a coding group. A message is also associated with a *protocol identifier*. The protocol identifier indicates how the receiving messaging application should handle an incoming message (normal case, ME data download, SIM data download, etc.). The protocol identifier values shown in Table 3.5 can be assigned to the TP-Protocol-Identifier (TP-PID) parameter.

3.8 Storage of Messages and SMS Settings in the SIM

SMS settings (default protocol identifier, service centre address, etc.), user preferences (validity period, default recipient address, etc.) and messages can be stored in the SIM. The internal architecture of a SIM card is organized around a processor and three different types of memory:

- Read Only Memory (ROM): this memory contains the card operating system along with one or more applications. The memory can only be read and cannot be modified by applications.

Table 3.5 Protocol identifiers

TP-PID value (hex)	Description
0x00	*Normal case:* this protocol identifier is used for a simple transfer from the SME to the SMSC and from the SMSC to the SME. The message is handled by the handset according to its class and presented accordingly to the subscriber
0x01...0x1F	*No telematic interworking, but SME-to-SME protocol*
0x20...0x3F	*Telematic interworking:* telematic interworking means that the message is sent from or to a telematic device such as a telefax, a teletex or an Internet email system. All telematic interworking devices that can be identified with these protocol identifiers are listed in Appendix A
0x40	*Short message type 0:* the ME acknowledges the receipt of such a message but discards its content. The common use case for using this message type is to page a mobile to check if the mobile is active without the user being aware that the mobile station has been paged
0x41...0x47	*Replace short message type n (n from 1...7):* the replacement of short messages is not supported by all handsets. If supported, a handset replaces the previously received 'replace short message type *n*' message by the last message received with the same protocol identifier (same replace type code) and same originating address
0x48...0x5D	*Reserved* (22 values)
0x5E	*Obsolete:* the intention was to use this value to indicate that the message contains some EMS content. However, the use of this value was incompatible with several SMSC implementations. Consequently, this value was later made obsolete in order to avoid any interoperability problems
0x5F	*Return call message:* a return call short message is used to indicate to the user that a call can be established to the originator address. As for the protocol identifier 'replace short message type *n*', a return call message overwrites the one which had been previously stored in the mobile station (only if it has the same TP-PID and same originator address)
0x60...0x7B	*Reserved* (28 values)
0x7C	*ANSI-136 R-DATA:* messages with such a protocol identifier are transmitted to the (U)SIM
0x7D	*ME data download:* this protocol identifier indicates that the associated message is to be handled directly by the ME. Messages with this protocol identifier belong to class 1
0x7E	*ME de-personalization short message:* messages with this protocol identifier instruct the ME to de-personalize according to the content of the message. Messages associated with this protocol identifier belong to class 1
0x7F	*(U)SIM data download:* messages with such a protocol identifier are transmitted to the (U)SIM
0x80...0xBF	*Reserved* (64 values)
0xC0...0xFF	*SMSC specific use:* this range of protocol identifier values is left for service centre specific use

- Electrically Erasable Programmable Read Only Memory (EEPROM): this memory contains all parameters defined by the GSM/3GPP technical specifications and data manipulated by applications. Information saved in this memory is persistently stored even if the mobile station has been powered off.

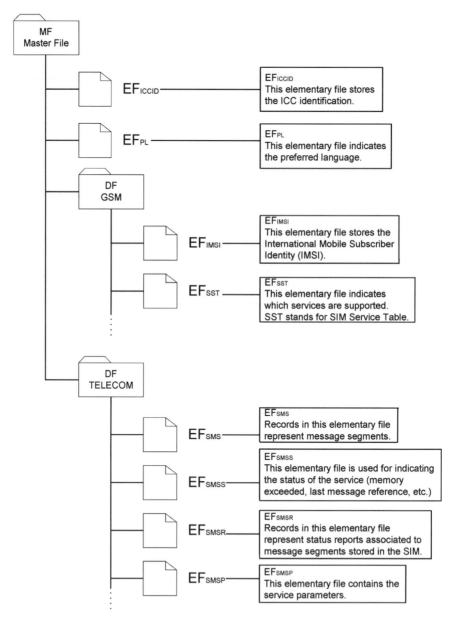

Figure 3.11   SIM storage

Table 3.6 SMS-related SIM elementary files

Elementary file	Description
EF$_{SMS}$	*Storage of a message segment:* this file can contain several records, each representing a message segment. A record contains the SMSC address followed by the message segment TPDU. Each record also indicates one of the following message status: • message received by the mobile station from network, message read • message received by the mobile station from network, message not read • message originated by the mobile station, message to be sent • message originated by the mobile station, message sent If the message is originated by the mobile station and has been sent, then the corresponding record also indicates whether or not a status report was requested. If a status report was requested, then the record indicates whether or not the status report has been received and optionally refers to the corresponding EF$_{SMSR}$ record
EF$_{SMSS}$	*Storage of the status of the service:* this elementary file indicates the status of the service. In this file, a flag is set if a message has been rejected because the SIM capacity for storing messages has exceeded. The file also contains the last message reference used for uniquely identifying messages sent by the mobile station
EF$_{SMSR}$	*Storage of a status report for a message segment:* records in this elementary file represent status reports corresponding to message segments stored in the SIM. Note that this elementary file is seldom supported in existing commercial implementations
EF$_{SMSP}$	*Storage of SMS parameters:* this elementary file is used for storing the default values for the following parameters: • recipient address • SMSC address • message protocol identifier • message data coding scheme • message validity period

• Random Access Memory (RAM): this memory contains data manipulated by applications. Information stored in this memory is lost when the mobile station is powered off.

The storage structure of a SIM card is based on a hierarchy of folders and files. The root folder is known as the Master File (MF), a normal folder is known as a Dedicated File (DF) and a file is known as an Elementary File (EF).

Four elementary files are used for storing SMS settings, user preferences and messages in the SIM. These SMS related elementary files are stored in the folder DF$_{TELECOM}$ as shown in Figure 3.11. The four elementary files, as defined in [3GPP-51.11], are described in Table 3.6.

The partial SIM structure for the storage of SMS user settings and messages is shown in Figure 3.11. Note that not all service parameters can be stored using the elementary file EF$_{SMSP}$. For instance, the default value for the Email gateway and the default setting for the request for a status report cannot be saved in standard SIM elementary files. To cope with this limitation, default values to be assigned to these parameters are sometimes stored directly in the ME memory (e.g. flash memory). This means that these parameters cannot be automatically retrieved if the subscriber inserts his/her SIM in another handset.

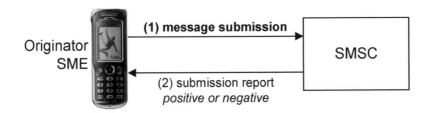

Figure 3.12   Message submission

3.9 Message Submission

In the context of SMS, the term *submission* refers to the transfer of a message segment from the originator SME to the serving SMSC (Figure 3.12).

Once the message segment has been successfully received by the serving SMSC, the originator SMSC queries the HLR in order to route forward the message segment towards the recipient SME.

At the transfer layer, a message segment is conveyed as part of a TPDU of type `SMS-SUBMIT`. The TPDU can contain the following parameters:

- Message type (`SMS-SUBMIT`)
- Request for rejecting duplicated messages
- Message validity period
- Request for reply path
- Request for a status report
- Message reference
- Address of the recipient SME
- Protocol identifier
- Data coding scheme
- User data header
- User data (with associated length).

Upon receipt of the message by the serving SMSC, the SMSC provides a submission report to the originator SME. Two types of reports can be provided: a *positive submission report* for a successful submission or a *negative submission report* for a failed submission. If the submission report is not received after a given period of time, then the originator SME concludes that the message submission has failed.

3.9.1 TPDU Layout

A the transfer layer, the TPDU of type `SMS-SUBMIT` has the layout shown in Figure 3.13. In this chapter, a specific graphical convention is used for the representation of TPDU layouts. Mandatory parameters are represented by grey-shaded boxes and optional parameters are represented by white boxes.

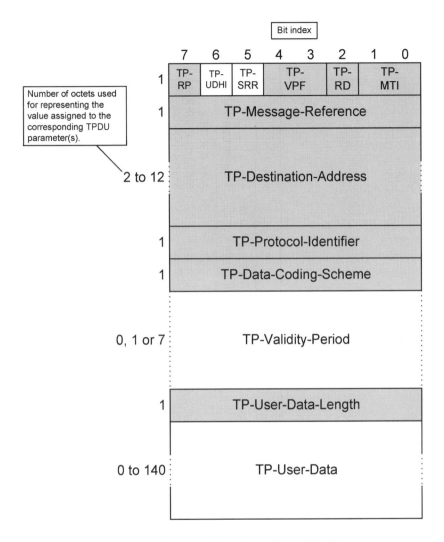

Figure 3.13 TPDU layout/type SMS-SUBMIT

3.9.2 TPDU Parameters

The TPDU of type SMS-SUBMIT contains the parameters listed in Table 3.7. The SMS standard defines three representations for numeric values and alphanumeric values assigned to TPDU parameters. These representations (integer representation, octet representation and semi-octet representation) are defined in Appendix B.

3.9.3 Rejection of Duplicates

It sometimes happens that a submission report gets lost. In this case, the originator SME has no means to determine if the message, for which the submission report has been lost, has been successfully submitted to the serving SMSC or not. If the originator SME re-transmits the

Table 3.7 Message submission/TPDU parameters[a]

Abbreviation	Reference	P	R	Description
TP-MTI	TP-Message-Type-Indicator	●	2 bits	Message type (bits 0 and 1 of first octet) Bit 1 — Bit 0 — Message type 0 — 0 — SMS-DELIVER 0 — 0 — SMS-DELIVER-REP 1 — 0 — SMS-STATUS-REP 1 — 0 — SMS-COMMAND **0 — 1 — SMS-SUBMIT** 0 — 1 — SMS-SUBMIT-REP
TP-RD	TP-Reject-Duplicates	●	1 bit	Indication of whether the SMSC shall accept or reject duplicated message segments. A message segment is a duplicate of an original message segment if it has the same TP-MR, TP-DA and TP-OA.
TP-VPF	TP-Validity-Period-Format	●	2 bits	Presence and format of the TP-VP field (bits 3 and 4 of first octet). Bit 4 — Bit 3 0 — 0 — TP-VP not present. 0 — 1 — TP-VP – enhanced format 1 — 0 — TP-VP – relative format 1 — 1 — TP-VP – absolute format

Field	P	R	Description
TP-RP	●	1 bit	Reply path (bit 7 of first octet). bit 7 at 0: reply-path is not set. bit 7 at 1: reply-path is set.
TP-UDHI	○	1 bit	Presence of a user data header in the user data part (bit 6 of first octet). bit 6 at 0: no user data header. bit 6 at 1: a user data header is present
TP-SRR	○	1 bit	Request for a status report. bit 5 at 0: no status report requested. bit 5 at 1: a status report is requested.
TP-MR	●	1 octet Integer rep.	Message segment reference number in the range 0...255 (decimal values)
TP-DA	●	2–12 octets	The destination address identifies the originating SME. The address format is defined in Section 3.9.6
TP-PID	●	1 octet	Protocol identifier as defined in Section 3.7.7
TP-DCS	●	1 octet	Data coding scheme as defined in Section 3.7
TP-VP	○	1 octet or 7 octets	Validity period identifies the time from when the message is no longer valid
TP-UDL	●	1 octet Integer rep.	The user data length is expressed in septets (GSM 7-bit default alphabet) or octets (UCS2 or 8-bit encoding)
TP-UD	○	TP-DCS dependent	The user data and user data header are defined in Section 3.15

[a] In the table field names, P stands for Provision and R for Representation. In the table body, ● stands for Mandatory whereas ○ stands for Optional.

message, while the first submission attempt was successful, the message will be transmitted twice to the message recipient. To avoid this situation, the originator SME has the ability to inform the SMSC that a previous submission was attempted for the message being submitted. In this case, if the SMSC detects that the previous submission had been successful, then the message is automatically discarded and not transmitted to the recipient SME. This ensures that the recipient SME receives the message only once.

Two parameters are associated with this ability to reject duplicates. The first parameter is the TP-Reject-Duplicates. This Boolean flag is set to 1 (true) for the new submission attempt and set to 0 (false) otherwise. The second parameter is the TP-Message-Reference which allows the SMSC to identify that the message, for which the TP-Reject-Duplicates is set, has already been successfully submitted.

3.9.4 Validity Period

The validity period of a message indicates the time after which the message content is no longer valid. The value assigned to the TP-Validity-Period parameter can take three different forms (as indicated by the field TP-Validity-Period -Format):

* *Relative format* (1 octet, integer representation). The value assigned to the TP-Validity-Period parameter in a relative format defines the length of the validity period starting from the time the message was received by the serving SMSC. The representation of the value assigned to the TP-Validity-Period parameter is as follows:

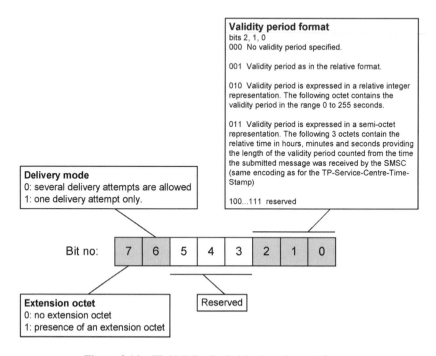

Figure 3.14   TP-Validity-Period in the enhanced format

TP-Validity-Period	Validity period value
0–143	(TP-Validity-Period) × 5 minutes
144–167	(12 hours + (TP-Validity-Period − 143) × 30 minutes)
168–196	(TP-Validity-Period − 166) × 1 day
197–255	(TP-Validity-Period − 192) × 1 week

- *Absolute format* (7 octets, semi-octet representation). The value assigned to the TP-Validity-Period parameter, in an absolute format, defines the date when the validity period terminates. The 7-octet value is an absolute time representation as defined in Section 3.9.5.
- *Enhanced format* (7 octets). The first octet of the 7-octet TP-Validity-Period, in the enhanced format, indicates how the following 6 octets are used. The presence of all octets is mandatory even if they are not all used. The first octet is structured as shown in Figure 3.14. Note that the value assigned to the TP-Validity-Period parameter is always expressed with either 1 octet or 7 octets, depending on its format. Any reserved or unused bit is set to 0.

3.9.5 Absolute Time Representation

Values assigned to several TPDU parameters represent an absolute time definition. This is the case for the TP-Validity-Period, TP-Service-Centre-Time-Stamp and the TP-Discharge-Time. For these parameters, the absolute time representation is decomposed into a sequence of time-related parameters, as described in Figure 3.15, which shows the absolute time 23rd December 01, 9:53:42 AM, GMT + 1 hour. Note that the time zone is expressed in quarters of an hour.

3.9.6 Destination Address

The value assigned to the TP-Destination-Address parameter represents the address of the recipient SME. This value is formatted as shown in the following section.

3.9.7 SME Addressing

Values assigned to the following parameters represent SME addresses:

- TP-Destination-Address
- TP-Recipient-Address
- TP-Originating-Address.

An SME address is decomposed into four sub-parameters:

- Address length (represents the number of useful semi-octets in the address value sub-parameter, the maximum length is 20 semi-octets)
- Type of number
- Numbering plan identification
- Address value.

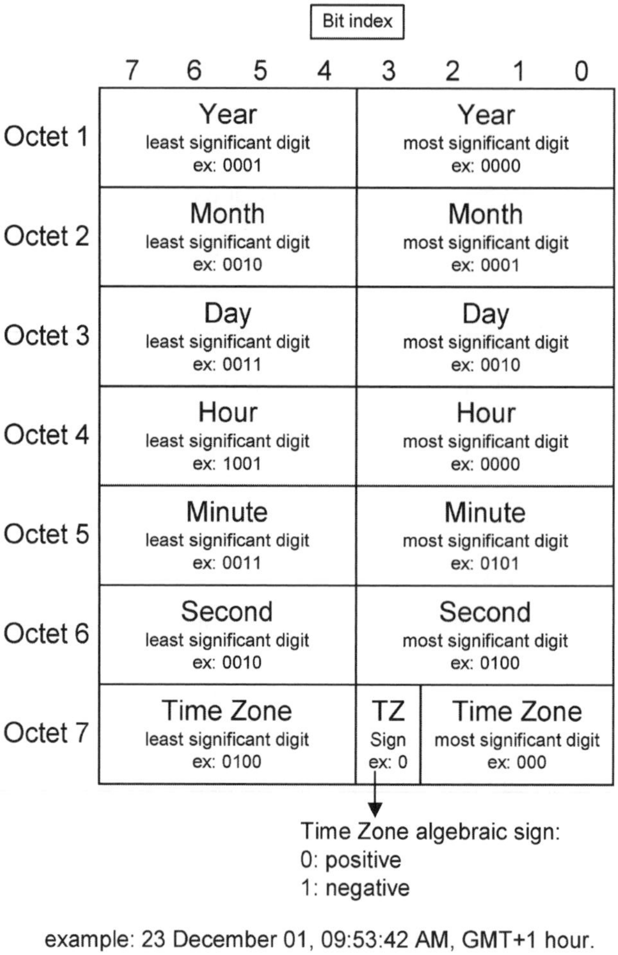

Bit index

	7 6 5 4	3	2 1 0
Octet 1	Year least significant digit ex: 0001		Year most significant digit ex: 0000
Octet 2	Month least significant digit ex: 0010		Month most significant digit ex: 0001
Octet 3	Day least significant digit ex: 0011		Day most significant digit ex: 0010
Octet 4	Hour least significant digit ex: 1001		Hour most significant digit ex: 0000
Octet 5	Minute least significant digit ex: 0011		Minute most significant digit ex: 0101
Octet 6	Second least significant digit ex: 0010		Second most significant digit ex: 0100
Octet 7	Time Zone least significant digit ex: 0100	TZ Sign ex: 0	Time Zone most significant digit ex: 000

Time Zone algebraic sign:
0: positive
1: negative

example: 23 December 01, 09:53:42 AM, GMT+1 hour.

Figure 3.15   Absolute time definition

The values assigned to the three addressing parameters are formatted as shown in Figure 3.16. The values listed in Table 3.8 can be assigned to the `type-of-number` sub-parameter. The values listed in Table 3.9 can be assigned to the `numbering-plan-identification` sub-parameter. At the transfer layer, SMS does not offer the group sending feature which consists of submitting one message addressed to several recipients by submitting one message only to the SMSC. However, this feature is sometimes emulated at the application layer at the cost of submitting one message to the SMSC for each recipient.

3.10 Message Submission Report

After the submission of a message segment from an originator SME to the serving SMSC, the SMSC acknowledges the submission by sending a report back to the originator SME. This report indicates the status of the submission. A positive submission report is sent back if the

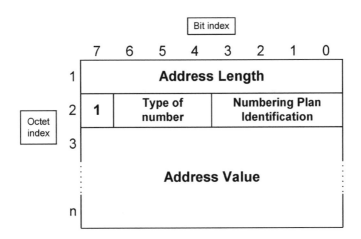

Figure 3.16   SMS addressing

Table 3.8   SMS addressing/type of number

Bit 6	Bit 5	Bit 4	Description
0	0	0	Unknown (address sub-parameters are organized according to the network dialling plan)
0	0	1	International number
0	1	0	National number
0	1	1	Network specific number (administration/service number specific to the serving network)
1	0	0	Subscriber number
1	0	1	Alphanumeric (coded in GSM 7-bit default alphabet)
1	1	0	Abbreviated number
1	1	1	Reserved

Table 3.9   SMS addressing/numbering plan identification[a]

Bit 3	Bit 2	Bit 1	Bit 0	Description
0	0	0	0	Unknown
0	0	0	1	ISDN/telephone numbering plan
0	0	1	1	Data numbering plan (X.121)
0	1	0	0	Telex numbering plan
0	1	0	1	SMSC specific plan (external SMEs attached to the service centre)
0	1	1	0	SMSC specific plan (external SMEs attached to the service centre)
1	0	0	0	National numbering plan
1	0	0	1	Private numbering plan
1	0	1	0	ERMES numbering plan

[a] All other values that can be assigned to this sub-parameter are reserved.

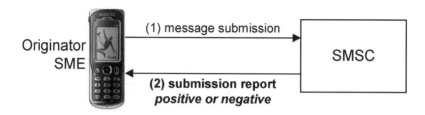

Figure 3.17 Submission report

submission was successful, otherwise a negative submission report is generated (Figure 3.17). Note that, with existing network configurations, submission reports are not always used. Instead, the acknowledgement of a message submission is often limited to a lower layer confirmation (relay layer). If provided, the submission report is conveyed in the form of a TPDU of type `SMS-SUBMIT-REPORT` at the transfer layer.

3.10.1 Positive Submission Report

The positive submission report can contain several of the following parameters:

- Message type (`SMS-SUBMIT-REPORT`)
- Parameter indicator (presence of protocol identifier, data coding scheme and user data length)
- Protocol identifier
- Data coding scheme
- Service centre time stamp (time at which the SMSC received the associated message)
- User data header
- User data (with associated length).

Upon receipt of the submission report, the originator SME may indicate to the subscriber whether or not the submission was successful. If the submission was not successful, then the originator may request the subscriber to modify the message in order to re-attempt the message submission.

After message submission, if the originator SME does not receive a submission report from the serving SMSC, then the SME can conclude that either:

- The message submission has failed, or
- The submission report has been lost.

In this situation, the originator SME may attempt another message submission. For this purpose, the originator SME can indicate for the new message submission that the message had already been submitted previously. This notice allows the serving SMSC to discard the newly submitted message if the first submission attempt was successful. The rejection of duplicate messages is described in Section 3.9.3.

The positive submission report TPDU has layout shown in Figure 3.18. The positive submission report TPDU can contain several of the parameters listed in Table 3.10.

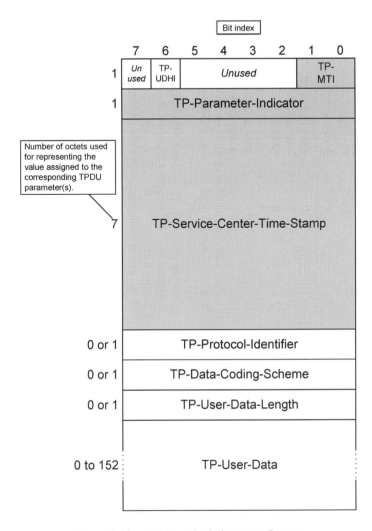

Figure 3.18 Positive submission report/layout

3.10.2 Negative Submission Report

Upon reception of a message segment, the serving SMSC may not be able to route forward the message (message badly formatted, SMSC busy, etc.). In this situation, the SMSC sends a negative submission report back to the originator SME. At the transfer layer, a negative submission report is transported as a TPDU of type SMS-SUBMIT-REPORT. This TPDU can contain several of the following parameters:

- Message type (SMS-SUBMIT-REPORT)
- Parameter indicator (presence of protocol identifier, data coding scheme and user data length)
- Protocol identifier
- Failure cause

Table 3.10 Positive submission report/TPDU parameters[a]

Abbreviation	Reference	P	R	Description
TP-MTI	TP-Message-Type-Indicator	●	2 bits	Message type (bits 0 and 1 of first octet)

Bit 1	Bit 0	Message type
0	0	SMS-DELIVER
0	0	SMS-DELIVER-REP
1	0	SMS-STATUS-REP
1	0	SMS-COMMAND
0	1	SMS-SUBMIT
0	1	**SMS-SUBMIT-REP**

Abbreviation	Reference	P	R	Description
TP-UDHI	TP-User-Data-Header-Indicator	○	1 bit	Presence of a user data header in the user data part (bit 6 of first octet). bit 6 at 0: no user data header. bit 6 at 1: a user data header is present
TP-PI	TP-Parameter-Indicator	●	1 octet	Presence of TP-PID, TP-DCS and TP-UDL fields. The format of the TP-Parameter-Indicator is defined in Section 3.10.3
TP-SCTS	TP-Service-Centre-Time-Stamp	●	7 octets	Service centre time stamp represents the time the SMSC received the message. The structure of the SMSC time stamp is defined in Section 3.10.3
TP-PID	TP-Protocol-Identifier	○	1 octet	Protocol identifier as defined in Section 3.7.7
TP-DCS	TP-Data-Coding-Scheme	○	1 octet	Data coding scheme as defined in Section 3.7
TP-UDL	TP-User-Data-Length	○	1 octet Integer rep.	The user data length is expressed in septets (GSM 7-bit default alphabet) or octets (UCS2 or 8-bit encoding)
TP-UD	TP-User-Data	○	TP-DCS dependent	The user data and user data header are defined in Section 3.15

[a] In the table field names, P stands for Provision and R for Representation. In the table body, ● stands for Mandatory whereas ○ stands for Optional.

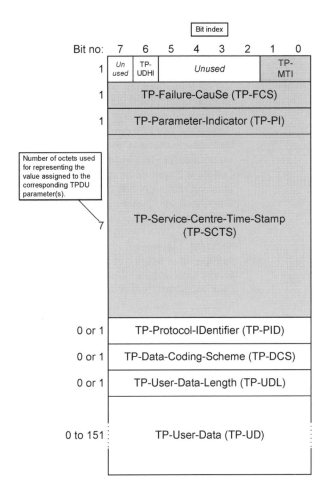

Figure 3.19   Negative submission report/layout

- Data coding scheme
- Service centre time stamp (time at which the SMSC received the associated message)
- User data header
- User data (with associated length).

The negative submission report TPDU has the layout shown in Figure 3.19.

The different reasons for which the serving SMSC can acknowledge a message submission negatively are described in Table 3.11 (corresponding reason identification to be assigned to the TP-Failure-Cause parameter. The negative submission report TPDU can contain several of the parameters listed in Table 3.12.

3.10.3 Parameter Indicator

The parameter indicator informs whether or not the following parameters are present in the

Table 3.11 Negative submission report/failure causes

Reason id (hex)	Description
0x80	Telematic interworking not supported
0x81	Short message Type 0 not supported
0x82	The short message cannot be replaced
0x8F	Unspecified TP-PID error
0x90	Data coding scheme (alphabet) not supported
0x9F	Unspecified TP-DCS error
0xA0	Command cannot be executed
0xA1	Command not supported
0xAF	Unspecified TP-Command error
0xB0	TPDU not supported
0xC0	SMSC busy
0xC1	No SMSC subscription
0xC2	SMSC system failure
0xC3	Invalid SME address
0xC4	Destination SME barred
0xC5	Message rejected – duplicate message
0xC6	TP-Validity-Period-Format not supported
0xC7	TP-Validity-Period not supported

TPDU: TP-Protocol-Identifier, TP-Data-Coding-Scheme and TP-User-Data-Length. The structure of this parameter is shown in Figure 3.20. If the TP-User-Data-Length is not present in the TPDU, then the TP-User-Data is not present either.

3.10.4 Service Centre Time Stamp

For a positive or negative report, the service centre time stamp parameter (TP-Service-Centre-Time-Stamp) indicates the time at which the associated message was received by the serving SMSC. The value assigned to this parameter is formatted in an absolute time representation as defined in Section 3.9.5.

3.11 Message Delivery

In the context of SMS, the term *delivery* refers to the transfer of a message segment from the serving SMSC to the recipient SME (Figure 3.21). If the recipient SME is not available for the delivery of the message segment, then the SMSC stores the message temporarily. The SMSC attempts to deliver the message until a delivery report is received from the recipient SME or until the message validity period expires. Upon receipt of a negative or positive report or upon message deletion, the serving SMSC may send a status report back to the originator SME (step 3). The status report is generated only if the originator SME requested it during message submission.

Table 3.12 Negative submission report/TPDU parameters[a]

Abbreviation	Reference	P	R	Description
TP-MTI	TP-Message-Type-Indicator	●	2 bits	Message type (bits 0 and 1 of first octet) Bit 1 Bit 0 Message type 0 0 SMS-DELIVER 0 0 SMS-DELIVER-REP 1 0 SMS-STATUS-REP 1 0 SMS-COMMAND 0 1 SMS-SUBMIT **0 1 SMS-SUBMIT-REP**
TP-UDHI	TP-User-Data-Header-Indicator	○	1 bit	Presence of a user data header in the user data part (bit 6 of first octet) Bit 6 at 0: no user data header Bit 6 at 1: a user data header is present
TP-FCS	TP-Failure-Cause	●	1 octet	Reason for submission failure (see table above). Values in the ranges 0x00...0x7F, 0x83...0x8E, 0x92...0x9E, 0xA2...0xAE, 0xB1...0xBF, 0xC8...0xCF, 0xD6...0xDF are reserved. The range 0xE0...0xFE contains values specific to an application. The value 0xFF is used for an unspecified cause
TP-PI	TP-Parameter-Indicator	●	1 octet	Presence of TP-PID, TP-DCS and TP-UDL fields. The format of the TP-Parameter-Indicator is defined in Section 3.10.3
TP-SCTS	TP-Service-Centre-Time-Stamp	●	7 octets	Service centre time stamp represents the time the SMSC received the message. The structure of the SMSC time stamp is defined in Section 3.10.3
TP-PID	TP-Protocol-Identifier	○	1 octet	Protocol identifier as defined in Section 3.7.7
TP-DCS	TP-Data-Coding-Scheme	○	1 octet	Data coding scheme as defined in Section 3.7
TP-UDL	TP-User-Data-Length	○	1 octet	The user data length is expressed in septets (GSM 7-bit default alphabet) or octets (UCS2 or 8-bit encoding)
TP-UD	TP-User-Data	○	TP-DCS dependent	The user data and user data header are defined in Section 3.15

[a] In the table field names, P stands for Provision and R for Representation. In the table body, ● stands for Mandatory whereas ○ stands for Optional.

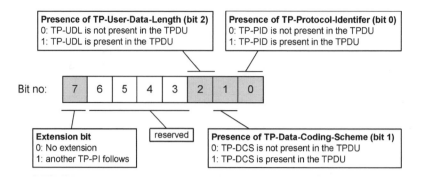

Figure 3.20   TP-Parameter-indicator structure

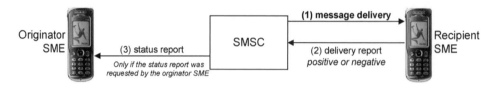

Figure 3.21   Message delivery

At the transfer layer, the message is delivered in the form of a TPDU of type SMS-DELIVER. The TPDU can contain several of the following parameters:

- Message type (SMS-DELIVER)
- Indication that there are more messages to be received
- Request for reply path
- Request for a status report
- Address of the originator SME
- Protocol identifier
- Data coding scheme
- Service centre time stamp (time at which the SMSC received the message)
- User data header
- User data (with associated length).

Upon receipt of the message, the recipient SME provides a delivery report back to the serving SMSC. The delivery report indicates the status of the message delivery. Two types of reports can be provided: a *positive delivery report* for a successful message delivery or a *negative delivery report* for a failed delivery.

If the delivery report is not received after a given period of time, then the serving SMSC concludes that the message delivery has failed and may try to retransmit the message later.

3.11.1 TPDU Layout

At the transfer layer, the message delivery TPDU has the layout shown in Figure 3.22.

3.11.2 TPDU Parameters

The message delivery TPDU can contain several of the parameters listed in Table 3.13.

3.11.3 Status Report Indicator

The status report indicator (TP-Status-Report-Indicator parameter) indicates whether or not the originator of the message requested a status report. The following values can be assigned to this 1-bit parameter:

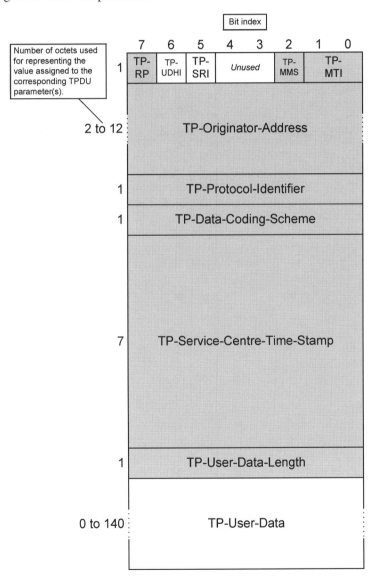

Figure 3.22 Message delivery/layout

Table 3.13 Message delivery/TPDU parameters[a]

Abbreviation	Reference	P	R	Description
TP-MTI	TP-Message-Type-Indicator	●	2 bits	Message type (bits 0 and 1 of first octet)
				Bit 1 Bit 0 Message type
				0 0 **SMS-DELIVER**
				0 0 SMS-DELIVER-REP
				0 0 SMS-DELIVER-REP
				1 0 SMS-STATUS-REP
				1 0 SMS-COMMAND
				0 1 SMS-SUBMIT
				0 1 SMS-SUBMIT-REP
TP-MMS	TP-More-Messages-to-Send	●	1 bit	Presence of more messages to send at the SMSC (bit 2 of first octet)
				Bit 2 at 0: more messages are waiting
				Bit 2 at 1: no more messages are waiting
TP-RP	TP-Reply-Path	●	1 bit	Reply path (bit 7 of first octet)
				Bit 7 at 0: reply-path is not set
				Bit 7 at 1: reply-path is set
TP-UDHI	TP-User-Data-Header-Indicator	○	1 bit	Presence of a user data header in the user data part (bit 6 of first octet)

Field	Name	P	R	Description
TP-SRI	TP-Status-Report-Indication	○	1 bit	Bit 6 at 0: no user data header. Bit 6 at 1: a user data header is present. Indication that a status report is requested (bit 5 of the first octet). Bit 5 at 0: no status report requested. Bit 5 at 1: a status report is requested
TP-OA	TP-Originating-Address	●	2 to 12 octets	The originating address identifies the originating SME. The address format is defined in Section 3.9.7
TP-PID	TP-Protocol-Identifier	●	1 octet	Protocol identifier as defined in Section 3.7.7
TP-DCS	TP-Data-Coding-Scheme	●	1 octet	Data coding scheme as defined in Section 3.7
TP-SCTS	TP-Service-Centre-Time-Stamp	●	7 octets	Service centre time stamp represents the time the SMSC received the message. The structure of the SMSC time stamp is defined in Section 3.10.3
TP-UDL	TP-User-Data-Length	●	1 octet Integer rep.	The user data length is expressed in septets (GSM 7-bit default alphabet) or octets (UCS2 or 8-bit encoding)
TP-UD	TP-User-Data	○	TP-DCS dependent	The user data and user data header are defined in Section 3.15

[a] In the table field names, P stands for Provision and R for Representation. In the table body, ● stands for Mandatory whereas ○ stands for Optional.

- Value 0: no status report requested
- Value 1: a status report is requested.

3.11.4 Service Centre Time Stamp

The service centre time stamp (`TP-Service-Centre-Time-Stamp` parameter) indicates the time at which the message has been received by the serving SMSC. The value assigned to this parameter is formatted in a time absolute representation as defined in Section 3.9.5.

3.12 Message Delivery Report

Upon delivery of a message from the serving SMSC to the recipient SME, the SME acknowledges the message delivery by sending back a delivery report to the serving SMSC. A positive delivery report is sent back if the delivery was successful, otherwise a negative delivery report is generated. The reception of the delivery report is necessary for the serving SMSC, which will then stop attempting to deliver the message to the recipient SME. If the originator SME requested a status report to be generated, then the serving SMSC generates the status report according to the delivery report received from the recipient SME (Figure 3.23).

Note that with existing network configurations, delivery reports are not always used. Instead, the acknowledgement of a message delivery is often limited to a lower layer confirmation (relay layer).

If provided, the delivery report is conveyed in the form of a TPDU of type SMS-DELIVER-REPORT at the transfer layer.

3.12.1 Positive Delivery Report

The positive delivery report TPDU can contain several of the following parameters:

- Message type (`SMS-DELIVER-REPORT`)
- Parameter indicator (presence of protocol identifier, data coding scheme and user data length)
- Protocol identifier
- Data coding scheme
- User data header
- User data (with associated length).

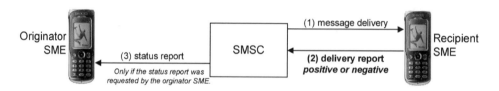

Figure 3.23   Delivery report

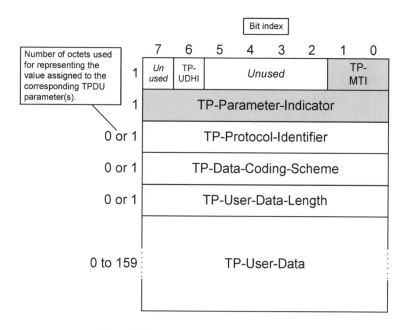

Figure 3.24 Positive delivery report/layout

At the transfer layer, a positive delivery report TPDU has the layout shown in Figure 3.24. A positive delivery report TPDU contains the parameters listed in Table 3.14.

3.12.2 Negative Delivery Report

In some situations, the recipient SME is not able to handle the message correctly (message badly formatted, storage capacity exceeded, etc.). In order to inform the serving SMSC that the message cannot be handled, the receiving SME generates a negative delivery report. At the transfer layer, the negative delivery report is transported in the form of a TPDU of type SMS-DELIVER-REPORT. The TPDU can contain several of the following parameters:

- Message type (SMS-DELIVER-REPORT)
- Parameter indicator (presence of protocol identifier, data coding scheme and user data length)
- Protocol identifier
- Failure cause
- Data coding scheme
- User data header
- User data (with associated length).

At the transfer layer, a negative delivery report TPDU has the layout shown in Figure 3.25. The different reasons for which the recipient SME can generate a negative delivery report are listed in Table 3.15 (corresponding reason identification to be assigned to the TP-Failure-Cause parameter). The parameters listed in Table 3.16 are included in the TPDU of a negative delivery report.

Table 3.14 Positive delivery report/TPDU parameters[a]

Abbreviation	Reference	P	R	Description
TP-MTI	TP-Message-Type-Indicator	●	2 bits	Message type (bits 0 and 1 of first octet) <table><tr><td>Bit 1</td><td>Bit 0</td><td>Message type</td></tr><tr><td>0</td><td>0</td><td>SMS-DELIVER</td></tr><tr><td>**0**</td><td>**0**</td><td>**SMS-DELIVER-REP**</td></tr><tr><td>1</td><td>0</td><td>SMS-STATUS-REP</td></tr><tr><td>1</td><td>0</td><td>SMS-COMMAND</td></tr><tr><td>0</td><td>1</td><td>SMS-SUBMIT</td></tr><tr><td>0</td><td>1</td><td>SMS-SUBMIT-REP</td></tr></table>
TP-UDHI	TP-User-Data-Header-Indicator	○	1 bit	Presence of a user data header in the user data part (bit 6 of first octet) Bit 6 at 0: no user data header Bit 6 at 1: a user data header is present
TP-PI	TP-Parameter-Indicator	●	1 octet	Presence of TP-PID, TP-DCS and TP-UDL fields. The format of the TP-Parameter-Indicator is defined in Section 3.10.3
TP-PID	TP-Protocol-Identifier	○	1 octet	Protocol identifier as defined in Section 3.7.7
TP-DCS	TP-Data-Coding-Scheme	○	1 octet	Data coding scheme as defined in Section 3.7.
TP-UDL	TP-User-Data-Length	○	1 octet Integer rep.	The user data length is expressed in septets (GSM 7-bit default alphabet) or octets (UCS2 or 8-bit encoding)
TP-UD	TP-User-Data	○	TP-DCS dependent	The user data and user data header are defined in Section 3.15

[a] In the table field names, P stands for Provision and R for Representation. In the table body, ● stands for Mandatory whereas ○ stands for Optional.

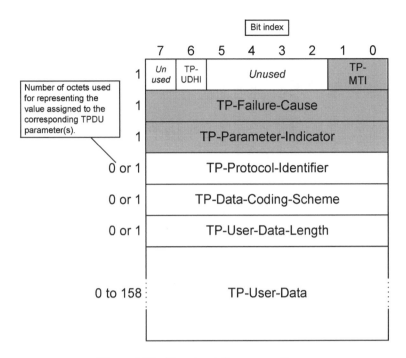

Figure 3.25   Negative delivery report/layout

Table 3.15   Negative delivery report/failure causes[a]

Reason id. (hex)	Description
0x81	Short message type 0 not supported
0x82	The short message cannot be replaced
0x8F	Unspecified TP-PID error
0x91	Message class not supported
0x9F	Unspecified TP-DCS error
0xB0	TPDU not supported
0xD0	(U)SIM SMS storage full
0xD1	No SMS storage capability in (U)SIM
0xD2	Error in MS
0xD3	Memory capacity exceeded
0xD4	(U)SIM application toolkit busy
0xD5	(U)SIM data download error
0xE0...0xFE	Application-specific errors
0xFF	Unspecified error cause

[a] Other values are reserved.

Table 3.16 Negative delivery report/TPDU parameters[a]

Abbreviation	Reference	P	R	Description
TP-MTI	TP-Message-Type-Indicator	●	2 bits	Message type (bits 0 and 1 of first octet)
				Bit 1 Bit 0 Message type
				0 0 SMS-DELIVER
				0 **0** **SMS-DELIVER-REP**
				1 0 SMS-STATUS-REP
				1 0 SMS-COMMAND
				0 1 SMS-SUBMIT
				0 1 SMS-SUBMIT-REP
TP-UDHI	TP-User-Data-Header-Indicator	○	1 bit	Presence of a user data header in the user data part (bit 6 of first octet)
				Bit 6 at 0: no user data header
				Bit 6 at 1: a user data header is present
TP-FCS	TP-Failure-Cause	●	1 octet	Reason for delivery failure (see table above)
TP-PI	TP-Parameter-Indicator	●	1 octet	Presence of TP-PID, TP-DCS and TP-UDL fields. The format of the TP-Parameter-Indicator is defined in Section 3.10.3
TP-PID	TP-Protocol-Identifier	○	1 octet	Protocol identifier as defined in Section 3.7.7
TP-DCS	TP-Data-Coding-Scheme	○	1 octet	Data coding scheme as defined in Section 3.7
TP-UDL	TP-User-Data-Length	○	1 octet Integer rep.	The user data length is expressed in septets (GSM 7-bit default alphabet) or octets (UCS2 or 8-bit encoding)
TP-UD	TP-User-Data	○	TP-DCS dependent	The user data and user data header are defined in Section 3.15

[a] In the table field names, P stands for Provision and R for Representation. In the table body, ● stands for Mandatory whereas ○ stands for Optional.

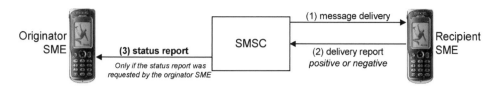

Figure 3.26   Status report

3.13 Status Report

Upon delivery of a message segment to a recipient SME, the serving SMSC may generate a status report and transfer it back to the originator SME. The status report is sent only if the originator SME requested it during the message submission. The serving SMSC generates the status report when the associated delivery report has been received from the recipient SME or when the message is discarded by the SMSC without delivery (e.g. validity period has expired). Figure 3.26 shows the three steps leading to the delivery of a status report.

A status report may also be generated upon receipt or execution of a message command (see next section). At the transfer layer, the status report is transported in the form of a TPDU of type `SMS-STATUS-REPORT`. The TPDU can contain several of the following parameters:

- Message type (`SMS-STATUS-REPORT`)
- Parameter indicator (presence of protocol identifier, data coding scheme and user data length)
- Indication that there are more messages to be received
- Protocol identifier
- Status report qualifier
- Delivery status
- Discharge time
- Message reference (from original message)
- Recipient address
- Service centre time stamp (from original message)
- Data coding scheme
- User data header
- User data (with associated length).

Upon receipt of the status report, the originator SME is able to identify the original message, to which the status report refers, by checking that the status report recipient address is equal to the destination address of the original message and that the status report message reference is equal to the original message reference. In addition, the originator SME may also check that the status report service centre time stamp corresponds to the service centre time stamp specified in the submit report of the original message. Note that the service centre time stamp is not always provided by the serving SMSC for a message submission. If the original message is stored in the SIM, then the status of the corresponding EF$_{SMS}$ file is updated (message originated by the mobile station, message sent, status report received). Additionally, a record may be created in the EF$_{SMSR}$ file containing the status report.

If the original message has been deleted, then the originator SME may discard the corresponding status report or may alternatively present the message to the subscriber as a normal message (as if it had been delivered as a TPDU of type SMS-DELIVER).

The status of the delivery, conveyed by the status report, can take one of the values listed in Table 3.17 (to be assigned to the TP-Status parameter of the status report TPDU).

Table 3.17 Status report codes[a]

Status id. (hex)	Description
Short message transaction completed	
0x00	Message successfully received by the recipient SME
0x01	Message forwarded by the SMSC to the SME but the SMSC is unable to confirm delivery
0x02	Message has been replaced by the SMSC
0x10...0x1F	Values specific to each SMSC
Temporary error, SC still trying to transfer the short message	
0x20	Congestion
0x21	SME busy
0x22	No response from SME
0x23	Service rejected
0x24	Quality of service not available
0x30...0x3F	Values specific to each SMSC
Permanent error, SMSC is no longer making delivery attempts	
0x40	Remote procedure error
0x41	Incompatible destination
0x42	Connection rejected by the SME
0x43	Not obtainable
0x44	Quality of service not available
0x45	No interworking available
0x46	Message validity period expired
0x47	Message deleted by originating SME
0x48	Message deleted by SMSC administration
0x49	Message does not exist (or the SMSC has no knowledge of the associated message)
0x50...0x5F	Values specific to each SMSC
Temporary error, SMSC is no longer making delivery attempts	
0x60	Congestion
0x61	SME busy
0x62	No response from SME
0x63	Service rejected
0x64	Quality of service not available
0x65	Error in SME
0x70...0x7F	Values specific to each SMSC

[a] Unused status index values are reserved.

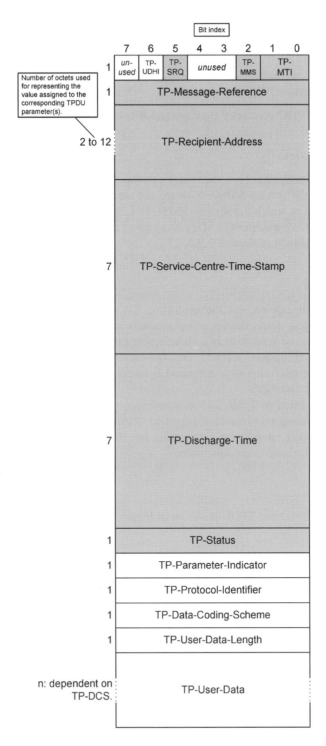

Figure 3.27 Status report/layout

3.13.1 TPDU Layout

The layout of the status report TPDU is shown in Figure 3.27.

3.13.2 TPDU Parameters

The status report TPDU contains the parameters listed in Table 3.18.

3.13.3 Discharge Time

If the associated message has been successfully delivered, then the discharge time represents the time at which the serving SMSC delivered the message. Otherwise, the discharge time represents the time at which the serving SMSC last attempted to deliver the message. The value assigned to this parameter (`TP-Discharge-Time`) is formatted as an absolute time representation as defined in Section 3.9.5.

3.14 Command

An originator SME can request the originator SMSC to execute a command. This is performed by sending a specific *command* message to the serving SMSC. The command can be a request for the generation of a status report by the serving SMSC, the deletion of a previously submitted message segment, etc. The command is usually not submitted from a mobile station. The execution of a command is usually requested by an application server (external SME) (Figure 3.28).

The commands listed in Table 3.19 can be executed by the serving SMSC (command identification to be assigned to the `TP-Command-Type` parameter of the command TPDU). The submission report (step 2) for a command has the same characteristics as those of a submission report for a message submission.

3.14.1 TPDU Layout

The layout of the command TPDU is shown in Figure 3.29.

3.14.2 TPDU Parameters

The command TPDU can contain several of the parameters listed in Table 3.20.

3.15 User Data Header and User Data

As shown in previous sections, the `TP-User-Data` (`TP-UD`) contains the text part of a message segment. Optionally, the `TP-UD` may also contain an 8-bit aligned User-Data-Header (UDH). The User-Data-Header is composed of a User-Data-Header-Length followed by a sequence of information elements. Information elements are used for the following purposes:

Table 3.18 Status report/TPDU parameters[a]

Abbreviation	Reference	P	R	Description
TP-MTI	TP-Message-Type-Indicator	●	2 bits	Message type (bits 0 and 1 of first octet) Bit 1 Bit 0 Message type 0 0 SMS-DELIVER 0 0 SMS-DELIVER-REP **1 0 SMS-STATUS-REP** 1 0 SMS-COMMAND 0 1 SMS-SUBMIT 0 1 SMS-SUBMIT-REP
TP-MMS	TP-More-Messages-to-Send	●	1 bit	Presence of more messages to send at the SMSC (bit 2 of first octet) Bit 2 at 0: more messages are waiting Bit 2 at 1: no more messages are waiting
TP-SRQ	TP-Status-Report-Qualifier	●	1 bit	Indication whether the associated TPDU was a submitted SM or a command Bit 5 at 0: associated TPDU was a submitted SM Bit 5 at 1: associated TPDU was a command
TP-UDHI	TP-User-Data-Header-Indicator	○	1 bit	Presence of a user data header in the user data part (bit 6 of first octet) Bit 6 at 0: no user data header Bit 6 at 1: a user data header is present
TP-MR	TP-Message-Reference	●	1 octet Integer rep.	Message segment reference number in the range 0...255
TP-RA	TP-Recipient-Address	●	2-12 octets	The recipient address identifies the SME which was identified by the TP-Destination-Address of the associated submitted message. The value assigned to this parameter is formatted as defined in Section 3.9.7
TP-SCTS	TP-Service-Centre-Time-Stamp	●	7 octets	Service centre time stamp represents the time the SMSC received the message. The structure of the SMSC time stamp is defined below
TP-DT	TP-Discharge-Time	●	7 octets	See description in Section 3.13.3

Table 3.18 (*continued*)

Abbreviation	Reference	P	R	Description
TP-ST	TP-Status	●	1 octet	Status of the associated message or command as defined in table. Values in the ranges 0x03...0x0F, 0x26...0x2F, 0x4A...0x4F, 0x66...0x6F, 0x80...0xFF are reserved
TP-PI	TP-Parameter-Indicator[b]	○	1 octet	Presence of TP-PID, TP-DCS and TP-UDL fields. The format of the TP-Parameter-Indicator is defined in Section 3.10.3
TP-PID	TP-Protocol-Identifier	○	1 octet	Protocol identifier as defined in Section 3.7.7
TP-DCS	TP-Data-Coding-Scheme	○	1 octet	Data coding scheme as defined in Section 3.7
TP-UDL	TP-User-Data-Length	○	Integer rep.	The user data length is expressed in septets (GSM 7-bit default alphabet) or octets (UCS2 or 8-bit encoding)
TP-UD	TP-User-Data	○	TP-DCS dependent	The user data and user data header are defined in Section 3.15

[a] In the table field names, P stands for Provision and R for Representation. In the table body, ● stands for Mandatory whereas ○ stands for Optional.
[b] Mandatory if any of the optional parameters following TP-Parameter-Indicator is present, otherwise optional.

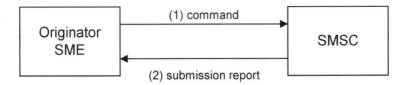

Figure 3.28   Command and submission report

Table 3.19   Command identifiers

Cmd. id. (hex)	Description
0x00	Enquiry relating to a previously submitted message. With this command, a request is made for the generation of a status report
0x01	Cancel status report request relating to a previously submitted message. No status report is to be generated for this command execution
0x02	Delete a previously submitted message. No status report is to be generated for this command execution
0x03	Enable status report request relating to a previously submitted message. No status report is to be generated for this command execution
0xE0...0xFF	Values specific for each SMSC

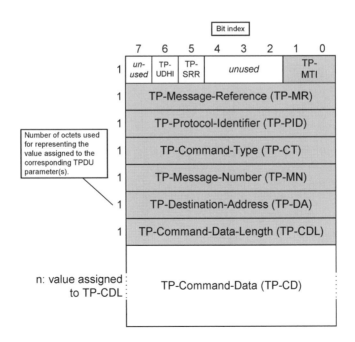

Figure 3.29   Command/layout

Table 3.20 SMS command/TPDU parameters[a]

Abbreviation	Reference	P	R	Description
TP-MTI	TP-Message-Type-Indicator	●	2 bits	Message type (bits 0 and 1 of first octet)
				Bit 1 Bit 0 Message type
				0 0 SMS-DELIVER
				0 0 SMS-DELIVER-REP
				1 0 SMS-STATUS-REP
				1 0 SMS-COMMAND
				0 1 SMS-SUBMIT
				0 1 SMS-SUBMIT-REP
TP-SRR	TP-Status-Report-Request	○	1 bit	Request for a status report Bit 5 at 0: no status report requested Bit 5 at 1: a status report is requested
TP-UDHI	TP-User-Data-Header-Indicator	○	1 bit	Presence of a user data header in the user data part (bit 6 of first octet) Bit 6 at 0: no user data header Bit 6 at 1: a user data header is present
TP-MR	TP-Message-Reference	●	1 octet Integer rep.	Message segment reference number in the range 0...255
TP-PID	TP-Protocol-Identifier	●	1 octet	Protocol identifier as defined in Section 3.7.7
TP-CT	TP-Command-Type	●	1 octet	Type of command to be executed as defined in table. Values in range 0x04...0x1F are reserved
TP-MN	TP-Message-Number	●	1 octet	Message number of the message on which the command is to be executed
TP-DA	TP-Destination-Address	●	2–12 octets	The destination address identifies the originating SME. The address format is defined in Section 3.9.7
TP-CDL	TP-Command-Data-Length	●	1 octet Integer rep.	Length of the TP-Command-Data in octets
TP-CD	TP-Command-Data	○	Value of TP-CDL (in octets)	This parameter contains the user data

[a] In the table field names, P stands for Provision and R for Representation. In the table body, ● stands for Mandatory whereas ○ stands for Optional.

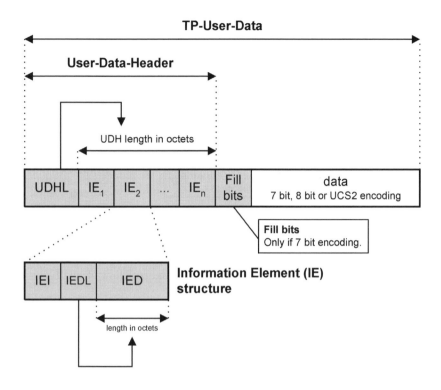

Figure 3.30 TP-User-Data structure

- *SMS control:* in this category, information elements contain some SMS control instructions such as concatenation information, application ports, SMSC control parameters, etc.
- *Basic and extended EMS objects:* in this category, information elements contain the definition of EMS objects such as melodies, pictures, animations, etc.

The structure of the TP-User-Data parameter is shown in Figure 3.30. The first octet of the User-Data-Header, called the User-Data-Header-Length (UDHL), indicates the length of the User Data Header. If the text is 7-bit encoded, then fill bits may be necessary between the User-Data-Header and the remaining part of the TP-User-Data. These fill bits ensure that the text, which follows the User Data Header, always starts on a septet boundary. This is important to allow the oldest SMEs, that do not support the concept of User-Data-Header, to still be able to interpret the text part of the message.

3.15.1 Information Elements

An *Information Element* (IE), if recognized, is decoded and interpreted by the recipient SME. An IE can be an EMS element such as a command for text formatting, a melody, an animation or a picture. It can also be an information element for SMS control. The structure of an IE is designed to allow easy introduction of new IEs in the standard along with maintaining

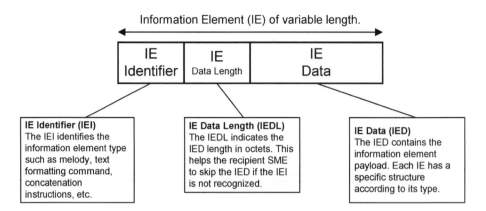

Figure 3.31   Information element structure

forward compatibility between heterogeneous SMEs. The structure of an IE is shown in Figure 3.31.

The *IE Identifier* (IEI) indicates the information element type (see Table 3.21). The *IE Data Length* (IEDL) indicates the length of the *IE Data* (IED) in octets. The IE data represents the information element payload. For an IE representing a bitmap picture, the payload is the picture position in the text along with the picture dimensions and picture bitmap. For an IE representing concatenation instructions, the payload consists of the concatenation segment index, message reference number and number of segments in the concatenated message.

Note that one single information element needs to fit into one message segment. An information element cannot be segmented and spread over several message segments.

If the receiving SME does not recognize a particular IE identifier, then the SME interprets the IEDL and skips the entire IE and continues the processing of the immediately following IE. For instance, this can occur if an SME was designed before the inclusion of an IE in the standard. Regarding this method of handling IEs, it has to be noted that older SMEs may completely ignore certain IEs. This can become the source of awkward message presentation or application misbehaviour.

SMS control information elements are presented in this chapter. EMS information elements are presented in Chapters 4 and 5.

An IE is *segment repeatable* if several IEs with the same identifier can appear more than once in a message segment. Additionally, an IE is *message repeatable* if several IEs with the same identifier can appear more than once in a message.

In the situation where several IEs with identical identifiers are present in a message segment and are not segment repeatable, then the last occurrence of the IE is used. If two IEs contain mutually exclusive instructions (e.g. an 8-bit port address and a 16-bit port address), then the last occurring IE is used.

The values listed in Table 3.21 can be assigned to the 1-octet IE identifier parameter. For each information element, a table summarizes the structure of the information element, as described in Figure 3.32.

Table 3.21 Information element identifiers

IEI value	Description	From release[a]	Type[b]	Segment repeat	Message repeat	See Section
0x00	Concatenated short message, 8 bit reference number	99	SMS	No	Yes	3.15.2
0x01	Special SMS message indication	99	SMS	Yes	Yes	3.15.3
0x02...0x03	Reserved	N/A	N/A	N/A	N/A	–
0x04	Application port addressing scheme 8 bit address	99	SMS	No	Yes	3.15.4
0x05	Application port addressing scheme 16 bit address	99	SMS	No	Yes	3.15.4
0x06	service centre control parameters	99	SMS	No	Yes	3.15.5
0x07	User-Data-Header source indicator	99	SMS	Yes	Yes	3.15.6
0x08	Concatenated short message 16 bit reference number	99	SMS	No	Yes	3.15.2
0x09	Wireless control message protocol	99	SMS	No	Yes	3.15.8
0x0A	Text formatting	99	EMS 1	Yes	Yes	4.3 and 5.20
0x0B	Predefined sound	99	EMS 1	Yes	Yes	4.5.1
0x0C	User defined sound[c]	99	EMS 1	Yes	Yes	4.5.2
0x0D	Predefined animation	99	EMS 1	Yes	Yes	4.6.1
0x0E	Large animation	99	EMS 1	Yes	Yes	4.6.2
0x0F	Small animation	99	EMS 1	Yes	Yes	4.6.2
0x10	Large picture	99	EMS 1	Yes	Yes	4.4.1
0x11	Small picture	99	EMS 1	Yes	Yes	4.4.2
0x12	Variable-size picture	99	EMS 1	Yes	Yes	4.4.3
0x13	User prompt indicator	4	EMS 1	Yes	Yes	4.7
0x14	Extended object	5	EMS 2	Yes	Yes	5.3
0x15	Reused extended object	5	EMS 2	Yes	Yes	5.4
0x16	Compression control	5	EMS 2	Yes	Yes	5.5
0x17	Object distribution indicator	5	EMS 1	Yes	Yes	4.8
0x18	Configurable-size WVG object	5	EMS 2	Yes	Yes	5.19.2
0x19	Character size WVG object	5	EMS 2	Yes	Yes	5.19.1
0x1A	Extended object data request cmd.	5	EMS 2	No	No	5.22
0x1B...0x1F	Reserved for future EMS features	N/A	N/A	N/A	N/A	–
0x20	RFC 822 Email header	99	SMS	No	Yes	3.20.2
0x21	Hyperlink format element	5	SMS	Yes	Yes	5.21
0x22	Reply address element	5	SMS	No	Yes	3.15.9
0x23...0x6F	Reserved for future use	99	N/A	N/A	N/A	–
0x70...0x7F	(U)SIM toolkit security header	99	SMS	No	No	3.15.7
0x80...0x9F	SME to SME specific use	N/A	N/A	N/A	N/A	–
0xA0...0xBF	Reserved for future use	N/A	N/A	N/A	N/A	–
0xC0...0xDF	SMSC specific use	N/A	N/A	N/A	N/A	–
0xE0...0xFF	Reserved for future use	N/A	N/A	N/A	N/A	–

[a] This table field indicates the release of the technical specification in which the information element was introduced.

[b] This table field indicates whether the information element is related to SMS control (SMS), basic EMS (EMS 1) or extended EMS (EMS 2).

[c] iMelody format with a maximum size of 128 octets.

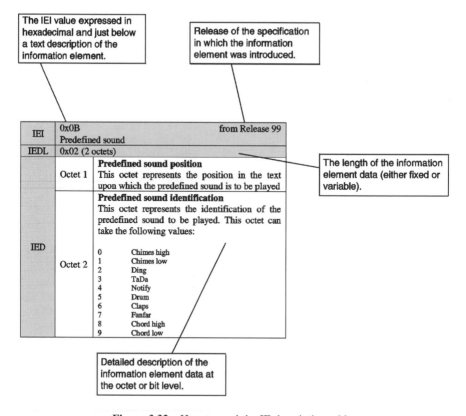

The IEI value expressed in hexadecimal and just below a text description of the information element.

Release of the specification in which the information element was introduced.

The length of the information element data (either fixed or variable).

Detailed description of the information element data at the octet or bit level.

Figure 3.32   How to read the IE description table

3.15.2 Concatenation of Message Segments

The concatenation information element indicates that a message segment represents one segment from a concatenated message. Parameters conveyed in this information element are the concatenated message reference number (not to be confused with the value assigned to the TP-Message-Reference parameter[4]), the number of segments in the concatenated message and the sequence number of the segment in the concatenated message. Two distinct information elements have been defined. The two concatenation information elements differ in the number of bits that are used for representing the message reference number.

- Information element identifier with IE identifier 0x00 allows the reference number to be represented with 8 bits (range 0...255, decimal values).
- Information element identifier with IE identifier 0x08 allows the reference number to be represented with 16 bits (range 0...65535, decimal values).

The information element, for an 8-bit-reference-number concatenation, is structured as

----

[4] All message segments composing a concatenated message have distinct values assigned to the TP-Message-Reference parameter. However, they have the same value assigned to the concatenated message reference number in the concatenation information element.

Table 3.22   IE/concatenation with 8-bit reference number

IEI	0x00 from release 99 Concatenation with 8-bit reference number		
IEDL	0x03 (3 octets)		
IED	Octet 1	**Concatenated message reference number.** This is a modulo 256 number which remains the same for all segments composing a concatenated message.	
	Octet 2	**Number of segments in the concatenated message.** This number represents the number of segments composing the concatenated message.	
	Octet 3	**Sequence number of this segment in the concatenated message.** The first segment of a concatenated message has a sequence number of 1. Value 0 is reserved.	

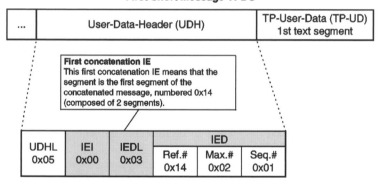

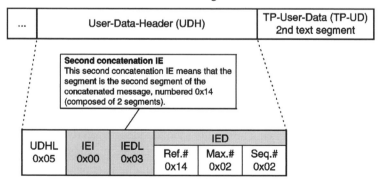

Figure 3.33   IE concatenation/example

Table 3.23 IE/concatenation with 16-bit reference number

IEI	0x08 from release 99		
IEI	Concatenation with 16-bit reference number		
IEDL	0x04 (4 octets)		
IED	Octet 1 and Octet 2	**Concatenated short message reference number.** This is a modulo 65536 number which remains the same for all segments composing a single concatenated message
IED	Octet 3	**Maximum number of short messages in the concatenated message.** This number represents the number of segments composing the concatenated message
IED	Octet 4	**Sequence number of this segment in the concatenated message.** The first segment of a concatenated message has a sequence number of 1. Value 0 is reserved

shown in Table 3.22. Note that if the concatenation sequence number of a segment is 0 or is greater than the number of segments in the concatenated message, then the concatenation information element is usually ignored by the recipient SME. In this case, message segments forming the concatenated message are presented independently as one-segment messages.

Figure 3.33 shows the coding of two message segments composing a concatenated message. In this example, the first message segment has a concatenation reference number of 0x14 and a sequence number equal to 1. The second message segment has the same concatenation reference number and a sequence number equal to 2. Before reconstruction, the receiving SME checks that the two segments have the same originator address.

Similarly, the information element, for a 16-bit reference number concatenation, is structured as shown in Table 3.23.

Box 3.2 Concatenated message identification

The triplet composed of the concatenation reference number, the message originator address and the number of message segments represents the key for uniquely identifying a concatenated message. The 16-bit reference number is preferred since it reduces the probability of encountering the conflicting situation where two different segments in the receiving SME have the same message key and the same segment sequence number. In this conflicting situation, the SME usually encounters difficulties in reconstructing the two distinct messages. However, the usage of the 16-bit reference number means that one additional octet is used per segment of a concatenated message. It is not advisable to mix concatenation elements with 8-bit and 16-bit reference numbers in the same concatenated message.

Table 3.24 IE/special SMS message indication

IEI	0x01 from release 99	
	Special SMS Message Indication	
IEDL	0x02 (2 octets)	
IED	Octet 1	**Message indication type and storage**
		Message storage:
		Bit 7 of octet 1 indicates whether the message shall be stored or not in the MS. Bit 7 is set to 1 if the short message is to be stored and set to 0 otherwise
		Indication type: Bits 6…0 of octet 1 indicate which indicator is to be updated. These bits can take the following values:
		Bit 6…0 — Indicator
		000 0000 — Voice Message Waiting
		000 0001 — Fax Message Waiting
		000 0010 — Email Message Waiting
		000 0011 — Other Message Waiting
	Octet 2	**Number of waiting messages** This number represents the number of waiting messages. It can range from 0 to 255. The value 255 for this octet means than 255 or more messages are waiting

3.15.3 Special SMS Message Indication

This information element is used for updating waiting message indicators in the receiving handset. These indicators inform the subscriber whether or not messages are awaiting retrieval from a server (voice mail, fax, Email, etc.). Depending on the handset capabilities, indicators may additionally inform the subscriber on the number of messages that are waiting.

The information element has the structure shown in Table 3.24. If several indicators are to be updated, then several information elements of this type can be inserted in one short message or in several message segments composing a concatenated message.

In addition to the *information element method*, three other methods exist for updating message waiting indicators.

- *TP-Data-Coding-Scheme method:* A message waiting indicator can also be updated with the delivery of a message for which a special value has been assigned to the TP-Data-Coding-Scheme parameter. With this method, a message waiting indicator can be activated or deactivated on the receiving handset. Unlike the information element method, the number of waiting messages cannot be specified. If the value assigned to the TP-Data-Coding-Scheme indicates that the message is to be stored (see also coding groups in Section 3.7.6), then the message may contain some additional text (GSM

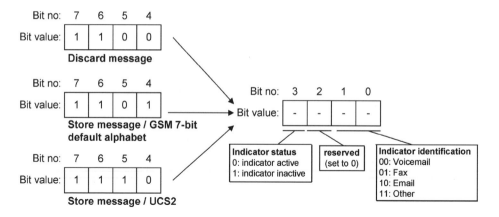

Figure 3.34   Message waiting indicator/TP-Data-Coding-Scheme method

7-bit default alphabet or UCS2). Special values to be assigned to the `TP-Data-Coding-Scheme` parameter are structured as shown in Figure 3.34.

- *TP-Originating-Address method:* Another method is commonly supported by SMEs. This method consists of assigning a special value to the `TP-Originating-Address` parameter of the message to be delivered. Note that this method has been defined by a group of operators (Orange and T-Mobile (UK), formerly One2one) and has not yet been published by any recognized standardization development organization. This method is defined in the Common PCN Handset Specification (CPHS). 3GPP members can download this specification from the following location: http://www.3gpp.org/ftp/tsg_t/WG3_USIM/ TSGT3_00_old_meetings/TSGT3_15/docs/T3-000450.zip.

- *Return call message:* A fourth method, to indicate to the subscriber that one or more messages are waiting, consists of sending a return call message (see `TP-Protocol-Identification` description in Section 3.7.7). In this situation, the message contains a text description indicating that one or more messages are waiting retrieval. Note that with this method, message waiting indicators are usually not updated automatically.

3.15.4 Application Port Addressing

Application port addressing is a feature allowing the routing of a received message to the port of an identified application running in the mobile station. Applications that benefit from such a feature are personal information managers, programs handling over-the-air provisioning of configuration parameters, etc. Application port addressing can be realized using two distinct information elements. The first information element is used for 8-bit address ports whereas the second information element is used for 16-bit address ports. If a concatenated message is to be routed to an application, then the associated application port addressing IE is incorporated in all message segments composing the concatenated message.

For applications with 8-bit address ports, the information element shown in Table 3.25 is used. In the 8-bit address range, the values listed in Table 3.26 are used. For applications with 16-bit address ports, the information element shown in Table 3.27 is used. In the 16-bit address range, the values listed in Table 3.28 are used.

Table 3.25 IE/application port addressing scheme, 8-bit address

IEI	0x04 from release 99	
	Application port addressing scheme, 8 bit address	
IEDL	0x02 (2 octets)	
IED	Octet 1	**Destination port** This octet indicates the 8-bit address of the receiving port
	Octet 2	**Originator port** This octet indicates the 8-bit address of the sending port

Table 3.26 Port number range/8-bit addresses

Port number range		Description
From	To	
0x00 0 (decimal)	0xEF 239 (decimal)	Reserved
0x0F 240 (decimal)	0xFF 255 (decimal)	Available for allocation by applications

Table 3.27 IE/application port addressing scheme, 16-bit address

IEI	0x05 from release 99	
	Application port addressing scheme, 16 bit address	
IEDL	0x04 (4 octets)	
IED	Octet 1 and Octet 2	**Destination port** This octet indicates the 16-bit address of the receiving port
	Octet 3 and Octet 4	**Originator port** This octet indicates the 16-bit address of the sending port

Table 3.28 Port number range/16-bit addresses

Port number range		Description
From	to	
0x0000 0 (decimal)	0x3E7F 15999 (decimal)	See default address ports as defined by IANA[a]
0x3E80 16000 (decimal)	0x4267 16999 (decimal)	Available for allocation by applications
0x4268 17000 (decimal)	0xFFFF 65535 (decimal)	Reserved

[a] IANA Port Numbers are available at http://www.iana.org/assignments/port-numbers.

3.15.5 Service Centre Control Parameters

This information element is used to convey control instructions in a flexible way. These instructions may be interpreted by the SMSC or the recipient SME. In particular, as part of a message, this information element identifies the set of events for which a status report should be generated by the SMSC. The possible events which can be identified for the generation of a status report are:

- A transaction has been completed
- A permanent error when the SMSC is not making anymore transfer attempts

Table 3.29   IE/service centre control parameters

IEI	0x06 from release 99 Service Centre Control Parameters	
IEDL	0x01 (1 octet)	
IED	Octet 1	**Selective Status Report (SR)** The four least significant bits (bits 3…0) indicate whether or not a specific status report is to be generated as shown below: **Bit 0**/SR for transaction completed 0: no status report to be generated 1: status report to be generated **Bit 1**/SR for permanent error when SMSC is not making anymore transfer attempts 0: no status report to be generated 1: status report to be generated **Bit 2**/ SR for temporary error when SMSC is not making anymore transfer attempts 0: no status report to be generated 1: status report to be generated **Bit 3**/SR for temporary error when SMSC is still trying to transfer the short message 0: no status report to be generated 1: status report to be generated **Bits 4 and 5** are reserved for future use **Bit 7**/Original UDH This bit indicates whether or not the original user data header is to be included in the status report. The bit is set to 0 if the original UDH is to be included, it is set to 1 otherwise

- A temporary error when the SMSC is not making anymore transfer attempts
- A temporary error when the SMSC is still trying to transfer the message.

For the successful generation of the status report by the SMSC or by the receiving entity, the originator SME must also set the `TP-Status-Report-Request` parameter in the submitted message segment. The information element has the structure shown in Table 3.29.

If the original user data header is to be included in the status report (see bit 7 description), then the status report normal UDH is differentiated from the original UDH with the UDH source indicator information element as described in the following section.

If a concatenated message includes service centre control parameters, then the service centre control parameter information element must be included in each message segment composing the concatenated message.

3.15.6 User-Data-Header Source Indicator

The User-Data-Header (UDH) source indicator information element is used to combine several User-Data-Headers provided by sources such as the originator SME, the recipient SME or the SMSC in a single message segment. A message segment can contain multi-source User-Data-Header information in status reports or in messages used for updating message waiting indicators. The information element has the structure shown in Table 3.30.

Figure 3.35 shows how the UDH Source Indicator can be used for separating various sequences of information elements in the User-Data-Header of a status report.

Table 3.30 IE/user data header source indicator

IEI	0x07 from release 99	
	User-Data-Header Source Indicator	
IEDL	0x01 (1 octet)	
IED	Octet 1	**Source Indicator** Value 0x01 indicates that the following part has been generated by the original sender (valid in case of a status report) Value 0x02 indicates that the following part has been generated by the original receiver (valid in case of a status report) Value 0x03 indicates that the following part has been generated by the SMSC (valid in any message or status report) Value 0x00 and values ranging from 0x04 to 0xFF are reserved values

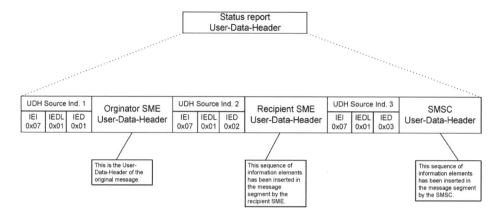

Figure 3.35 UDH source indicator/example

3.15.7 (U)SIM Toolkit Security Header

This group of 16 information elements is used for indicating the presence of a (U)SIM toolkit security header in the `TP-User-Data` after the User-Data-Header. Such an information element is only inserted in the first message segment of a concatenated message. One characteristic of these information elements is that they do not have any payload (value 0 assigned to IEDL and there is no IED). These information elements are structured as shown in Table 3.31.

3.15.8 Wireless Control Message Protocol

This information element is used for exchanging states of applications or protocols between SMEs. This method avoids the unnecessary retransmission of packets in the WAP environment when an application is not available for receiving data. The information element is structured as shown in Table 3.32.

Table 3.31 IE/(U)SIM toolkit security header

IEI	0x70 to 0x7F from release 99
	(U)SIM Toolkit Security Header
IEDL	0x00 (0 octet - no IED)

Table 3.32 IE/wireless control message protocol

IEI	0x09 from release 99
	Wireless Control Message Protocol
IEDL	variable
IED	Octet 1 to Octet n	These octets contain Wireless Control Message Protocol information

Table 3.33 IE/alternate reply address

IEI	0x22 from release 5	
	Alternate reply address	
IEDL	variable	
IED	Octet 1 to Octet n	These octets represent the alternate reply address. This address is formatted as defined in Section 3.9.7

3.15.9 Alternate Reply Address

In the context of SMS, a message reply is typically sent to the address of the message originator. Alternatively, during message submission, the message originator has the capability to indicate that message replies should be sent to an alternate reply address. This is performed by inserting, in the submitted message, a dedicated information element containing the alternate reply address. This dedicated information element is structured as shown in Table 3.33.

Note that for external SMEs, an alternative method is sometimes supported by SMSCs. This method is called the `TP-Originating-Address` substitution method. It consists of replacing the value, assigned by the external SME to the `TP-Originating-Address`, by an alternate reply address specified by the external SME during message submission.

3.16 Network Features for Message Delivery

It may happen that the SMSC is unable to deliver a message to a recipient SME due to some temporary or permanent error conditions. This can occur, for instance, if the mobile device has been powered off, if the device has no more storage capacity to handle the message or if the recipient subscriber is unknown. For failures due to permanent error conditions, the SMSC does not attempt to deliver the message anymore. For failures due to temporary

Table 3.34 SMS failure causes

Failure cause	Status
Unknown subscribed	Permanent
Teleservice not provisioned	Permanent
Call barred	Temporary
Facility not supported	Temporary
Absent subscriber	Temporary
MS busy for MT SMS	Temporary
SMS lower layers capabilities not provisioned	Temporary
Error in MS	Temporary
Illegal subscriber	Permanent
Illegal equipment	Permanent
System failure	Temporary
Memory capacity exceeded	Temporary

error conditions, the SMSC may store the message temporarily in a queue and may attempt to deliver the message again. Table 3.34 shows the list of temporary and permanent errors that can be returned to an SMSC for a failed message delivery.

To allow the retransmission of messages by the SMSC, the network which is serving the recipient SME can maintain a Message Waiting Indication (MWI). This indication enables the serving network to know when the recipient SME has retrieved the capabilities for handling messages. In such a situation, the serving network alerts SMSCs for which a previous message delivery to the recipient SME has failed due to some temporary error conditions. Upon alerting, SMSCs can appropriately re-attempt the transmission of queued messages. The MWI is composed of the following set of parameters, maintained by various network elements (HLR, VLR and GGSN):

- Address of the recipient SME (MSISDN-Alert) and associated addresses of SMSCs that unsuccessfully attempted to deliver messages to the recipient SME. This information is maintained by the HLR.
- The Mobile-station-memory-Capacity-Exceeded-Flag (MCEF) Boolean flag indicates whether or not a message delivery has failed because the storage capacity of the recipient SME has exceeded. This flag is maintained by the HLR.
- In the context of GPRS, the Mobile-station-Not-Reachable-for-GPRS (MNRG) Boolean flag indicates whether or not a message delivery has failed because the recipient SME is absent from the network. This flag is maintained by the HLR or by the SGSN.
- The Mobile-station-Not-Reachable-Flag (MNRF) Boolean flag indicates whether or not a message delivery has failed because the recipient SME is absent from the network. This flag is maintained by the HLR or the VLR.
- The Mobile-station-Not-Reachable-Reason (MNRR) indicates the reason why the recipient SME is absent from the network. This indicator, maintained by the HLR, is either cleared or contains one of the following values: no paging response via the MSC, no paging response via the SGSN, IMSI detached or GPRS detached.

The MWI is updated on signalling events such as network attach, location update, call set-up, etc. After update, if the HLR detects that the associated recipient SME has retrieved the ability to handle messages, then it send an 'Alert-SC' message to all SMSCs that unsuccessfully attempted to deliver messages to the recipient SME. Upon alerting, SMSCs can retry to deliver the message(s) for this particular recipient SME, without delay.

In the situation where the recipient SME rejected a message delivery because of a storage capacity failure, the SME can alert the HLR when it has retrieved the ability to store messages. This is performed by sending a specific message (RP-SM-MEMORY-AVAILABLE), at the relay layer, to the HLR. Upon receipt of such a message, the HLR alerts the SMSCs which unsuccessfully attempted to deliver one or more messages to the associated SME.

In addition to the possibility of an alert from the serving network that the recipient SME is available for handling message deliveries, an SMSC can also independently re-attempt the message delivery after an appropriate period of time. A well designed SMSC usually supports separate configurable retry algorithms for each delivery failure cause. With such SMSCs, the operator can configure the delay between two successive delivery attempts and the number of attempts performed by the SMSC. Typically, the delay between successive delivery attempts depends on the cause of the delivery failure and will increase if the causes for message delivery failures remain the same. Overall, the operator configures the SMSC retry procedure in order to achieve the best balance between subscriber service quality and network loading.

3.17 SMSC Access Protocols

SMSC access protocols enable interactions between two SMSCs or interactions between an external SME and an SMSC. The 3GPP, in technical report [3GPP-23.039], recognizes five main SMSC access protocols:

- Short Message Peer to Peer (SMPP) interface specification from the SMS Forum. Technical specifications from the SMS Forum (SMAP and SMPP) can be downloaded from the SMS Forum website at http://www.smsforum.net
- Short Message Service Centre external machine interface from the Computer Management Group (CMG). Technical specifications from CMG are available from http://www.cmgtelecom.com
- SMSC to SME interface specification from Nokia Networks. Technical specifications from Nokia Networks available at cimd.support@nokia.com
- SMSC open interface specification from SEMA Group. Technical specifications from SEMA group available from http://www.semagroup.com/m&t/telecoms.htm
- SMSC computer access service and protocol manual from Ericsson.

These protocols are proprietary and are usually binary protocols operating over TCP/IP or X.25. Recently, a non-profit organization was established by companies willing to promote SMS in the wireless industry. This organization is known as the SMS Forum (formerly SMPP Forum). The SMS Forum has adopted SMPP as its recommended binary access protocol for service centres. In addition, the SMS Forum is developing a text-based protocol operating over protocols such as HTTP where messages are represented in XML. This protocol is known as the Short Message Application Protocol (SMAP).

> **Box 3.3 Commercial availability of SMSCs**
>
> Several manufacturers provide SMSC solutions. The most well known are Logica, Nokia, CMG, SEMA and Ericsson.

This chapter outlines two binary protocols: SMPP from the SMS Forum and the SMSC open interface specification from SEMA Group. In addition, a description of the text-based SMAP protocol is also provided.

3.17.1 SMPP from SMS Forum

The Short Message Peer to Peer (SMPP) was originally developed by Logica and the protocol has now been adopted by the SMS Forum as an open binary access protocol enabling interactions between external SMEs and service centres developed by different manufacturers.

In order to interact with an SMSC via the SMPP protocol, an external SME first establishes a session. The transport of operation requests over this session is usually performed over TCP/IP or X.25 connections. For TCP/IP, application port 2775 is used for SMPP (this may, however, vary in implementations from different manufacturers). Operations over SMPP sessions can be categorized into four groups:

- *Session management:* these operations enable the establishment of SMPP sessions between an external SME and the SMSC. In this category, operations also provide a means of handling unexpected errors.
- *Message submission:* these operations allow external SMEs to submit messages to the SMSC.
- *Message delivery:* these operations enable the SMSC to deliver messages to external SMEs.
- *Ancillary operations:* these operations provide a set of features such as cancellation, query or replacement of messages.

SMPP is an asynchronous protocol. This means that the external SME may send several instructions to the SMSC without waiting for the result of previously submitted instructions. SMPP itself does not define any encryption mechanism for the exchange of messages between the external SME and the SMSC. However, two methods are recommended for avoiding confidentiality issues. The first method consists of using the well known Secure Socket Layer (SSL) and its standardized form, Transport Layer Security (TLS). The second method consists of allowing an SMPP session to operate over a secure tunnel. In this configuration, the connection is not made directly between the external SME and the SMSC but is relayed between two secure tunnel servers.

3.17.2 SMSC Open Interface Specification from Sema Group

Sema Group Telecoms provides the SMSC open interface specification along with its SMSCs. An external SME interacts with the SMSC with a protocol designed to operate over a variety of interfaces such as X.25, DECnet and SS7. This SMSC is usually accessed via the general X.25 gateway (either by using a radio Packet Assembler Disassembler or a dedicated link to the SMSC). The SME connected to the SMSC invokes an operation by sending a request to the SMSC. When the SMSC has completed the request, it sends the operation result back to the SME. Alternatively, the SMSC may invoke an operation by sending a request to the SME. When the SME has completed the request, it sends the operation result back to the SMSC. Possible operations are shown in Figure 3.36.

An external SME can request the following operations from the service centre:

- *Submit message segment:* this operation is for sending a message segment to an SME.
- *Delete message segment:* this command is used for deleting a previously submitted message segment.
- *Replace message segment:* this operation allows the replacement of a previously submitted message segment (which has not been delivered yet) by a new message segment.
- *Delete all message segments:* delete all previously submitted message segments which have not been delivered yet.
- *Enquire message segment:* request the status of a previously submitted message segment.
- *Cancel status report request:* this operation is for cancelling a request for the generation of a status report related to the previous submission of a message segment.
- *Alert SME request:* this command allows alerts when a specified SME has registered.

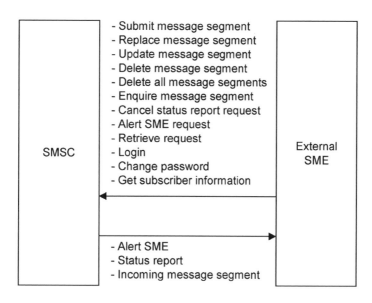

Figure 3.36 SEMA Group service centre configuration

- *Login:* this operation allows users to access to the SMSC remotely.
- *Change password:* this operation allows users to change their password.
- *Get subscriber information:* this operation allows an SME to query the network HLR to determine if a network node is currently serving a mobile station.

The SMSC can request the following operations from the external SME:

- *Alert SME:* this operation is used by the SMSC to indicate to the SME that a mobile station has registered in the network.
- *Status report:* this operation is used by the SMSC to provide a status report to the external SME. The status report indicates the status of the corresponding message segment delivery.
- *Incoming message segment:* this operation is used by the SMSC to provide an incoming message segment addressed to the SME.

3.17.3 SMAP

As previously indicated, access protocols initially developed to allow interactions between SMSCs and external SMEs are binary protocols. The SMS Forum is in the process of developing a text-based protocol that it expects to become the de facto access protocol for SMSCs in the future. An external SME may communicate directly with an SMSC via the SMAP protocol (only if the SMSC has native support for SMAP). Alternatively, a SMS gateway can fit between the external SME and the SMSC. This latter solution allows an easier evolution path from previous proprietary configurations. Such a configuration is shown in Figure 3.37.

One of SMS Forum objectives is to design SMAP (version 1.0) as functionally equivalent to SMPP (version 3.4). For this purpose, SMAP operations are categorized into the four SMPP groups of operations. SMAP is an application protocol independent from underlying

Figure 3.37 SMAP configuration with SMS gateway

transport protocols. However, HTTP represents a suitable candidate for transporting SMAP operation requests and results. These operation requests and results are formatted in XML. Figure 3.38 shows an example of a SMAP operation request formatted in XML. This operation corresponds to a message submission request from an external SME.

Four operational modes are available with SMAP:

- *Immediate mode:* with this mode, the external SME does not maintain a session with the gateway. Each operation contains the application context. This mode is used for message submissions only.
- *Client session mode:* with this mode, the external SME first establishes a session with the gateway prior to requesting operations to be processed by the SMSC. The gateway may also establish such a session for message delivery to an external SME.

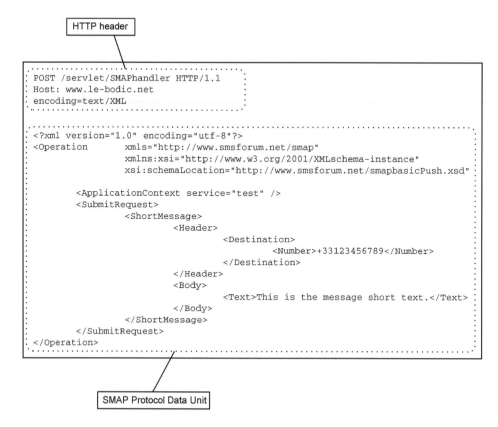

Figure 3.38 SMAP protocol data unit conveyed over HTTP (immediate mode)

- *Peer-to-peer session mode:* this mode allows a bi-directional session to be established between the external SME and the SMSC. Message submissions and deliveries can be performed over a single bi-directional session.
- *Batch mode:* with this mode, the gateway receives a set of SMAP operations to be processed, from the external SME. The gateway processes each operation in turn and builds a set of results. The set of results is also provided in a batch to the external SME. The batch mode is usually used when an interactive session is not required or would be unsuitable due to timeout issues.

3.18 SIM Application Toolkit

The SIM Application Toolkit (SAT), defined in [3GPP-31.111], defines mechanisms for allowing SIM-hosted applications to interact with the mobile equipment. This includes the following mechanisms:

- *Profile download:* this mechanism allows the mobile equipment to inform the SIM about its capabilities.
- *Proactive SIM:* a proactive SIM can issue commands to the mobile equipment. Section 3.18.1 provides a description of features available to proactive SIMs.
- *Data download to SIM:* it was shown in this chapter that a particular TP-Protocol-Identifier value could be used to download data to the SIM. This mechanism is further detailed in Section 3.18.2.
- *Menu selection:* this mechanism allows the (U)SIM to define menu items and to be notified by the mobile equipment when the subscriber has selected one of the menu items.
- *Call control by SIM:* with this mechanism, the SIM performs a control prior to the establishment of calls by the mobile equipment. This allows the SIM to authorize or reject the call establishment or to modify the parameters of the call to be established.
- *Control of outgoing messages by SIM:* with this mechanism, the SIM performs a control prior to the sending of messages by the mobile equipment. This allows the SIM to authorize or reject the sending of a message.
- *Event download:* this mechanism allows the SIM to provide a set of events to be monitored by the mobile equipment. If an event occurs then the mobile equipment notifies the SIM.
- *Security:* this mechanism ensures data confidentiality, data integrity and data sender validation.
- *Timer expiration:* the SIM can manage a set of timers running physically in the mobile equipment.
- *Bearer independent protocol:* this mechanism enables the SIM to establish a data connection between the SIM and the mobile equipment and between the mobile equipment and a remote server.

3.18.1 Proactive SIM

Technical specification [3GPP-11.11] defines the protocol of communications between the SIM and the mobile equipment. The protocol is known as the $T = 0$ protocol (or $T = 1$ for the USIM). A characteristic of this protocol is that the mobile equipment remains the 'master' during the communications and is the element which initiates all commands to the SIM. In this protocol, there is no means for the SIM to issue commands to the mobile equipment. In order to cope with this limitation, the concept of a *proactive command* was introduced. A SIM, making use of proactive commands, is known as a *proactive SIM*. With proactive commands, the SIM is able to issue a command to the mobile equipment by specifying the proactive command in the response to a normal command previously submitted by the mobile equipment to the SIM. Upon receipt of such a response, the mobile equipment executes the specified command and provides the execution results to the SIM as part of a normal command.

3.18.2 SIM Data Download

As shown in Section 3.7.7, two specific values can be assigned to the `TP-Protocol-Identifier` parameter (0x7C and 0x7F) for SIM data download. Upon receipt of a message containing one of these two values, the receiving mobile equipment provides the message to the SIM. The message is provided to the SIM by the use of a SAT command known as an `ENVELOPE (SMS-PP DOWNLOAD)` command. The subscriber is not notified of the receipt of the message by the mobile equipment.

3.18.3 SIM Interactions: Example

In order to illustrate the use of SIM proactive commands and SIM data download messages, a simple scenario is described in this section. A SIM application maintains a SIM elementary file in which the subscriber geographical location is regularly updated. For the purpose of collecting statistics on subscriber moves, an application hosted in a remote server (external SME) regularly queries the mobile station for the subscriber location. For this purpose, the external SME submits a 'querying' message to the SMSC. The SMSC delivers the message to the recipient mobile station. Upon receipt, the mobile equipment detects that the message is a SIM data download message and therefore provides the message to the SIM as part of an `ENVELOPE` SAT command. The SIM analyses the message payload and generates an additional message that contains the results of the transaction (subscriber location, e.g. retrieved with a GPS module connected to the mobile equipment). The SIM issues the message to the mobile equipment via the `SEND SHORT MESSAGE` proactive command. Upon receipt of the proactive command, the mobile equipment submits the message to the SMSC which in turn delivers the message to the external SME. Finally, the external SME analyses the payload of the 'result' message and updates a database. The flow of interactions for such a scenario is depicted in Figure 3.39.

3.19 SMS Control via a Connected Terminal Equipment

Technical specification [3GPP-27.005] defines interface protocols for control of SMS functions between the MS and an external Terminal Equipment (TE) via an asynchronous interface. The MS and the TE are connected with a serial cable, an infrared link or any other similar link as shown in Figure 3.40. Communications between the MS and the TE can be carried out in three different modes:

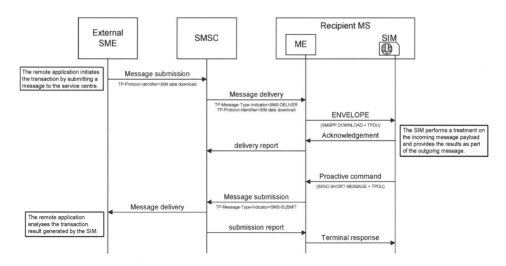

Figure 3.39   Example of interactions between an external SME and a SIM

- *Block mode:* a binary communications protocol including error protection. It is advisable to use this mode if the link between the MS and the TE is not reliable.
- *Text mode:* a character-based protocol suitable for high-level software applications. This protocol is based on AT commands. AT stands for ATtention. This two-character abbreviation is always used to start a command line to be sent from TE to MS in the text mode.
- *Protocol Data Unit (PDU) mode:* a character-based protocol with hexadecimal-encoded binary transfer of commands between the MS and the TE. This mode is suitable for low-level software drivers that do not understand the content of commands.

Regardless of the mode, the TE is always in control of SMS transactions. The TE operates as the 'master' and the MS as the 'slave'. The block mode is a self-contained mode whereas the text and PDU modes are just sets of commands operated from the V.25ter command state or online command state.

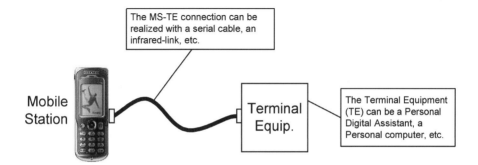

Figure 3.40   MS connection with terminal equipment

3.19.1 AT Commands in Text Mode

This section provides an outline of commands available in the text mode only. In this mode, SMS-related AT commands are categorized in four groups:

- *General configuration commands* allow the terminal equipment to configure the way it wishes to communicate with the mobile station.
- *Message configuration commands* enable the terminal equipment to consult and update the mobile station SMS settings (service centre address, etc).
- *Message receiving and reading commands* allow the terminal equipment to read messages locally stored in the mobile station and to be notified of incoming messages.
- *Message sending and writing commands* enable the terminal equipment to create, send and delete messages in the mobile station.

The list of SMS-related AT commands is given in Table 3.35.

Table 3.35   SMS related AT commands

AT Command	Description
General configuration commands	
+ CSMS	Select message service
+ CPMS	Preferred message storage
+ CMGF	Message format
+ CESP	Enter SMS block mode protocol
+ CMS ERROR	Message service failure code
Message configuration commands	
+ CSCA	Service centre address
+ CSMP	Set text mode parameters
+ CSDH	Show text mode parameters
+ CSCB	Select cell broadcast message types
+ CSAS	Save settings
+ CRES	Restore settings
Message receiving and reading commands	
+ CNMI	New message indications to TE
+ CMGL	List messages
+ CMRG	Read message
+ CNMA	New message acknowledgement
Message sending and writing commands	
+ CMGS	Send message
+ CMSS	Send message from storage
+ CMGW	Write message to memory
+ CMGD	Delete message
+ CMGC	Send command
+ CMMS	More message to send

Figure 3.41   AT command usage example

3.19.2 AT Command Usage: Example

The example in Figure 3.41 shows how a message can be created in the ME message local store and sent to a recipient.

3.20 SMS and Email Interworking

Interworking between SMS and Email is enabled by allowing the conversion of an SMS message into an Email message, and vice versa. This conversion is performed by the Email gateway as shown in the architecture depicted in Figure 3.2. An originator SME can indicate that a message has to be delivered to an Internet recipient by setting specific values for the parameters listed in Table 3.36.

The process of sending an SMS message to the Internet domain is summarized in Figure 3.42. The Email gateway may also convert an Email message to an SMS message. The conversion process consists of incorporating the Email message content in the `TP-User-Data` parameter of the SMS message TPDU. For this purpose, two methods have been developed. The first method is a *text-based method* consisting of inserting the Email message (RFC 822 header and footer) directly into the text section of the `TP-User-Data` parameter. The second method, called the *information element-based method*, consists of using a specific information element for separating the Email header from the Email body in the text part of the `TP-User-Data` parameter.

3.20.1 Text-based Method

With this method, the content of the Email message is directly incorporated in text form in the `TP-User-Data` parameter. The text part representing the Email message content shall comply with the grammar rules listed in Table 3.37.

Table 3.36   TPDU parameters for Internet interworking

TPDU parameter	Value
`TP-Protocol-Identifier`	Internet electronic mail (0x32)
`TP-Destination-Address`	Gateway address

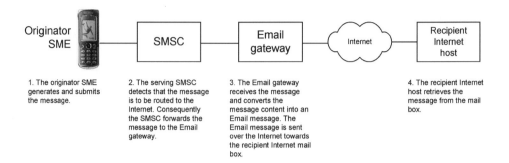

1. The originator SME generates and submits the message.

2. The serving SMSC detects that the message is to be routed to the Internet. Consequently the SMSC forwards the message to the Email gateway.

3. The Email gateway receives the message and converts the message content into an Email message. The Email message is sent over the Internet towards the recipient Internet mail box.

4. The recipient Internet host retrieves the message from the mail box.

Figure 3.42   Process of sending a message to an Internet host

Fields <to-address> and <from-address> can take the two following forms: user@domain or User Name <user@domain>. In the latter form, angle brackets are part of the address and are conveyed in the message. A message can contain multiple recipient addresses for the <to-address> field. In this case, addresses are separated by a comma: user1@domain1,user2@domain2,user3@domain3. According to

Table 3.37   Email text format[a]

Email type	Message Content
Internet to SMS – Without subject – Without real name	[<from-address> <space>] <message>
SMS to Internet – Without subject – Without real name	[<to-address> <space>] <message>
Internet to SMS – With subject – Without real name	[<from-address>] (<subject>) <message> or [<from-address>] ## <subject> # <message>
SMS to Internet – With subject – Without real name	[<to-address>] (<subject>) <message> or [<to-address>] ## <subject> # <message>
Internet to SMS – With subject – With real name	[<from-address>]#<real-name>##[<subject>]#<message>
SMS to Internet – With subject – With real name	[<to-address>]#<real-name>##[<subject>]#<message>

[a] In this table, the following notation is used: [ ] denotes optional fields; < > delimit fields; <space> denote a single space character.

user@domain.com This is the text of the message.	user@domain.com,user 2@domain2.com,user3 @domain3.com This is the text of the message.
user@domain.com##Me ssage Subject#This is the text of the message.	user@domain.com#My real name goes here##Message Subject#This is the text of the message.

Figure 3.43   Examples of SMS Email messages

the grammar rules, the examples shown in Figure 3.43 are valid. In SMS messages, the character '@' can be replaced by the character '*' and the character '_' (underscore) can be replaced by the character '$'.

If the content of the Email message does not fit into one short message, then concatenation may be used. It is advisable to concatenate message segments with one of the concatenation information elements as described in Section 3.15.2. Alternatively, a text-based concatenation mechanism consists of adding the symbol '+' in specific positions in the SMS message. The first message segment contains the Email header as described above and ends with '+'. Subsequent message segments start with '+' and end with '+'. The last segment starts with a '+' but does not end with a '+'. The Email message header is only inserted once in a concatenated message. The example in Figure 3.44 shows three message segments composing an Email message with text-based concatenation.

3.20.2 Information Element-based Method

Another method for representing the content of an Email message in the TP-User-Data parameter consists of using a dedicated information element structured as shown in Table 3.38.

The presence of this information element indicates that the text part of the TP-User-Data parameter contains an Email header and an optional Email body. The Email header and

user@domain.com,user 2@domain2.com,user3 @domain3.com This first short message contains the first segment of the Internet email.+	+ And this second message (without header) contains the second segment of the Internet email. It is followed by a third short message.+	+ The last short message contains the last segment of the Internet Email.

Figure 3.44   Example of a concatenated SMS Email message

Table 3.38   IE/RFC 822 Email header

IEI	0x20 from release 99	
	RFC 822 Email Header	
IEDL	0x01 (1 octet)	
IED	Octet 1	This octet represents the length of the Email header (or the length of the Email header fraction if used in a concatenated message). This allows to differentiate the Email header from the Email body in the text part of the `TP-User-Data` parameter The length is expressed in terms of: • number of septets for GSM 7-bit default alphabet. • number of 16-bit symbols for UCS2. • number of octets for 8-bit encoding

body composing the Email message are formatted according to the conventions published by the IETF in [RFC-822].

The Email header shall always precedes the Email body and both the header and body shall be encoded using the same character set (GSM 7 bit default alphabet, UCS2 or 8-bit data for ASCII).

If the Email message content does not fit into one message segment, then the concatenation mechanism defined in Section 3.15.2 can be used. In this situation, the information element 'RFC 822 Email header' is inserted into each message segment composing the concatenated message. Figure 3.45 shows how this information element can be used for separating a 45-septet Email header from the Email body.

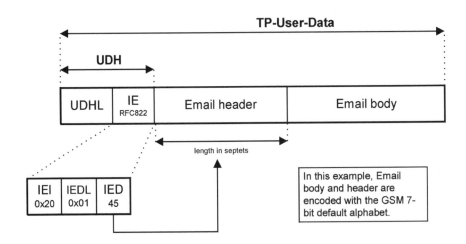

Figure 3.45   Information element RFC 822 Email header

Table 3.39 SMS TPDU parameter index

Abbreviation	Name	Short message submission From MS to SMSC	Submission report From SMSC to MS	Short message delivery From SMSC to MS	Delivery report From MS to SMSC	Status report From SMSC to MS	Command From MS to SMSC
TP-CD	TP-Command-Data Length 1 octet	No	No	No	No	No	Optional Variable position
TP-CDL	TP-Command-Data-Length Length 1 octet	No	No	No	No	No	Mandatory Variable position
TP-CT	TP-Command-Type Length 1 octet	No	No	No	No	No	Mandatory Octet 4
TP-DA	TP-Destination-Address Length 2 to 12 octets	Mandatory From octet 3	No	No	No	No	No
TP-DCS	TP-Data-Coding-Scheme	Mandatory Variable position	Optional Variable position	Mandatory Variable position	Optional Variable position	Optional Variable position	No
TP-DT	TP-Discharge-Time Length 7 octets	No	No	No	No	Mandatory Variable position	No
TP-FCS	TP-Failure-CauSe Length 1 octet (Integer)	No	Mandatory Octet 2	No	Mandatory Octet 2	No	No
TP-MMS	TP-More-Message-to-Send Length 1 bit	No	No	Mandatory Octet 1/Bit 2	Mandatory Octet 1/Bit 2	No	No
TP-MN	TP-Message-Number Length 1 octet	No	No	No	No	No	Mandatory Octet 5
TP-MR	TP-Message-Reference Length 1 octet (Integer)	Mandatory Octet 2	No	No	No	No	Mandatory Octet 2
TP-MTI	TP-Message-Type-Indicator Length 2 bits	Mandatory Octet 1/Bits 0 and 1	Mandatory Octet 1/Bits 0 and 1	Mandatory Octet 1/Bits 0 and 1	Mandatory Octet 1/Bits 0 and 1	Mandatory Octet 1/Bits 0 and 1	Mandatory Octet 1/Bits 0 and 1
TP-OA	TP-Originating-Address Length 2 to 12 octets	No	No	Mandatory From octet 2	No	No	No

Table 3.39 (*continued*)

Abbreviation	Name	Short message submission From MS to SMSC	Submission report From SMSC to MS Positive	Negative	Short message delivery From SMSC to MS	Delivery report From MS to SMSC Positive	Negative	Status report From SMSC to MS	Command From MS to SMSC
TP-PI	TP-Parameter-Indicator Length 1 octet	No	Mandatory Octet 2	Mandatory Octet 3	No	Mandatory Octet 2	Mandatory From octet 3	Optional Variable position	No
TP-PID	TP-Protocol-Identifier Length 1 octet	Mandatory Variable position	Mandatory Octet 10	Optional Octet 11	Mandatory Variable position	Optional Octet 11	Optional Octet 11	Optional Variable position	Mandatory Octet 3
TP-RA	TP-Recipient-Address Length 2 to 12 octets	No	No		No	No		Mandatory From octet 3	No
TP-RD	TP-Reject-Duplicates Length 1 bit	Mandatory Octet 1/Bit 2	No		No	No		No	No
TP-RP	TP-Reply-Path Length 1 bit	Mandatory Octet 1/Bit 7	No		Mandatory Octet 1/Bit 7	No		No	No
TP-SCTS	TP-Service-Centre-Time-Stamp Length 7 octets	Mandatory	Mandatory From octet 3	Mandatory From octet 4	No	No		Mandatory	No
TP-SRI	TP-Status-Report-Indication Length 1 bit	No	No		Optional Octet 1/Bit 5	No		No	No
TP-SRQ	TP-Status-Report-Qualifier Length 1 bit	No	No		No	No		Mandatory Octet 1/Bit 5	No
TP-SRR	TP-Status-Report-Request Length 1 bit	Optional Octet 1/Bit 5	No		No	No		No	No
TP-ST	TP-Status Length 1 octet	No	No		No	No		Mandatory Variable position	No

TP-UD TP-User-Data Variable length	Optional Variable position	Optional Variable position	Optional Variable position	Optional Variable position	Optional Variable position	No
TP-UDHI TP-User-Data-Header-Indicator Length 1 bit	Optional Octet 1/Bit 6	Optional Octet 1/Bit 6	Optional Octet 1/Bit 6	Optional Octet 1/Bit 6	Optional Octet 1/Bit 6	Optional Octet 1/Bit 6
TP-UDL TP-User-Data-Length Length 1 octet (Integer)	Mandatory Variable position	Optional Variable position	Mandatory Variable position	Optional Variable position	Optional Variable position	No
TP-VP TP-Validity-Period Length 1 octet or 7 octet	Optional Variable position	No	No	No	No	No
TP-VPF TP-Validity-Period-Format Length 2 bits	Mandatory Octet 1/Bit 3	No	No	No	No	No

3.21 Index of TPDU Parameters

Table 3.39 provides a list of all TPDU parameters and indicates whether or not the parameter is supported according to the TPDU type. If the parameter is supported, then an indication is given whether the parameter is mandatory or optional.

3.22 Pros and Cons of SMS

The incontestable advantage of the Short Message Service is that is has become an ubiquitous service in most GSM networks. One hundred percent of GSM handsets support the Short Message Service. A message can be sent from almost any GSM network and delivered to any other GSM subscriber attached to the same network, to another network in the same country or even to a network in another country.

The main drawback of the Short Message Service is that only limited amounts of data can be exchanged between subscribers. In its simplest form, SMS allows 140 octets of data to be exchanged. Concatenation has been introduced to allow longer messages to be transmitted. Another drawback is that only text can be included in messages and this does not allow the creation of messages with content more compelling than text. Furthermore, the lack of content support for SMS prevents the development of commercial applications based on SMS. To cope with these limitations, an application-level extension of SMS has been introduced in the standard. This extension, known as the Enhanced Messaging Service (EMS), supersedes SMS by allowing subscribers to exchange messages containing elements such as melodies, pictures and animations. At the transfer layer, EMS messages are transported in the same way as SMS messages.

The next two chapters provide an in-depth description of the Enhanced Messaging Service.

Further Reading

G. Peersman and S. Cvetkovic, The Global System for Mobile Communications Short Message Service, IEEE Personal Communications Magazine, June, 2000.

4

Basic EMS

Without any doubt, the Short Message Service has been a tremendous commercial success. However, SMS traffic growth started to slow down. In 2000, to prevent this slowdown, several mobile manufacturers, mobile operators and third party vendors decided to give a further breath to SMS by collaborating on the development of an application-level extension of SMS: the Enhanced Messaging Service (EMS). EMS supersedes SMS capabilities by allowing the exchange of rich-media messages containing text with pictures, melodies, animations, etc. Standardization work went on for almost two years to define and finalize EMS features. A close analysis of the standardization work and availability of EMS handsets on the market leads to the identification of two EMS features sets. The first set of features is defined in 3GPP technical specifications release 99 (with several updates in release 4 and release 5) whereas the second set of features is defined in 3GPP technical specifications release 5. In this book, the first features set is described as *basic EMS* (this chapter) and the second features set is described as *extended EMS* (Chapter 5).

This chapter first describes how text can be formatted (alignment, font size and style). It also presents how to compose messages with text, small pictures, animations and monophonic melodies. The realization of a standardized download service with EMS content is outlined. The last section concludes by identifying the pros and cons of basic EMS. Extended EMS is presented in Chapter 5.

4.1 Service Description

Basic EMS allows the exchange of rich-media messages. Note that basic EMS features were mainly introduced in [3GPP-23.040] release 99, with several updates in release 4 and release 5. EMS messages can contain several of the following elements:

- Text, with or without formatting (alignment, font size and style);
- Black and white bitmap pictures;
- Black and white bitmap-based animations;
- Monophonic melodies.

One of the characteristics of EMS is that graphical elements (pictures and animations) and melodies are always placed in the text at a specific position: that of a character. This method looks appropriate for graphical elements. However, the applicability of this positioning method is less appropriate for melodies. Indeed, a melody is rendered by the receiving device when its associated character position in the message text becomes visible to the subscriber.

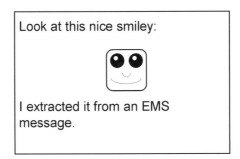

Figure 4.1   Example of rich-media EMS message

In basic EMS, an element position is always expressed with a character position in the text of the message segment in which it is contained (relative positioning). In basic EMS, an element has to fit into a single message segment and cannot be positioned in the text part of another message segment. Such an element cannot be segmented and spread over several message segments. Consequently, its maximum size is equal to the maximum size of a message segment payload (<140 octets).

Figure 4.1 shows an example of rich-media message formatted with basic EMS. In addition, for the machine-to-person scenario, basic EMS includes the possibility of realizing a download service with the capability to limit the distribution of downloaded elements.

EMS is an application-level extension of SMS using the concept of information element, as defined in the previous chapter, to convey EMS elements in messages. Each EMS element is associated with a dedicated information element. The payload of a dedicated information element is specifically structured to represent the corresponding EMS element (sequence of notes for a melody, bitmap for a picture, etc.). Information elements dedicated to basic EMS are listed in Section 3.15.1 and fully described in this chapter.

4.2 Basic EMS Compatibility with SMS

Two forms of compatibility with SMS were considered for the design of basic EMS: *forward compatibility* and *backward compatibility*. In this book, a device is said to be EMS-enabled if it supports at least one EMS feature (basic or extended). An EMS-enabled device is said to be backward compatible with an SMS-only device if it is capable of interpreting messages sent from SMS-only devices. All EMS-enabled devices are backward compatible with SMS-only devices. The other way round, an SMS-only device is said to be forward compatible with an EMS-enabled device if it supports messages sent from EMS-enabled devices. Existing implementations of SMS-only devices are able to correctly interpret the text part of EMS messages and simply ignore EMS-specific elements such as images, animations and sounds. It can therefore be said that SMS-only devices are forward-compatible with EMS-enabled devices.

Note that some EMS-enabled devices have partial support for the whole set of EMS features. For instance, an EMS-enabled device may support sounds only while another may support sounds, images and animations. If an EMS-enabled device receives a message containing an element which it is not capable of rendering, then the element is simply ignored by the device. Only the remaining part of the message is presented to the subscriber.

4.3 Formatted Text

Basic EMS allows text formatting instructions to be conveyed as part of the message. The appearance of the text can be formatted on the following aspects:

- Alignment (language dependent – default, align left, centre, align right);
- Font size (normal, small, large);

Table 4.1   IE/text formatting

IEI	0x0A from release 99		
	Text formatting		
IEDL	0x03 (3 octets)		
IED	Octet 1	**Start position** This octet represents the position of the first character of the message text to which the text formatting instructions are to be applied.	
	Octet 2	**Text formatting length** This octet represents the number of characters to which the text formatting instructions are to be applied.	
	Octet 3	**Formatting mode** This octet specifies how the associated text is to be formatted. The structure of the octet is as follows:	

Bit 1	Bit 0	Alignment
0	1	Align left
0	1	Centre
1	0	Align right
1	1	Language dependent (default)

Bit 3	Bit 2	Font size
0	0	Normal (default)
0	1	Large
1	0	Small
1	1	reserved

Bit 4	Bold
0	Bold off
1	Bold on

Bit 5	Italic
0	Italic off
1	Italic on

Bit 6	Underlined
0	Underlined off
1	Underlined on

Bit 7	Strikethrough
0	Strikethrough off
1	Strikethrough on

This message contains some **bold text**, *italic text*, <u>underlined text</u>, ~~strikethrough text~~ with various

font sizes.

Figure 4.2   Example of message with text formatting instructions

- Style (bold, italic, strikethrough and underlined).

The information element dedicated to text formatting is structured as shown in Table 4.1. If the text formatting length is set to 0, then text formatting instructions apply to all text characters from the start position. However, this may be overridden for a text section if a subsequent text formatting information element follows.

In the conflicting situation where several text formatting information elements apply to the same text section, text formatting instructions are applied in the sequential order of occurrence of corresponding information elements.

Note that this information element has been enhanced in extended EMS with the support of text foreground and background colours. This enhanced information element has an additional octet (octet 4 of IED, value 0x04 is assigned to IEDL). The description of the enhanced text formatting information element is given in Section 5.20. Figure 4.2 shows a message containing some formatted text. Figure 4.3 presents the content of the associated TPDU.

4.4 Pictures

In basic EMS, three types of black-and-white bitmap pictures can be included in messages as listed in Table 4.2. The first two types are used for the representation of pictures with predefined dimensions whereas the last type is for the representation of pictures for which dimensions can be configured by the message originator. Each picture format is associated with a dedicated information element for which the identifier is provided in Table 4.2. The bitmap of each picture is encoded in the data part of the associated information element.

4.4.1 Large Picture

The information element for a large picture (32×32 pixels) is structured as given in Table 4.3. The bitmap structure for a large picture is shown in Figure 4.4. The size for a large picture (16×16) is 129 octets (including position and bitmap, excluding IEI and IEDL fields).

4.4.2 Small Picture

Another type of picture that can be used in basic EMS is for the representation pictures of $16 \times$

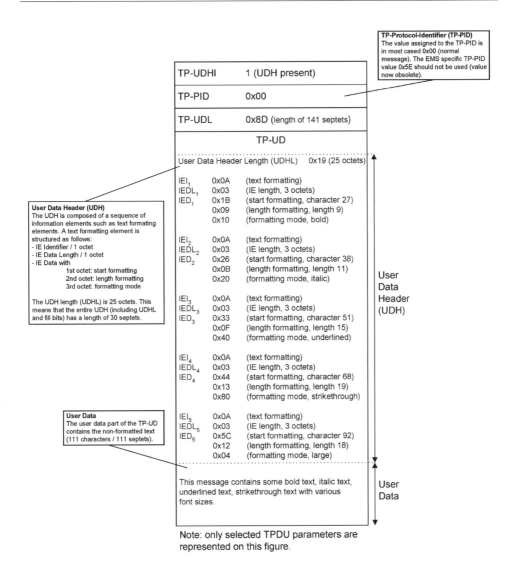

Figure 4.3 Text formatting information element/example

Table 4.2 Basic EMS bitmap pictures

IE Identifier	Description	Dimensions	Bitmap size
0x10	Large picture	32 × 32 pixels	128 octets
0x11	Small picture	16 × 16 pixels	32 octets
0x12	Variable-size picture	Variable	Variable

Table 4.3 IE/large bitmap picture

IEI	0x10 Large picture (32x32)	From release 99
IEDL	0x81 (129 octets)	
IED	Octet 1	**Large picture position** This octet represents the position in the text where the large picture is to be placed.
	Octets 2...129	**Large picture bitmap** This sequence of octets represents the bitmap of the large picture. The four first octets are used to encode the first row (top picture row), the four following octets are used to encode the immediately following picture row. For a 32 pixel-row, the 8 bits of the first octet represent respectively the 8 first pixels (pixels at the extreme left of the picture), the 8 bits of the second octet represent respectively the next 8 pixels and so on. The most significant bit of an octet represents the pixel which is on the left side whereas the least significant bit represents the pixel on the right side. A bit at 0 means that the pixel is white and a bit at 1 means that the pixel is black. The mapping between the sequence of octets and the pixels is shown in Figure 4.4.

16 pixels. The associated information element is structured as given in Table 4.4. The bitmap structure for a small picture is as shown in Figure 4.5. The size for a small picture (16×16) is 33 octets (including position and bitmap, excluding IED and IEDL fields).

Figure 4.6 shows how a picture of 16×16 pixels is encoded. The picture can be inserted in the text part of a message as shown in Figure 4.7.

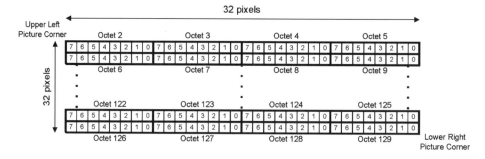

Figure 4.4 Large picture/bitmap representation

Table 4.4   IE/small bitmap picture

IEI	0x11 From release 99	
	Small picture (16x16)	
IEDL	0x21 (33 octets)	
IED	Octet 1	**Small picture position** This octet represents the position in the text where the small picture is to be placed.
	Octets 2...33	**Small picture bitmap** This sequence of octets represents the bitmap of the small picture. The two first octets are used to encode the first row (top picture row), the two following octets are used to encode the immediately following picture row. For a 16 pixel-row, the 8 bits of the first octet represent respectively the 8 first pixels (pixels at the extreme left of the picture) and the 8 bits of the second octet represent respectively the next 8 pixels. The most significant bit of an octet represents the pixel which is on the left side whereas the least significant bit represents the pixel on the right side. A bit at 0 means that the pixel is white and a bit at 1 means that the pixel is black. The mapping between the sequence of octets and the pixels is shown in Figure 4.5.

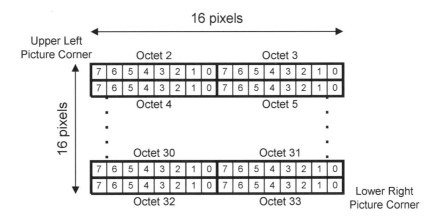

Figure 4.5   Small picture/bitmap representation

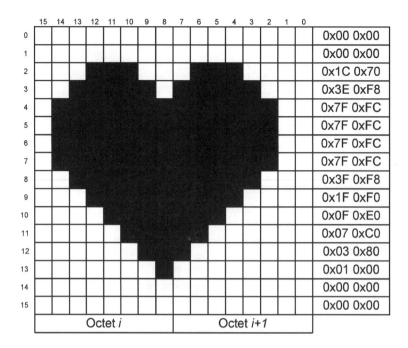

Figure 4.6 Example/small bitmap picture

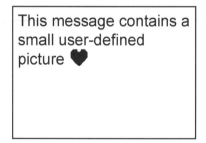

Figure 4.7 Message with a small bitmap picture

4.4.3 Variable-size Picture

In addition to small and large pictures (predefined dimensions), a picture with customizable dimensions can also be conveyed as part of a message. The information element dedicated to *variable-size pictures* (width × height) is structured as given in Table 4.5.

The picture size (including picture position, picture dimensions and picture bitmap, excluding IEI and IEDL fields), expressed in octets, is calculated as follows:

$$\text{Picture size (octets)} = \text{width (in 8 pixels)} \times \text{height (in pixels)} + 3$$

In basic EMS, a picture can only fit into one message segment at most and considering this limitation, the following conditions must be fulfilled:

Table 4.5 IE/variable-size bitmap picture

IEI	0x12 From release 99 Variable-size picture (width x height)	
IEDL	variable	
IED	Octet 1	**Variable-size picture position** This octet represents the position in the text where the variable-size picture is to be placed.
	Octet 2	**Horizontal dimension (width)** This octet represents the picture width. The unit is 8 pixels (for instance 2 represents a width of 16 pixels).
	Octet 3	**Vertical dimension (height)** This octet represents the picture height. The unit is 1 pixel.
	Octets 2...n	**Variable-size picture bitmap** This sequence of octets represents the bitmap of the variable-size picture. The bitmap is coded row per row from top left to bottom right. The first octets represent the first pixel row of the picture and so on.

Picture size ≤ 137 octets   for a one segment message.
Picture size ≤ 132 octets   for a concatenated message (8-bit reference number).
Picture size ≤ 131 octets   for a concatenated message (16-bit reference number).

Figure 4.8 presents a few examples of pictures available with existing handsets.

Box 4.1 Obsolete method for stitching pictures

3GPP technical specifications release 99 and release 4 define a method for segmenting a large picture (larger than a message segment) to cope with the size limitations of basic EMS. With this method, if multiple pictures with the same width are received side by side, then they are stitched together horizontally to form a larger picture. Similarly, if multiple series of stitched pictures of the same resulting width are received, separated by a carriage return only, then the multiple series of pictures are stitched together vertically to form an even larger picture. It has to be noted that very few handsets on the market support this stitching method, therefore it is not recommended to generate content based on this method. The User Prompt Indicator defined in Section 4.7 should be used instead. Another alternative is to use extended EMS in which elements do not have the size limitations of basic EMS. Extended EMS is presented in the next chapter.

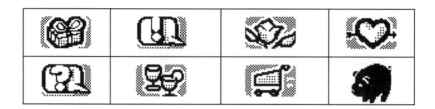

Figure 4.8   Examples of bitmap pictures. Reproduced by permission of Alcatel Business Systems

4.5 Sounds

With basic EMS, two types of sounds can be represented in messages: predefined and user-defined sounds.

A *predefined sound* is a commonly used sound for which the sound representation (sequence of notes) is included by default in the EMS-enabled device supporting this feature. Consequently, the sequence of notes of the predefined sound does not require to be included as part of the message. The sound identification only is sufficient. The advantage of using predefined sounds is that the associated information element requires a very limited amount of message space. Available predefined sounds are given below:

- Chime high
- Chime low
- Ding
- Claps
- Tada

- Notify
- Drum
- Fanfar
- Chord high
- Chord low

A *user-defined sound* is a sound for which the sequence of notes is entirely inserted in a message segment. With basic EMS, a user-defined sound is represented in the iMelody format and has a maximum size of 128 octets. The iMelody format is used for representing monophonic sounds/melodies (several notes cannot be rendered at the same time). It has to be noted that 128 octets are usually not enough for defining a sound of more than a few seconds (this explains why the word 'sound' is employed here instead of 'melody'). Extended EMS allows the definition of longer monophonic melodies or polyphonic melodies.

4.5.1 Predefined Sounds

The information element dedicated to predefined sounds is structured as shown in Table 4.6. Predefined sounds are implemented in a manufacturer-specific manner. This means that these sounds may be rendered differently on devices produced by different manufacturers (polyphonic or monophonic sounds, short or long sounds, etc).

4.5.2 User-defined Sound

The information element dedicated to user-defined sounds is structured as shown in Table 4.7. A user-defined sound consists of a sequence of instructions. Instructions include the execution of notes and the definition of additional information such as composer name, sound

Table 4.6   IE/predefined sounds

IEI	0x0B from release 99	
	Predefined sound	
IEDL	0x02 (2 octets)	
IED	Octet 1	**Predefined sound position** This octet represents the position in the text where the predefined sound is to be placed.
	Octet 2	**Predefined sound identification** This octet represents the identification of the predefined sound to be played. The following values can be assigned to this octet: **Id** **Description** 0 Chimes high 1 Chimes low 2 Ding 3 TaDa 4 Notify 5 Drum 6 Claps 7 Fanfare 8 Chord high 9 Chord low *other* *reserved*

Table 4.7   IE/user-defined sound

IEI	0x0C from release 99	
	User-defined sound	
IEDL	variable length	
IED	Octet 1	**User-defined sound position** This octet represents the position in the text where the user-defined sound is to be placed.
	Octets 2…n	**User-defined sound code** Octets 2…n represent the iMelody representation of the user-defined sound.

name, volume, etc. Such a sequence of instructions is represented in the iMelody format. The BNF grammar of the iMelody format is provided in Appendix D. The iMelody format is a monophonic sound format originally developed by the IrDA Forum in [IRDA-iMelody]. An iMelody sound is composed of an iMelody sound header, an iMelody sound body and an iMelody sound footer as shown in Figure 4.9.

iMelody Sound Header: the default alphabet for coding an iMelody is UTF8 (as defined in Appendix C). Each header parameter is coded over one line. The value assigned to a parameter is separated with a column from the parameter name and is placed on its right. The presence of a parameter in a sound header is either mandatory or optional. Table 4.8 provides an exhaustive list of sound header parameters.

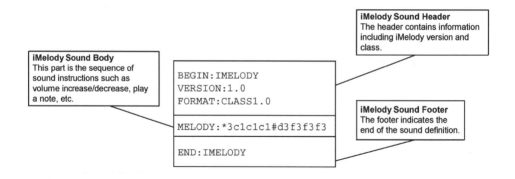

Figure 4.9   iMelody example

Table 4.8   iMelody commands[a]

	Parameter name	Description	Status	Example
1	BEGIN	Tag identifying the type of melody; in this case, the iMelody format	●	**BEGIN**: IMELODY
2	VERSION	The version of the melody in the format 'x.y'	●	**VERSION**: 1.2
3	FORMAT	The class of the melody; the class is represented by concatenating the word 'CLASS' with a version number in the format 'x.y'.	●	**FORMAT**: CLASS1.0
4	NAME	The name of the melody as a string of characters	○	**NAME**: MyMelody
5	COMPOSER	The name of the melody's composer as a string of characters	○	**COMPOSER**: MyName
6	BEAT	The beat value is expressed as a number of beats per minute (bpm). The value to be assigned to this parameter ranges from 25 to 900 (decimal values). This parameter is optional, if not specified, then notes are usually played at 120 bpm	○	**BEAT**: 63 (means that a 1/4 note has a duration of 60/63 = 0.95 seconds)
7	STYLE	Three styles can be specified for playing the melody: 'S0': natural style (with rests between notes) 'S1': continuous style (no rest between notes) 'S2': staccato style (shorter notes and longer rest period) If no style is defined in the melody definition, then the natural style is used as the default style	○	**STYLE**: S1
8	VOLUME	A volume level on a scale of 16 levels can be specified as part of the sound header. Volume level identifiers range from 'V0' to 'V15'. 'V0' means volume off whereas 'V15' is the maximum volume level. All other values refer to intermediary volume levels and 'V7' is the default volume level if this parameter has been omitted from the melody definition	○	**VOLUME**: V10

[a] Note: status is ● when the associated parameter is mandatory in the melody, otherwise the parameter is optional and the associated status is marked ○.

> ### Box 4.2 Recommendation for the creation of iMelody sounds
>
> In order to ensure that the melody is properly interpreted by the largest number of EMS-enabled devices, it is recommended to specify the header parameters in the melody in the order indicated in the first column of Table 4.8.

iMelody Sound Body: the melody body is composed of at least one command. Five command types can be mixed in order to build the melody. These command types shown in Tables 4.9–4.13. In addition to the five commands defined in the tables, a sequence of commands can be repeated more than once. For this purpose, the sequence to be repeated is inserted in the iMelody body between a pair of brackets:

`"("<sequence-of-commands> "@" <repeat-count> [<volume-modifier>] ")"`

where `<sequence-of-commands>` is the sequence of commands to be repeated, `<repeat-count>` is the number of times the sequence is to be repeated. It can take values such as '0' for repeat for ever, '1' for play once, '2' for play twice, and so on. `<volume-modifier>` can be assigned values 'V+' to increase the volume by one level each time the sequence of notes is repeated, or 'V−' to decrease the volume by one level each time the sequence of notes is repeated. This parameter is optional. With iMelody class 1.0, a sequence of commands to be repeated cannot be nested into a sequence of commands which is already designed to be repeated.

Table 4.9   iMelody/note command

Command	Note command
Description	The note command is used to play a note
Format	The note command is structured as follows:
	`<note><duration>[<duration-specifier>]`
	where `<note>` can be either:
	a basic note such as 'c', 'd', 'e', 'f', 'g', 'a' or 'b'.
	a flat note such as '&d', '&e', '&g', '&a' or '&b'.
	a sharp note such as '#c', '#d', '#f', '#g', '#a'
	where `<duration>` can take the following values:
	'0' for full note '1' for 1/2 note '2' for 1/4 note
	'3' for 1/8 note '4' for 1/16 note '5' for 1/32 note
	where `<duration-specifier>` can take the following values:
	'.' for dotted not ':' for double-dotted note ';' for 2/3 length
	The `<duration-specifier>` is an optional parameter in the note command

Table 4.10   iMelody/octave command-prefix

Command	Octave command
Description	The octave command indicates that the immediately following note is associated with a specific octave
Format	The octave command is a two-character command which takes one of the following values:

'*0' for 55 Hz	'*1' for 110 Hz	'*2' for 220 Hz
'*3' for 440 Hz	'*4' for 880 Hz	'*5' for 1760 Hz
'*6' for 3520 Hz	'*7' for 7040 Hz	'*8' for 14080 Hz

Note	At least one note command is to be inserted between two octave commands. If the octave level has not been specified, notes are played with the default octave level of 880 Hz (value '*4')

Table 4.11   iMelody/silence command

Command	Silence command
Description	The silence command is used to insert a silence between two notes
Format	The silence command is structured as follows: 'r' <duration>[<duration-specifier>] where <duration> and <duration-specifier> can take the same values as the note command <duration> and <duration-specifier>
Example	'r1:' means a rest for a half dotted note

Table 4.12   iMelody/volume command

Command	Volume command
Description	The volume command is used to change the volume level at which the following notes are going to be played
Format	The silence command can be assigned one of the following values: 'V1' to 'V15' to indicate that the following notes are to be played with respectively a low volume level to a high volume level. The silence command can also be assigned the values 'V+' or 'V−' to indicate respectively one volume level up or one volume level down

Table 4.13   iMelody/special effect command

Command	Special effect commands
Description	A special effect command is used to light on/off a led on the mobile handset, activate/deactivate the vibrator or light on/off the screen backlight
Format	Below are the list of special effect commands: 'ledon' for lighting on a mobile handset led 'ledoff' for lighting off a mobile handset led 'vibeon' for activating the vibrator 'vibeoff' for deactivating the vibrator 'backon' for lighting on the screen backlight 'backoff' for lighting off the screen backlight

Table 4.14 Differences between versions of the iMelody format

iMelody version	Supported commands
iMelody v1.0	Definition of header information, body commands and footer
iMelody v1.1	Corrections for relative and absolute volume commands in the iMelody body
	Corrections for the support of commands for lighting on and off the screen backlight
iMelody v1.2	Correction for the command for executing more than once a group of commands

iMelody Sound Footer: the iMelody sound is always ended by the sequence of characters: 'END:IMELODY'.

At the time of writing, three versions of the iMelody format have been publicly released. Table 4.14 provides a list of the differences between the three iMelody format versions.

Box 4.3 Short iMelody sound (without header and footer)

Headers and footers are mandatory in the iMelody format. However, in the context of EMS, they usually convey information which is not always required by the receiving device. This can be seen as a waste of message space. Devices can sometimes interpret correctly an iMelody composed of the iMelody body only (iMelody header and footer are omitted). Note that such a melody does not conform to the iMelody grammar and may not be interpreted correctly by all EMS-enabled devices. It is therefore not recommended to use this shorter form of iMelody sounds unless one can ensure that the receiving device supports it.

Box 4.4 Ambiguity in the iMelody grammar definition

The iMelody specification does not clearly define the signification of the octave prefix/command. It is unclear whether the octave prefix/command should apply only to the immediately following note or to all the notes until another octave prefix/command is encountered. After lengthy discussions between device manufacturers regarding the potential interoperability issue, it was decided to apply the octave prefix only to the immediately following note. Unfortunately, some handsets have been released on the market with the wrong interpretation of the iMelody format.

4.6 Animations

With basic EMS, several types of animations can be included in messages as shown in Table 4.15 which lists the size of each animation type (including animation position, predefined animation identification or animation bitmap pictures; excluding IEI and IEDL fields).

Table 4.15 Animations in basic EMS

IE Identifier	Description	Dimensions	Animation Size
0x0D	Predefined animation	Not applicable	2 octet
0x0E	Large user-defined animation	4 pictures (16 × 16 pixels)	129 octets
0x0F	Small user-defined animation	4 pictures (8 × 8 pixels)	33 octets

4.6.1 Predefined Animations

A *predefined animation* is a commonly known animation whose definition is supported by all EMS-enabled handsets. Consequently, the full representation (animated sequence of pictures) of such predefined animations is not transferred as part of messages. Only the identification of the predefined animation is included in messages. As with predefined sounds, the main advantage of using a predefined animation is its low requirement in terms of message space (four octets for the information element including one octet for the identification of the predefined animation).

The information element used to insert a predefined animation in a message is structured as shown in Table 4.16. The six first predefined animations (id 0–5) are defined in EMS release

Table 4.16 IE/predefined animation

IEI	0x0D Predefined animation	from release 99, new predefined animations added in release 4	
IEDL	0x02 (2 octets)		
IED	Octet 1	**Predefined animation position** This octet represents the position in the text where the predefined animation shall be displayed.	
	Octet 2	**Predefined animation identification** This octet represents the identification of the predefined animation to be displayed. The following values can be assigned to this octet:<table><tr><td>**Id**</td><td>**Description**</td><td>**Introduced in:**</td></tr><tr><td>0</td><td>I am ironic, flirty</td><td>Release 99</td></tr><tr><td>1</td><td>I am glad</td><td>Release 99</td></tr><tr><td>2</td><td>I am sceptic</td><td>Release 99</td></tr><tr><td>3</td><td>I am sad</td><td>Release 99</td></tr><tr><td>4</td><td>Wow !!!</td><td>Release 99</td></tr><tr><td>5</td><td>I am crying</td><td>Release 99</td></tr><tr><td>6</td><td>I am winking</td><td>Release 4</td></tr><tr><td>7</td><td>I am laughing</td><td>Release 4</td></tr><tr><td>8</td><td>I am indifferent</td><td>Release 4</td></tr><tr><td>9</td><td>In love / kissing</td><td>Release 4</td></tr><tr><td>10</td><td>I am confused</td><td>Release 4</td></tr><tr><td>11</td><td>Tongue hanging out</td><td>Release 4</td></tr><tr><td>12</td><td>I am angry</td><td>Release 4</td></tr><tr><td>13</td><td>Wearing glasses</td><td>Release 4</td></tr><tr><td>14</td><td>Devil</td><td>Release 4</td></tr><tr><td>*other*</td><td>*reserved*</td><td></td></tr></table>	

99. The nine additional animations (id 6–14) are defined in EMS release 4. If an EMS-enabled device conforms to EMS release 99 only, then only the six first predefined animations are rendered.

Predefined animations are implemented in a manufacturer-specific manner. This means that animations may be rendered differently on devices produced by different manufacturers.

4.6.2 User-defined Animations

Unlike predefined animations, *user-defined animations* are animations for which the representation of animated sequences of pictures is included in messages. A user-defined animation consists of a sequence of four pictures with identical dimensions. No timing information can be specified to define the display duration of each picture. Consequently, a handset usually represents an animation by looping over the four pictures. The display duration for each picture is often arbitrarily set to a few hundreds of milliseconds by the device manufacturer. User-defined animations have a maximum size of 128 octets. Furthermore, two types of user-defined animations can be inserted in messages: small user-defined animations (4 pictures 16 × 16 pixels) and large user-defined animations (4 pictures 8 × 8 pixels). Examples of user-defined animations are given in Table 4.17.

Table 4.17   Examples of user-defined animations[a]

Animation n°1	Animation n°2
(1) (2) (3) (4)	(1) (2) (3) (4)

[a] Reproduced by permission of Alcatel Business Systems.

Table 4.18   IE/large user-defined animation

IEI	0x0E from release 99		
	Large animation (4 pictures 16x16 pixels)		
IEDL	0x81 (129 octets)		
	Octet 1	**Large animation position** This octet represents the position in the text where the large animation is to be displayed.	
IEDL	Octets 2…129	**Large animation definition** These octets represent a sequence of four bitmap pictures as defined in Section 4.4. The picture dimensions are 16x16 pixels.	
		Octets 2..33	First picture (16x16 pixels).
		Octets 34..65	Second picture (16x16 pixels).
		Octets 66..97	Third picture (16x16 pixels).
		Octets 98..129	Fourth picture (16x16 pixels).

Table 4.19   IE/small user-defined animation

IEI	0x0E from release 99	
	Small animation (4 pictures 8x8 pixels)	
IEDL	0x21 (33 octets)	
IED	Octet 1	**Small animation position**
		This octet represents the position in the text where the small animation is to be displayed.
	Octets 2…33	**Small animation definition**
		These octets represent a sequence of four bitmap pictures as defined in Section 4.4. The picture dimensions are 8x8 pixels.
		Octets 2..9 First picture (8x8 pixels).
		Octets 10..17 Second picture (8x8 pixels).
		Octets 18..25 Third picture (8x8 pixels).
		Octets 26..33 Fourth picture (8x8 pixels).

The information element used to insert a large user-defined animation in a message is structured given in Table 4.18. The information element used to insert a small user-defined animation in a message is structured as shown in Table 4.19.

In basic EMS, there is no means for defining an animation whose dimensions are configured by the message originator.

4.7 User Prompt Indicator

As indicated in the introduction of this book, the download service has become very popular. This service allows subscribers to download objects, mainly to customize handsets. This includes the download of ringtones, switch-on and switch-off animations, etc. For this purpose, the object being downloaded is contained in a *download message*. A simplified network configuration for the support of a download service, and the necessary steps to download a ringtone, are shown in Figure 4.10.

Before the introduction of EMS, each manufacturer willing to support a download service had to specify its own proprietary object formats and its own proprietary transport method. Such proprietary download services presented a major drawback: the content of the download message had to be specifically generated according to the characteristics of the receiving device. This made the deployment of generic and widely available download services impossible.

With EMS, a standard transport protocol is used in the form of SMS. Additionally, a set of standard object formats are defined as part of basic EMS. Manufacturers, operators and content providers quickly realized that EMS can be used appropriately as the building block for the realization of a generic and standard download service. However, basic EMS presents one limitation: it can transport only small objects. This is a strong limitation for a download service which requires the download of potentially large objects (up to 512 octets). In order to cope with this EMS limitation, a basic application-level object segmentation and reconstruction method has been introduced in basic EMS. This method, allowing the trans-

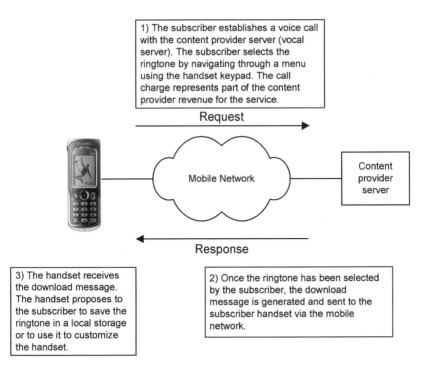

1) The subscriber establishes a voice call with the content provider server (vocal server). The subscriber selects the ringtone by navigating through a menu using the handset keypad. The call charge represents part of the content provider revenue for the service.

Request

Mobile Network

Content provider server

Response

3) The handset receives the download message. The handset proposes to the subscriber to save the ringtone in a local storage or to use it to customize the handset.

2) Once the ringtone has been selected by the subscriber, the download message is generated and sent to the subscriber handset via the mobile network.

Figure 4.10 Typical architecture for the download service

port of large objects, is based on the use of an additional information element: the *User Prompt Indicator* (UPI). The purpose of this information element is twofold:

1. The UPI information element is used to indicate that a message (short message or concatenated message) contains an object being downloaded for the customization of the receiving handset.
2. The UPI information element is also used to indicate that the object being downloaded has been segmented over several message segments and needs to be reconstructed by the receiving device.

The structure of the UPI information element is as shown in Table 4.20

Table 4.20 IE/user prompt indicator

IEI	0x13	from release 4
	User Prompt Indicator	
IEDL	0x01 (1 octet)	
IED	Octet 1	**Number of associated Information Elements** This octet represents the number of object segments to be reconstructed to build the downloaded object.

4.7.1 UPI Management

Upon receipt of the download message, the receiving device can propose to the subscriber:

- to save the downloaded object locally in order to be used later for composing new messages;
- to customize the device (change the default ringtone, change switch-on and switch-off animations, etc.).

The set of actions that may be performed on a downloaded object obviously depends on the object format and on the capabilities of the receiving device. The UPI information element is used in the machine-to-person scenario. This information element is usually not supported in the person-to-person scenario. The text that may be placed in a download message is usually not presented to the subscriber. It is therefore not recommended to add text as part of a download message.

4.7.2 UPI Segmentation and Reconstruction

The UPI reconstruction method can only be applied to melodies (user-defined) and pictures (variable, large and small).

- *Melody Reconstruction:* melody segments are stitched together according to the concatenation index number. Only the header and footer of the first melody segment are kept in the reconstructed melody.
- *Picture reconstruction:* the height of all picture segments shall be identical, otherwise the UPI is ignored. Picture segments are stitched vertically in order to build a larger picture.

Figure 4.11 shows how a large picture can be segmented into three picture slices in a download message.

4.8 Independent Object Distribution Indicator

The *independent Object Distribution Indicator* (ODI) is used to control the way the receiving SME can redistribute one or more objects contained in an EMS message. The major use case for the independent ODI is to control the redistribution of copyrighted content once it has been received by the subscriber. In the situation where the subscriber purchases copyrighted content, the content provider is able to indicate that one or more objects contained in the message cannot be redistributed via SMS. This means that the message cannot be forwarded by the receiving SME. This also means that objects can be extracted from the message but cannot be reused to compose new messages. For this purpose, information elements representing basic EMS objects, which must be limited for redistribution, are preceded by an independent ODI information element. One single ODI information element can be associated with one or more objects. The structure of the ODI information element is as shown in Table 4.21

If an object is not associated with any object distribution indicator, then the object is considered as not being limited in its distribution. Although the independent ODI information element was not introduced before release 5, this information element is considered, in this book, as a basic EMS feature. This is explained by the fact that this object distribution

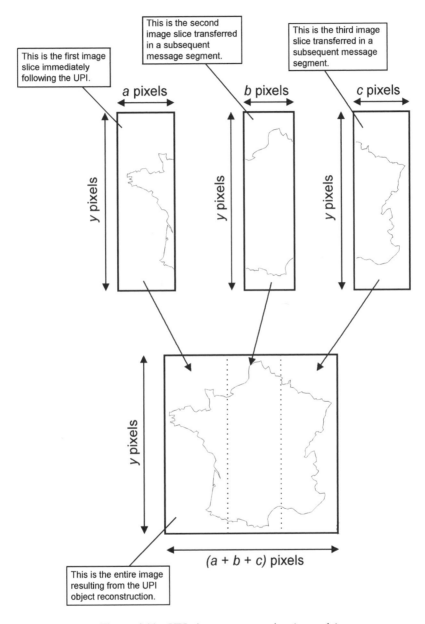

Figure 4.11   UPI picture segmentation (example)

indicator is used to limit the distribution of basic EMS objects only. The extended EMS defines another method for limiting the distribution of extended EMS objects. In this book, this method is known as the *integrated ODI* and is presented in the next chapter.

Figure 4.12 shows how the redistribution of two objects can be limited with the use of one independent ODI information element.

Table 4.21   IE/object distribution indicator

IEI	0x17 from release 5	
	Object Distribution Indicator	
IEDL	0x02 (2 octets)	
IED	Octet 1	**Number of associated information elements** This octet represents the number of information elements associated with the distribution indication. If set to 0, then the distribution indication applies to all information elements until the end of the message or until another ODI is encountered.
	Octet 2	**Distribution status** The distribution status octet is coded as follows: **Bit 0 Distribution attribute** 0 associated object(s) may be forwarded. 1 associated object(s) shall not be forwarded by SMS. **Bit 1..7 Reserved for future use** Unused bits shall be set to 0.

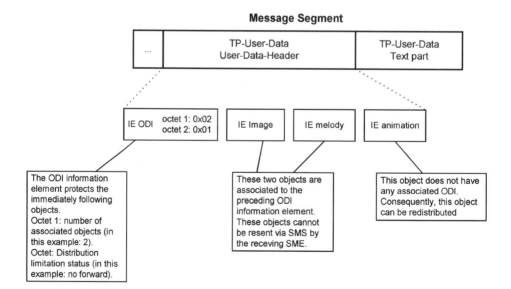

Figure 4.12   ODI (example)

4.9 EMS Features Supported by Existing Handsets

The first EMS-capable devices started to appear on the market in 2001. New models are released regularly with various levels of support for EMS features. Table 4.22 provides a list of supported basic EMS features for a selection of commercially available devices. This list is not exhaustive.

4.10 Content Authoring Tools

Mobile manufacturers and service providers have developed a number of tools to allow content providers to easily generate EMS messages. Several of these tools are presented in this section.

4.10.1 Alcatel Multimedia Conversion Studio

Alcatel provides a tool called the Multimedia Conversion Studio. Features of this tool include:

- conversions of widely used formats (BMP, MIDI, etc) into EMS formats (iMelody, bitmap pictures, etc.) with preview of converted elements;
- creation of message TPDUs, ready to be sent to remote mobile devices;
- batch facilities for converting a set of elements.

Figure 4.13 is a screenshot showing the process of converting a MIDI sound into an iMelody EMS sound. The Multimedia Conversion Studio can convert the following formats into EMS formats:

- BMP and PNG for images;
- WAVE and MIDI for sounds/melodies.

After registration, this tool can be downloaded from http://www.alcatel.com/wap.

4.10.2 Miscellaneous

Sony-Ericsson provides a tool which converts existing Nokia Smart Messaging logos and ringtones into EMS formats. After registration, this tool can be downloaded from http://www.ericsson.com/mobilityworld.

4.11 Pros and Cons of Basic EMS

Compared to SMS, basic EMS is a substantial evolution for the person-to-person messaging scenario. With basic EMS, messages containing formatted text pictures, sounds and animations can be exchanged between subscribers.

As far as the network SMSC supports message concatenation, then the network operator only needs to provide EMS-enabled devices to subscribers to activate the EMS service. In other words, the availability of an EMS-enabled device is the only requirement for activating the service. In particular, the service becomes automatically available without any network

Table 4.22 EMS features supported by commercial devices

	Alcatel			Motorola				Siemens			Sony-Ericsson							
	311/511	511/512	715	T192i	T280i-m	V66i-m	V60i-m	C45	S45	ME45	T20	T29	R520	T39	T65	T66	T68	R600
Text formatting			●					●	●					●				●
Sound																		
Predefined	●	●	●	●	●	●	●		●	●			●	●	●	●	●	●
User defined	●	●	●	●	●	●	●		●	●			●	●		●	●	●
UPI																		
Picture																		
Small	●	●	●	●	●	●	●		●	●	●	●	●	●	●	●	●	●
Large	●	●	●	●	●	●	●		●	●	●	●	●	●	●	●	●	●
Variable	●	●	●	●	●	●	●		●	●	●	●	●	●	●	●	●	●
UPI																		
Animation																		
Predefined	●	●	●	●	●	●	●		●	●			●	●		●	●	●
Small	●	●	●	●	●	●	●		●	●			●	●		●	●	●
Large	●	●	●	●	●	●	●		●	●			●	●		●	●	●
Variable	●	●	●	●	●	●	●		●	●			●	●		●	●	●
Concatenation[a]	10	10	10							3	3	3	6	6	3	6	6	4

[a] Available handsets support the concatenation of message segments up to a given number of segments. The last row of the table indicates the maximum number of segments supported for each handset, where known.

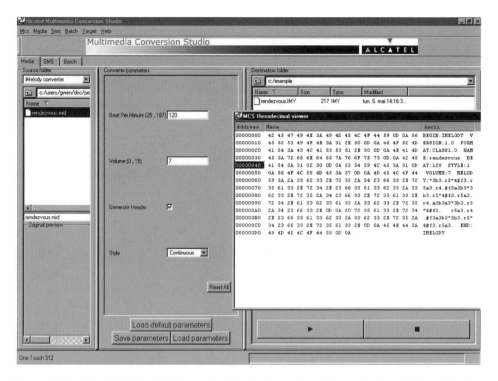

Figure 4.13 Alcatel Multimedia Conversion Studio. Reproduced by permission of Alcatel Business Systems

infrastructure upgrade. This makes EMS a service easy to deploy without additional network elements, other than capacity due to the potential increase in traffic volumes.

However, basic EMS suffers two major limitations:

- *Object size limitation:* the SMS concatenation mechanism can be used to build potentially large messages. However, in such large messages, each object (picture, animation or sound) has to fit into one message segment. This does not allow the exchange of pictures with large dimensions and melodies/sounds have to be very short. A high-level method for segmentation and reconstruction of objects has been defined in basic EMS in the form of the User Prompt Indicator concept. However, this concept is used only for the download service and does not apply to the person-to-person messaging scenario.
- *Minimal set of supported objects:* the number of object formats supported by basic EMS is limited. For instance, only black-and-white pictures and animations are supported. There is no support for greyscale or colour. Sounds are short and monophonic.

It can be said that basic EMS is a significant improvement for the person-to-person scenario. However, this service has many limitations that do not make the service very attractive for most professional uses. The next chapter presents the extended EMS, which copes with most of the limitations of basic EMS.

5

Extended EMS

The previous chapter presented basic EMS, an application-level extension of SMS. Basic EMS enables elements such as black-and-white bitmap pictures and animations, and monophonic sounds to be inserted in messages. It was shown that basic EMS has many limitations preventing the development of attractive services, in particular for commercial and professional uses.

To cope with these limitations, standardization work has been carried out on the development of an additional set of EMS features. This evolutionary step is designated in this book as extended EMS. Like basic EMS, extended EMS is also an application-level extension of SMS that builds on basic EMS by enabling the inclusion of large objects in messages. In addition to elements already supported in basic EMS, extended EMS also supports greyscale and colour bitmap pictures and animations, monophonic and polyphonic melodies, vector graphics, etc. For this purpose, a framework has been designed to cope with the object size limitation of basic EMS. Compression of objects is also supported in extended EMS to allow the development of cost-effective services.

This chapter, dedicated to extended EMS, first outlines features of extended EMS and describes the principle ensuring backward compatibility with existing SMS and basic EMS-enabled devices. The extended EMS framework is also presented along with objects that can be included as part of extended EMS messages. Compression and decompression of extended EMS objects are also described and illustrated.

5.1 Service Description

In order to break the limitations of the basic EMS, an extended version of EMS has been developed. Extended EMS features were mainly introduced in [3GPP-23.040] release 5. In addition to all the features of basic EMS, extended EMS-enabled devices support the following features:

- *A framework for the support of extended objects:* in the framework, an *extended object* is either an object with a type supported in basic EMS (which no longer has the basic EMS size limitation) or an object of a new type defined in extended EMS. One of the main limitations of basic EMS resides in the impossibility of including large objects in messages. With basic EMS, concatenation can be used for building large messages with many objects but each single object is limited to fit into one message segment. With extended EMS, objects are no longer limited to the size of one message segment but may be segmented and spread over more than one message segment.

- *Compression of objects:* since the extended EMS framework allows large objects to be included in messages, the number of segments per message can become significantly high. In order to allow the development of cost-effective services, a method for compressing extended objects has been introduced in extended EMS.
- *Integrated Object Distribution Indicator (ODI):* in the previous chapter, it was shown that an independent ODI could be used for limiting the distribution of basic EMS objects. Similarly, an ODI tag has been integrated into the definition of each extended object. This tag is known, in this book, as the integrated ODI.
- *A new set of objects:* additional object formats are supported for the construction of extended EMS messages. The entire set of supported object formats is given below:
  - black-and-white bitmap pictures (also supported in basic EMS)
  - 4-level greyscale bitmap pictures
  - 64-colour bitmap pictures
  - black-and-white bitmap animations (also supported in basic EMS)
  - 4-level greyscale bitmap animations
  - 64-colour bitmap animations
  - vCard data streams (used to define business cards)
  - vCalendar data streams (used to define appointments, reminders, etc.)
  - monophonic (iMelody) melodies (also supported in basic EMS)
  - polyphonic (MIDI) melodies
  - vector graphics.

- *Colour formatting for text:* in basic EMS, text formatting is limited to changing the text alignment (left, right and centre), font style (bold, italic, underlined, strikethrough) and font size (small, normal or large). In addition to these basic features, text background and foreground can also be coloured with extended EMS.
- *Hyperlink:* the hyperlink feature allows the association of some text and/or graphical elements (pictures, animations, etc) with a Uniform Resource Identifier (URI).
- *Capability profile:* many EMS-enabled devices have partial support for basic and extended EMS features. A mechanism in extended EMS allows SMEs to exchange their extended EMS capabilities. This enables SMEs (e.g. an application server) to format the content of messages according to what a specific recipient device is capable of rendering.

5.2 Extended EMS Compatibility with SMS and Basic EMS

Two forms of compatibility, forward compatibility and backward compatibility, were introduced in Section 4.2. At the time of writing this book, no extended EMS device is available on the market. However, considering design principles of SMS and EMS, it can be said that extended EMS devices will be backward compatible with SMS-only devices. It is expected that extended EMS devices will also be backward compatible with basic EMS devices. However, a manufacturer could design an extended EMS device that does not understand EMS messages generated by basic EMS devices, but such a device would be a bit awkward to introduce into the market. SMS-only devices and basic EMS devices will correctly interpret the text part of messages generated by extended EMS-enabled devices. Specific extended

EMS content is simply ignored by SMS-only devices and basic EMS devices. It is possible to format a message with a combination of both extended and basic EMS objects.

5.3 Extended Object Framework

The extended EMS framework allows a large object to be segmented and spread over more than one contiguous message segment. For this purpose, User-Data-Header information elements, introduced in Section 3.15, are used to avoid potential impacts on the network infrastructure for the deployment of extended EMS. It also enables compatibility with devices already available on the market.

With this framework, a large object is segmented into several parts and each part is included in the payload of a dedicated information element. Each information element, known as an *extended object information element*, is conveyed in one message segment with other information elements (SMS, basic and extended EMS) and optional text.

The representation of a large bitmap picture with three extended object IEs is shown in Figure 5.1. In the example, the first part of the picture is encoded in the first extended object IE. For a concatenated message, the maximum length of the picture segment in the first

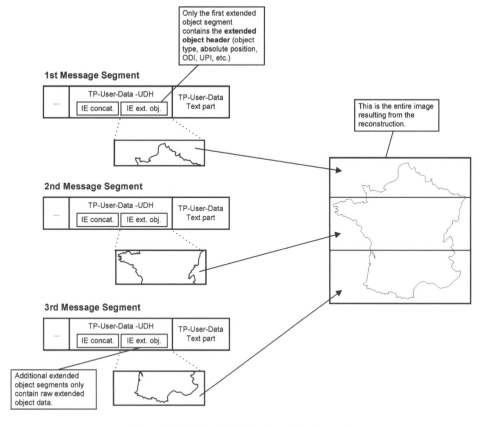

Figure 5.1 Extended object encoding/example

extended object is 124 octets.[1] In addition to containing the first picture segment, this first IE extended object also contains the *extended object header*. This extended object header is composed of the following parameters:

- *Object length:* this is the length of the entire extended object, expressed in octets. The part in the first extended object IE (excluding the extended object header) plus parts in additional extended object IEs are taken into consideration for the calculation of the length. The knowledge of the object length is required by the receiving SME to be able to correlate extended objects with corresponding information elements. This information is required because additional extended object IEs do not refer to any particular object reference number (or object index number).
- *Object handling status:* the object handling status indicates (1) whether or not the associated object is part of a download message and (2) whether or not the object can be redistributed by SMS.
- *Object type:* the object type indicates which type of object is being conveyed (picture, animation, etc.).
- *Object position:* because an extended object can be segmented over several message segments, this object position indicates the absolute message text position where the object should be placed. Note that in basic EMS, only relative positions (positions in the text of the message segment where the object is inserted) can be specified.

If the large object cannot be conveyed using one single extended object IE, then additional extended object IEs may be used. For each additional extended object IE, the extended object header is omitted. Consequently, any additional extended object IE contains only raw data from the extended object remaining parts (parts that could not be included in the first extended object IE). Due to this, an additional extended object IE can contain an extended object segment for which the size can reach 131 octets.[2]

The structure of the first extended object IE (including the extended object header) is shown in Table 5.1. Octets 1 to 7 of this extended object IE represent the extended object header and are not taken into consideration for the object length calculation (see description of octets 2 and 3).

In the situation where more than one extended object IE has to be used to convey an extended object, the concatenation IE 16-bit reference number shall be used. The use of this concatenation IE, compared with the 8-bit reference number concatenation IE, helps reduce the probability of receiving conflicting concatenated message segments (see Box 3.2).

If the extended object is too large to fit into one message segment, then the first part of the object is inserted into a first extended object IE as defined in Table 5.1. The remaining part(s) of the extended object is/are inserted into subsequent message segments in additional extended object IEs as defined in Table 5.2. Note that the first and additional extended object information elements have the same IE identifier (value 0x14).

---

[1] Calculation: 140 octets (TP-UD size) − 1 octet (UDHL) − 6 octets (concat. IE, 16 bit reference number) − 9 octets (first extended object IE without payload) = 124 octets.
[2] Calculation: 140 octets (TP-DU size) − 1 octet (UDHL) − 6 octets (concat. IE, 16 bit reference number) − 2 octets (extended object IE without payload and without extended object header) = 131 octets.

Table 5.1   IE/extended object (first)

IEI	0x14 from release 5
	Extended Object (first)
IEDL	variable

IED	Extended Object Header	Octet 1	**Reference number** This octet represents the reference number of the extended object. This reference number may be used for re-inserting the object in another part of the message. Each extended object, in a message, shall have a unique reference number.
		Octets 2 and 3	**Object length** These octets represent the length of the object, expressed in octets. This length comprises the part of the object which is included in the first extended object information element (excluding extended object header) plus object parts contained in additional extended object information elements.
		Octet 4	**Object handling status** This octet indicates how the object should be handled by the receiving SME. **Bit 0 Integrated ODI** This bit indicates whether or not the message can be redistributed by SMS. 0 Object may be redistributed. 1 Object shall not be redistributed by SMS. If this bit is set to 1, then the associated object is handled as shown in Section 4.8 **Bit 1 User Prompt Indicator** This bit indicates whether or not the object is part of a download message. 0 Part of a download message. 1 Not part of a download message. If this bit is set to 1, then the object is handled as shown in Section 4.7 Unused bits for this octet shall be set to 0.
		Octet 5	**Extended Object Type** This octet indicates which type of object is being conveyed as part of this information element. Available object formats are: (dec.) (hex.) Object type See Section 0 0x00 Predefined sound 5.7 1 0x01 iMelody melody 5.8 2 0x02 Black and white picture 5.9 3 0x03 4-level greyscale picture 5.10 4 0x04 64-colour picture 5.11 5 0x05 Predefined animation 5.12 6 0x06 Black and white animation 5.13 7 0x07 4-level greyscale animation 5.14 8 0x08 64-colour animation 5.15 9 0x09 vCard object 5.16 10 0x0A vCalendar object 5.17 11 0x0B Vector graphic (WVG) 5.19 12 0x0C Polyphonic melody (MIDI) 5.18 255 0xFF Capability return 5.22 other reserved
		Octets 6 and 7	**Extended Object Position** These octets represent the absolute character position in the message where the object is placed.
		Octets 8 to n	**Extended Object Definition** These octets represent the entire object definition or the first part of the object definition. If more than one message segment are required to convey the object, then the first object part is inserted in this information element and the remaining part(s) of the object is/are inserted in subsequent message segments (inserted as part of other extended object information elements 0x14).

Table 5.2   IE/extended object (additional)

IEI	0x14 from release 5	
	Extended Object (additional)	
IEDL	variable	
IED	Octets 1..n	These octets represent an additional extended object segment. Unlike the first extended object IE, an additional extended object IE does not contain an extended object header (reference number, object length, etc.)

Box 5.1 Recommendation for a Maximum Size for Extended EMS Messages

A receiving SME is capable of interpreting a message composed of up to eight message segments. A message composed of more than eight message segments might not be interpreted correctly by all receiving SMEs. It is therefore highly recommended to limit the number of segments per message to eight.

Box 5.2 Recommendation for the Use of Integrated and Independent ODIs

Two ODIs are available: an independent ODI (basic EMS) and an integrated ODI (extended EMS). The integrated ODI is part of the extended object header and the independent ODI is managed via an independent IE which is associated with one or more objects. If the object for which the distribution is limited is an extended object, then it is recommended to use the integrated ODI (resource gain). If the object for which the distribution is limited is not an extended object (basic EMS objects), then the use of the independent ODI is necessary. In this latter case, SMEs that do not support the ODI concept just ignore the independent ODI IE but interpret associated objects. In this situation, SMEs are able to freely distribute objects for which the message originator requested a limited distribution.

Figure 5.2 shows the encoding of a large black-and-white picture conveyed as part of a concatenated message composed of two segments.

5.4 Extended Object Reuse

In the extended object framework, each extended object is associated with a unique[3] *extended object reference number* (see extended object header). This reference number can be used for reinserting the associated extended object in other part(s) of the message. In order to perform this, an additional IE, known as the Reused Extended Object IE, has been introduced in extended EMS. Figure 5.3 shows a message where a first picture is defined once and reused

---

[3] The extended object reference number is unique in the message only.

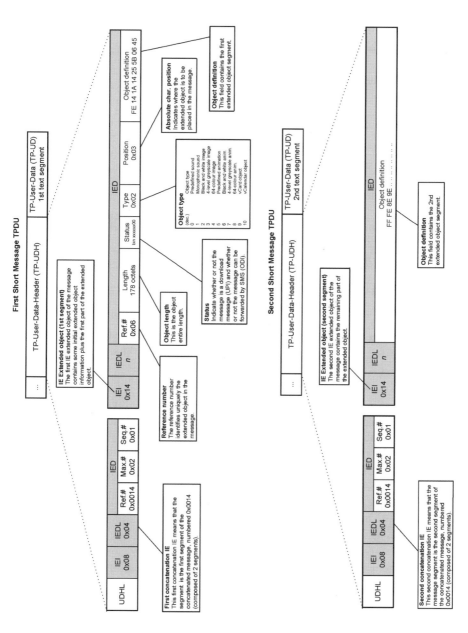

Figure 5.2 Extended object encoding

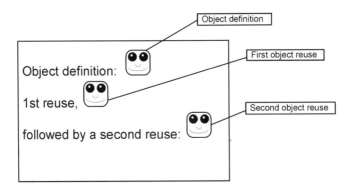

Figure 5.3   Example/object reuse

Table 5.3   IE/reused extended object

IEI	0x15 from release 5	
	Reused Extended Object	
IEDL	0x03 (3 octets)	
IED	Octet 1	**Reference number** This octet represents the reference number of the already-defined object.
	Octets 2 and 3	**Absolute character position** These octets represent the absolute character position in the message where the reused extended object shall be placed.

twice subsequently in other positions in the message. The Reused Extended Object IE has the structure shown in Table 5.3. The absolute character position of the original extended object is overwritten by the position indicated as part of the Reused Extended Object IE.

Box 5.3 Recommendations for the Reuse of Extended Objects

It is recommended to define the object to be reused in the first segment(s) of the message with the extended object IE and to reuse the defined extended object in the same or next message segments (according to the message concatenation index).

An extended object defined in a message can be reused in any part of the message in which it has been defined. However, an extended object cannot be reused, with the reused extended object IE, in other messages.

5.5 Compression of Extended Objects

Extended objects can be significantly larger than objects supported in basic EMS. In order to avoid an explosion of the number of message segments exchanged between SMEs, a support of compression/decompression has been introduced in extended EMS. Note that compression applies to extended objects only. Basic EMS elements cannot be decompressed/compressed unless they are encoded as extended objects. Several extended objects may be compressed together into a single compressed stream. Compressing several objects together usually achieves a better compression ratio.

5.5.1 Compressed Stream Structure

The compression mechanism introduced in extended EMS has similarities with the way uncompressed extended objects are encoded using the extended object IE. Compression is managed by a dedicated information element: the *compression control IE*. If a compressed stream is too large to fit into one message segment, then the compressed stream may be segmented in several parts and each part is conveyed into a single compression control IE. Each compression IE is inserted in a message segment along with other information elements and optional text. If several compression control IEs are used to convey a large compressed stream, then only the first compression IE contains the *compression control header*. Additional compression control IEs only contain raw compression data. The compression control header contains the following information:

- *Compression method:* this field identifies the compression method which has been used to compress extended objects, along with optional compression settings. At the time of writing, the only compression/decompression method supported in extended EMS is based on LZSS compression principles.
- *Compressed bytestream length:* this is the length of the entire compressed bytestream (including the part in the first compression control IE plus part(s) in additional compression control IEs). This compressed bytestream length is required for the receiving SME to be able to correlate compressed streams with corresponding compression control IEs (more than one compressed bytestream may be present in a message). This information is necessary because additional compression control IEs do not refer to a particular compressed stream reference/index number.

Octets 1–3 of the compression control IE represent the compression control header and are not taken into consideration for the calculation of the compressed bytestream length (see description of octets 2 and 3).

Note that the compression/decompression method introduced in extended EMS is not used to compress the text part of the message: an SME which does not support the extended EMS compression/decompression method is still able to interpret the text part of a message containing one or more compressed bytestreams.

The first compression control IE is structured as shown in Table 5.4. If the compressed stream is too large to fit into one message segment, then the first part of the compressed stream is conveyed in the first message segment and remaining parts are conveyed in subsequent message segments with the information element shown in Table 5.5.

Table 5.4   IE/compression control (first)

IEI	0x16 from release 5 Compression control (first)		
IEDL	variable		
IED	**Compression control header**	Octet 1	**Compression method** This octet informs on the method which has been used for compressing the bytestream. **Bit 3...0 compression method** 0000 LZSS compression 0001...1111 reserved **Bit 7...4 compression parameters** Value 0000 (binary) is assigned to this parameter for the LZSS compression scheme. Other values are reserved.
		Octets 2 and 3	**Compressed bytestream length** These octets represent the length of the compressed stream (excluding compression control header). The compressed stream may expand over several message segments.
		Octets 4 to n	**Compressed bytestream definition** These octets represent the entire compressed bytestream definition or the first part of the compressed bytestream definition. If more than one message segment are required to represent the compressed bytestream, then the first compressed bytestream part is inserted in this information element and remaining part(s) of the compressed bytestream is/are inserted in subsequent message segments (inserted in other compression control information elements 0x16). The uncompressed bytestream consists of a sequence of extended objects or reused extended objects (IEI and IED only excluding IEDL). Note that only extended objects and reused extended objects can be compressed using this method. Basic EMS objects cannot be compressed with this method.

Table 5.5 IE/compression control (additional)

IEI	0x16		from release 5
	Compression control (additional)		
IEDL	variable		
IED	Octets 1..n	These octets represent an additional part of a compressed stream. Unlike the first compression control IE, this IE does not contain a compression control header.	

Box 5.4 Recommendation for a Maximum Size for Messages Containing Compressed Extended Objects

A receiving SME is always able to interpret messages of an uncompressed size of up to eight message segments. A message with one or more compressed streams for which the uncompressed size is over eight message segments may not be interpreted correctly by all receiving SMEs. It is therefore highly recommended to limit the number of segments per message to eight (before compression). Note that decompression is mandatory for devices supporting extended EMS objects. On the other hand, compression is optional.

Figure 5.4 shows the encoding of a compressed stream containing three extended objects conveyed in a concatenated message composed of two message segments.

5.5.2 Compression and Decompression Methods

With extended EMS, the only compression method supported so far is derived from the LZSS compression principle.[4] LZSS-derived algorithms are often called dictionary-based compression methods. These methods compress a stream by adding, in the corresponding compressed stream, references to previously defined octet patterns rather than repeating them. The compression ratio for such dictionary-based methods is therefore proportional to the frequency of occurrences of octet patterns in the uncompressed stream.

After compression, a compressed bytestream is composed of two types of elements: data blocks and block references. A *data block* contains an uncompressed block of octets. On the other hand, a *block reference* is used to identify a sequence of octets in the uncompressed stream in order to repeat the identified sequence simply by referring to it. Figure 5.5 shows an example of a compressed bytestream structure. The data block is composed of a length and a payload. The length indicates the payload length in octets. The payload contains a sequence of up to 127 octets. The structure of a data block is shown in Figure 5.6. The block reference is composed of a repeated block length followed by a block location offset. The block reference is structured as shown in Figure 5.7.

---

[4] The Lempel–Ziv–Storer–Szymanski (LZSS) compression principle is a modified version of the LZ77 compression principle proposed by Storer and Szymanski.

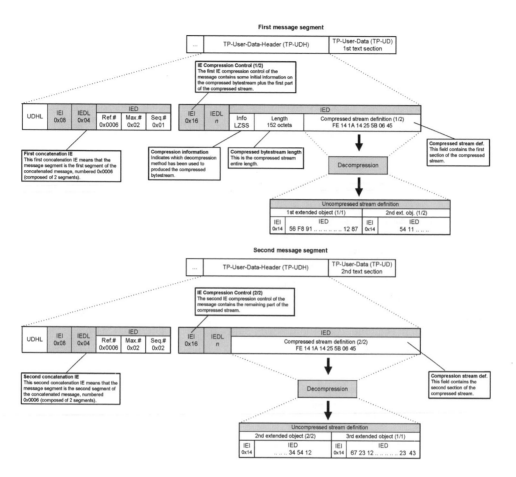

Figure 5.4   Compression control encoding

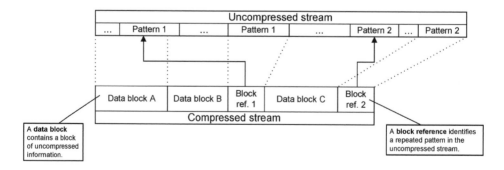

Figure 5.5   Structure of a compressed stream

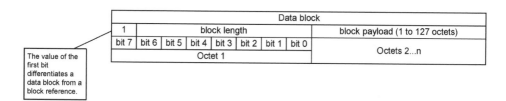

Figure 5.6   Structure of a data block

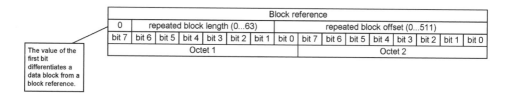

Figure 5.7   Structure of a block reference

Figure 5.8   Decompression process

5.5.3 Decompression Method

The decompression method consists of generating an output uncompressed stream from an input compressed stream (Figure 5.8). The input compressed stream is composed of a compression control header followed by a sequence of data and reference blocks. At the beginning of the decompression process, the output uncompressed stream is empty. A pointer starts from the beginning of the input compressed stream. All data blocks and block references are interpreted in their sequential order of occurrence in order to build the output uncompressed stream. Once a data block or block reference has been interpreted, then the pointer moves to the next data block or block reference until the end of the compressed stream is reached. If the pointer is located on a data block, then the block payload is extracted and appended at the end of the output uncompressed stream. If the pointer is located on a block reference, then a block of octets is identified in the output uncompressed stream and is appended at the end of the output uncompressed stream. The block to be repeated is the one which has the size specified in the reference block (repeated block length) and which is located at a specified offset from the end of the output uncompressed stream, as specified in the reference block (repeated block offset).

5.5.4 Compression Method

The compression method consists of identifying repeated patterns of octets in the input uncompressed stream and inserting associated reference blocks in the output compressed

Figure 5.9 Compression process

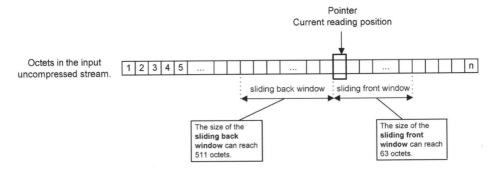

Figure 5.10 Compression process elements

stream (Figure 5.9). Octets which cannot be compressed in the input compressed stream are inserted as data blocks in the output compressed stream. The input uncompressed stream is scanned with a pointer from the start to the end, octet by octet. The current reading position in the input uncompressed stream is the octet designated by the pointer as shown in Figure 5.10. For each octet pointed by the pointer in the uncompressed stream, the following process is performed:

In the *sliding back window*, the system looks for the longest octet pattern matching the sequence of octets starting at the beginning of the *sliding front window*. Only patterns with a size over two octets and less than or equal to 63 octets are considered. At this stage two alternatives are possible:

1. If *no matching pattern is found*, then the octet in the current reading position in the input uncompressed stream is appended at the end of the output compressed stream. For this purpose, if a data block is available at the end of the output compression stream, then the octet is appended to the data block payload (only if the payload has not reach the maximum length of 127 octets). Otherwise a new data block is inserted at the end of the output compressed stream (with the octet as only payload). The pointer of the input uncompressed stream moves to the next octet and the process is iterated at the new reading position.

2. If *a matching pattern is found*, then a reference block is constructed according to the matching pattern characteristics. The reference block length is the length of the matching pattern and the reference block offset is equal to the number of octets between the current reading position and the beginning of the matching pattern in the input uncompressed stream. After construction, the reference block is appended at the end of the output compressed stream. The pointer of the input uncompressed stream moves to the octet which immediately follows the matching pattern in the input uncompressed stream and the process is iterated at the new reading position.

Table 5.6 Compression/test vectors

Test vector name	Uncompressed size (octets)	Compressed size (octets)	Compression ratio (%)
Black-and-white picture 64 × 62 pixels	498	44	10.04
Black-and-white picture 64 × 62 pixels	383	383	0
Greyscale picture 54 × 54 pixels	747	610	18.34
Colour picture 54 × 54 pixels	2205	1566	28.98
Black-and-white animation 64 × 64 pixels (4 frames)	5652	2223	60.67
Black-and-white animation 16 × 16 pixels (4 frames)	148	94	36.49

The compression process terminates when the pointer of the input uncompressed stream reaches the end of the stream.

The 3GPP technical specification [3GPP-23.040] defining the compression and decompression methods is provided with a set of test vectors. These test vectors enable implementers to test implementations.

It is difficult to estimate the compression ratio of the compression scheme. This ratio really depends on the frequency of occurrences of repeating patterns in the uncompressed stream. However, Table 5.6 shows the compression ratios for five of the test vectors provided with the [3GPP-23.040] technical specification.

5.6 Extended Objects

With the extended EMS framework, a new set of objects can be supported by extended EMS-enabled devices. The object type, in Table 5.7, is assigned to the 'extended object type'

Table 5.7 List of extended objects

Type	Description	
0x00	Predefined sound	
0x01	iMelody sound	
0x02	Black and white picture	
0x03	4-level greyscale picture	(new)
0x04	64-colour picture	(new)
0x05	Predefined animation	
0x06	Black and white animation	
0x07	4-level greyscale animation	(new)
0x08	64-colour animation	(new)
0x09	vCard object	(new)
0x0A	vCalendar object	(new)
0x0B	Vector graphics (WVG)	(new)
0x0C	Polyphonic melody (MIDI)	(new)
0x0B...0xFE	*Reserved for future use*	
0xFF	Capability information	(new)

parameter (octet 5) of the associated extended object information element (extended object header). In Table 5.7, an annotation indicates whether the format is new in extended EMS, or if it was already available in basic EMS.

5.7 Predefined Sound

Predefined sounds are already available in basic EMS. Another way of including a predefined sound in a message consists in inserting it as an extended object. The information element representing a predefined sound as an extended object is shown in Table 5.8.

User Prompt Indicator (UPI) and ODI flags do not apply to a predefined sound. Values assigned to these flags are ignored by receiving SMEs.

Box 5.5 Recommendations for the Use of Predefined Sounds

There are two ways of inserting a predefined sound in a message: as a basic EMS predefined sound or as an extended EMS predefined sound. With basic EMS, the associated information element has a length of 4 octets (including IEI, IEDL and IED). With extended EMS, the predefined sound information element has a length of 10 octets (without compression). There is a clear resource gain in using the basic EMS predefined sound rather than the extended predefined EMS sound. Furthermore, compared with the extended EMS predefined sound, a predefined sound in the basic EMS format is likely to be interpreted by a larger number of devices. Predefined sounds in the basic EMS format should therefore always be preferred.

5.8 iMelody Melody

iMelody sounds are already available in basic EMS. An alternative method for including such sounds in a message consists of inserting them as extended objects. The information element representing an iMelody sound as an extended object is shown in Table 5.9.

Unlike basic EMS, an iMelody sound inserted in a message as an extended object may have a length greater than 128 octets. Because sounds in extended EMS may have long length, they are known as melodies. An iMelody melody, in the extended EMS format, may also benefit from compression.

5.9 Black-and-white Bitmap Picture

Compared with black-and-white bitmap pictures in basic EMS, dimensions of extended EMS bitmap pictures can reach a maximum of 255 × 255 pixels.

The information element representing a black-and-white bitmap picture as an extended object is shown in Table 5.10. Octets 1–7 of this information element contain the generic extended object header. Octets 8–n contain the picture-specific information (dimensions and bitmap). If the picture cannot be encoded in one message segment, then additional extended object IEs are needed as shown in Section 5.3. A picture, in the extended EMS format, may be compressed.

Table 5.8   IE/predefined sound

IEI			0x14 (extended object) from release 5 Predefined sound	
IEDL			variable	
IED	Extended Object Header		*Octet 1*	*Reference number*
			Oct. 2-3	*Object length*
			Octet 4	*Object handling status (UPI+ODI)*
			Octet 5	*Extended object type* ***0x00 (hex) Predefined sound***
			Oct.6-7	*Extended object position*
	Extended Object Definition		Oct. 8	**Predefined sound number** This octet represents the predefined sound number as defined below: Id Description 0 Chimes high 1 Chimes low 2 Ding 3 TaDa 4 Notify 5 Drum 6 Claps 7 Fanfar 8 Chord high 9 Chord low *other* *reserved*

Table 5.9   IE/iMelody melody

IEI			0x14 (extended object) from release 5 iMelody melody	
IEDL			variable	
IED	Extended Object Header		*Octet 1*	*Reference number*
			Oct. 2-3	*Object length*
			Octet 4	*Object handling status (UPI+ODI)*
			Octet 5	*Extended object type* ***0x01 (hex) iMelody melody***
			Oct.6-7	*Extended object position*
	Extended Object Definition		Octets 8...n	**iMelody melody definition** Octets 8...n represent the melody in the iMelody format as described in Section 4.5.2.

Table 5.10   IE/black-and-white bitmap picture

IEI		0x14 (extended object) from release 5	
		Black-and-white bitmap picture	
IEDL		variable	
IED	Extended Object Header	*Octet 1*	*Reference number*
		Oct. 2-3	*Object length*
		Octet 4	*Object handling status (UPI+ODI)*
		Octet 5	*Extended object type*
			0x02 (hex) Black and white image
		Oct.6-7	*Extended object position*
	Extended Object Definition	Octet 8	**Horizontal dimension (width)** Octet representing the picture width. Range: 1…255, value 0 is reserved (decimal).
		Octet 9	**Vertical dimension (height)** Octet representing the picture height. Range: 1…255, value 0 is reserved (decimal).
		Octets 10 to n	**Black-and-white picture bitmap** These octets represent the bitmap of the picture from top left to bottom right. Unlike basic EMS, this bitmap does not necessary have an horizontal length which is a multiple of 8. There is no fill bits between the encoding of each row of pixels. Fill bits are inserted, if needed, in the last octet. The most significant bit of each pixel represents the leftmost pixel. Each pixel state is encoded with 1 bit: 0 Pixel is white. 1 Pixel is black.

Table 5.11   Black-and-white picture size

Dimensions (pixel × pixel)	Bitmap size (octets)	Number of message segments
16 × 16	32	1
32 × 32	128	2
64 × 64	512	4

Table 5.11 presents the bitmap size and the number of message segments required to convey black-and-white pictures (without compression).

5.10 4-Level Greyscale Bitmap Picture

As for black-and-white pictures, the dimensions of 4-level greyscale bitmap pictures can

Table 5.12   IE/4-level greyscale bitmap picture

IED				
IEI			0x14 (extended object) from release 5 4-level greyscale bitmap picture	
IEDL			variable	
IED	Extended Object Definition	Extended Object Header	*Octet 1*	*Reference number*
			Oct. 2-3	*Object length*
			Octet 4	*Object handling status (UPI+ODI)*
			Octet 5	*Extended object type* ***0x03 (hex)*** ***4-level greyscale picture***
			Oct.6-7	*Extended object position*
			Octet 8	**Horizontal dimension (width)** Octet representing the image width. Range: 1...255, value 0 is reserved (decimal).
			Octet 9	**Vertical dimension (height)** Octet representing the image height. Range: 1...255, value 0 is reserved (decimal).
			Octets 10 to n	**Greyscale picture bitmap** These octets represent the bitmap of the picture from top left to bottom right. Unlike basic EMS, this bitmap does not necessarily have an horizontal length which is a multiple of 8. There is no fill bits between the encoding of each row of pixels. Fill bits are inserted, if needed, in the last octet. Each pixel state is encoded with 2 bits: 00 Pixel is black 01 Pixel is dark grey 10 Pixel is light grey 11 Pixel is white

reach a maximum of 255 × 255 pixels. In the uncompressed form, the state of each pixel of the bitmap picture is represented with 2 bits. It is therefore highly recommended to compress greyscale pictures. The corresponding extended object IE is structured as shown in Table 5.12.

Octets 1–7 of this information element contain the generic extended object header. Octets 8–n contain the picture-specific information (dimensions and bitmap). If the picture cannot be encoded in one message segment, then additional extended object IEs are needed as shown in Section 5.3.

Table 5.13 presents the bitmap size and the number of message segments required to convey greyscale pictures (without compression):

5.11 64-Colour Bitmap Picture

As for black-and-white and greyscale pictures, dimensions of colour bitmap pictures can

Table 5.13 Greyscale picture size

Dimensions (pixel × pixel)	Bitmap size (octets)	Number of message segments
16 × 16	64	1
32 × 32	256	3
64 × 64	1024	8

reach a maximum of 255 × 255 pixels. In the uncompressed form, the state of each pixel of the bitmap picture is represented with 6 bits. It is therefore highly recommended to compress colour bitmap pictures. The corresponding extended object IE is structured as shown in Table 5.14.

Octets 1–7 of this information element contain the generic extended object header. Octets 8–n contain the picture-specific information (dimensions and bitmap). If the picture cannot be encoded in one message segment, then additional extended object IEs are needed as shown in Section 5.3.

Each pixel state of the colour picture is represented by three pairs of bits. These three pairs of bits represent the quantities of red, blue and green for each pixel as shown in Table 5.15.

Table 5.14 IE/64-colour bitmap picture

IEI		0x14 (extended object) from release 5 64-colour bitmap picture	
IEDL		variable	
IED	Extended Object Header	*Octet 1*	*Reference number*
		Oct. 2-3	*Object length*
		Octet 4	*Object handling status (UPI+ODI)*
		Octet 5	*Extended object type* ***0x04 (hex) 64-colour picture***
		Oct.6-7	*Extended object position*
	Extended Object Definition	Octet 8	**Horizontal dimension (width)** Octet representing the image width. Range: 1…255, value 0 is reserved (decimal).
		Octet 9	**Vertical dimension (height)** Octet representing the image height. Range: 1…255, value 0 is reserved (decimal).
		Octets 10 to n	**Greyscale picture bitmap** These octets represent the bitmap of the picture from top left to bottom right. Each pixel state is represented with 6 bits (64 colours) as shown in Table 5.15. As for other large object pictures, fill bits are only used on the last octet of the bitmap, if required.

Table 5.15   Colour picture/colour code – RGB(2,2,2)

Red		Green		Blue	
MSB	LSB	MSB	LSB	MSB	LSB
Bit 5	Bit 4	Bit 3	Bit 2	Bit 1	Bit 0

Table 5.16   Colour picture/size

Dimensions (pixel × pixel)	Bitmap size (octets)	Number of message segments
16 × 16	192	2
32 × 32	768	6
64 × 64	3072	24

Table 5.16 presents the bitmap size and the number of message segments required to convey colour pictures (without compression).

5.12 Predefined Animation

Predefined animation is already available in basic EMS. Another way of including a predefined animation in a message consists of inserting it as an extended object. The information element representing a predefined animation as an extended object is shown in Table 5.17.

UPI and ODI flags do not apply to predefined animations. Values assigned to these flags are ignored by receiving SMEs. The analysis, provided in Box 5.5, regarding the choice of formats (basic EMS or extended EMS) for inserting predefined sounds in a message, also applies to predefined animations. Consequently, predefined animations should rather be inserted in a message as basic EMS objects rather than as extended EMS objects.

5.13 Black-and-white Animation

Compared with basic EMS black-and-white animations (maximum dimensions: four frames of 16x16 pixels), black-and-white animations in extended EMS can take various forms (up to 255 frames with dimensions of up to 255 × 255 pixels). The frame display time can be configured (same frame display time for the entire animation). The frame display time ranges from 100 ms to 1.6 s. The number of animation repetitions can also be specified. The number of repetitions can be unlimited or specified in the range 1–15. The corresponding extended object IE is structured as shown in Table 5.18.

Octets 1–7 of this information element contain the generic extended object header. Octets 8–n contain the animation specific information. If the animation cannot be conveyed in one message segment, then additional extended object IEs are needed as shown in Section 5.3.

Table 5.17   IE/predefined animation

IEI		0x14 (extended object) from release 5	
		Predefined animation	
IEDL		variable	
IED	Extended Object Header	*Octet 1*	*Reference number*
		Oct. 2-3	*Object length*
		Octet 4	*Object handling status (UPI+ODI)*
		Octet 5	*Extended object type*
			0x05 (hex) Predefined animation
		Oct.6-7	*Extended object position*
	Extended Object Definition	Oct. 8	**Predefined animation number**

For Oct. 8:

Predefined animation number
This octet represents the predefined animation number as defined below:

Id	Description
0	I am ironic, flirty
1	I am glad
2	I am sceptic
3	I am sad
4	Wow !!!
5	I am crying
6	I am winking
7	I am laughing
8	I am indifferent
9	In love / kissing
10	I am confused
11	Tongue hanging out
12	I am angry
13	Wearing glasses
14	Devil
other	*reserved*

5.14 4-Level Greyscale Animation

Greyscale animations are not available in basic EMS. In extended EMS, the frame display time can be specified (same frame display time for the entire animation). The frame display time ranges from 100 ms to 1.6 s. The number of animation repetitions can also be specified. The number of repetitions can be unlimited or specified in the range 1–15. The corresponding extended object IE is structured as shown in Table 5.19.

Octets 1–7 of this information element contain the generic extended object header. Octets 8–n contain the animation specific information. If the animation cannot be encoded in one message segment, then additional extended object IEs are needed as shown in Section 5.3.

5.15 64-Colour Animation

Colour animations are not available in basic EMS. In extended EMS, the frame display time can be specified (same frame display time for the entire animation). The frame display time ranges from 100 ms to 1.6 s. The number of animation repetitions can also be specified. The

Table 5.18   IE/black-and-white animation

IEI			0x14 (extended object) from release 5 Black and white animation
IEDL			variable
IED	Extended Object Header	*Octet 1*	*Reference number*
		Oct. 2-3	*Object length*
		Octet 4	*Object handling status (UPI+ODI)*
		Octet 5	*Extended object type* **0x06 (hex) black-and-white animation**
		Oct.6-7	*Extended object position*
	Extended Object Definition	Octet 8	**Horizontal dimension (width)** Octet representing the animation width. Range: 1...255, value 0 is reserved (decimal).
		Octet 9	**Vertical dimension (height)** Octet representing the animation height. Range: 1...255, value 0 is reserved (decimal).
		Octet 10	**Number of frames in the animation** Range 1...255, value 0 is reserved.
		Octet 11	**Animation control** This octet indicates the number of times the animation is to be repeated (from 1 to 15 repetitions or unlimited repetition). It also indicates the display time for each frame. The octet is structured as follows: **Bits 7-4** **Frame display time** 0000 (0.1 s.) 0110 (0.7 s.) 1100 (1.3 s.) 0001 (0.2 s.) 0111 (0.8 s.) 1101 (1.4 s.) 0010 (0.3 s.) 1000 (0.9 s.) 1110 (1.5 s.) 0011 (0.4 s.) 1001 (1.0 s.) 1111 (1.6 s.) 0100 (0.5 s.) 1010 (1.1 s.) 0101 (0.6 s.) 1011 (1.2 s.) **Bits 3-0** **Repeat count** 0000 Unlimited repetition 0001 1 repetition ... n repetitions ($1<n<15$) 1111 15 repetitions
		Octets 12 to n	**Animation frame bitmap(s)** These octets represent n bitmaps in the format defined in Section 5.9. At the end of each frame representation, fill bits may be inserted, if required, to allow the following frame to start on an octet boundary.

Table 5.19 IE/greyscale animation

IEI	0x14 (extended object) from release 5 4-level greyscale animation		
IEDL	variable		
IED	Extended Object Header	*Octet 1*	*Reference number*
		Oct. 2-3	*Object length*
		Octet 4	*Object handling status (UPI+ODI)*
		Octet 5	*Extended object type* ***0x07 (hex) 4-level greyscale animation***
		Oct.6-7	*Extended object position*
	Extended Object Definition	Octet 8	**Horizontal dimension (width)** Octet representing the animation width. Range: 1…255, value 0 is reserved (decimal).
		Octet 9	**Vertical dimension (height)** Octet representing the animation height. Range: 1…255, value 0 is reserved (decimal).
		Octet 10	**Number of frames in the animation** Range 1…255, value 0 is reserved (decimal).
		Octet 11	**Animation control** This octet indicates the number of times the animation is to be repeated (from 1 to 15 repetitions or unlimited repetition). It also indicates the display time for each frame. The octet is structured as follows: **Bits 7-4** **Frame display time** 0000 (0.1 s.) 0110 (0.7 s.) 1100 (1.3 s.) 0001 (0.2 s.) 0111 (0.8 s.) 1101 (1.4 s.) 0010 (0.3 s.) 1000 (0.9 s.) 1110 (1.5 s.) 0011 (0.4 s.) 1001 (1.0 s.) 1111 (1.6 s.) 0100 (0.5 s.) 1010 (1.1 s.) 0101 (0.6 s.) 1011 (1.2 s.) **Bits 3-0** **Repeat count** 0000 Unlimited repetition 0001 1 repetition … n repetitions ($1<n<15$) 1111 15 repetitions
		Octets 12 to n	**Animation frame bitmap(s)** These octets represent n bitmaps in the format defined in Section 5.10. At the end of each frame representation, fill bits may be inserted, if needed, to allow the following frame starts on an octet boundary.

Table 5.20   IE/colour animation

IEI		0x14 (extended object) from release 5 64-colour animation	
IEDL		variable	
IED	**Extended Object Header**	*Octet 1*	*Reference number*
		Oct. 2-3	*Object length*
		Octet 4	*Object handling status (UPI+ODI)*
		Octet 5	*Extended object type* ***0x08 (hex) 64-colour animation***
		Oct.6-7	*Extended object position*
	Extended Object Definition	Octet 8	**Horizontal dimension (width)** Octet representing the animation width. Range: 1...255, value 0 is reserved (decimal).
		Octet 9	**Vertical dimension (height)** Octet representing the animation height. Range: 1...255, value 0 is reserved (decimal).
		Octet 10	**Number of frames in the animation** Range 1...255, value 0 is reserved (decimal).
		Octet 11	**Animation control** This octet indicates the number of times the animation is to be repeated (from 1 to 15 repetitions or unlimited repetition). It also indicates the display time for each frame. The octet is structured as follows: **Bits 7-4** **Frame display time** 0000 (0.1 s.) 0110 (0.7 s.) 1100 (1.3 s.) 0001 (0.2 s.) 0111 (0.8 s.) 1101 (1.4 s.) 0010 (0.3 s.) 1000 (0.9 s.) 1110 (1.5 s.) 0011 (0.4 s.) 1001 (1.0 s.) 1111 (1.6 s.) 0100 (0.5 s.) 1010 (1.1 s.) 0101 (0.6 s.) 1011 (1.2 s.) **Bits 3-0** **Repeat count** 0000 Unlimited repetition 0001 1 repetition ... n repetitions ($1 < n < 15$) 1111 15 repetitions
		Octets 12 to n	**Animation frame bitmap(s)** These octets represent n bitmaps in the format defined in Section 5.11. At the end of each frame representation, fill bits may be inserted, if required, to allow the following frame to start on an octet boundary.

Table 5.21 Animations/sizes

	Frame dimensions (pixel × pixel)	Bitmap size (4 frames) (octets)	Number of message segments
Black-and-white	16 × 16	128	2
	32 × 32	512	4
	64 × 64	2048	16
Greyscale	16 × 16	256	3
	32 × 32	1024	8
	64 × 64	4096	32
Colour	16 × 16	768	6
	32 × 32	3072	24
	64 × 64	12288	94

number of repetitions can be unlimited or limited in the range 1–15. The corresponding extended object IE is structured as shown in Table 5.20.

Octets 1–7 of this information element contains the generic extended object header. Octets 8–n contains the animation specific information. If the animation cannot be encoded in one message segment, then additional extended object IEs are needed as shown in Section 5.3.

Table 5.21 shows the bitmap size and the number of message segments required to insert various types of 4-frame animations in extended EMS messages.

> **Box 5.6 Recommendation for a Maximum Object Size**
>
> Although a message could theoretically be composed of 255 message segments, one can seldom find devices that can interpret messages composed of more than eight message segments. To ensure that a message can be rendered properly on the largest number of devices, the number of message segments should not exceed eight (uncompressed size).

5.16 vCard Data Stream

The vCard format is used for representing electronic business cards [VERSIT-vCard]. This format is already widely used with Personal Digital Assistants and is becoming the de facto format for the exchange of electronic business cards over infrared links. It is also becoming common to attach a vCard data stream, as a signature, to an Email message. A vCard data stream contains basic contact details such as last name, first name, postal and electronic addresses, phone, mobile and fax numbers, etc. It may also contain more complex data such as photographs, company logos or audio clips. The information element enabling the inclusion of a vCard data stream in a message is structured as shown in Table 5.22.

A vCard data stream is a collection of one or more properties. A property is composed of a property name and an assigned value. In a vCard data stream, a set of properties can be grouped to preserve the coupling of the properties' meaning. For instance, the properties for a

Table 5.22   IE/vCard

IEI			0x14 (extended object) from release 5	
			vCard data stream (version 2.1)	
IEDL			variable	
IED	Extended Object Header		*Octet 1*	*Reference number*
			Oct. 2-3	*Object length*
			Octet 4	*Object handling status (UPI+ODI)*
			Octet 5	*Extended object type*
				0x09 (hex) vCard object
			Oct.6-7	*Extended object position*
	Extended Object Definition		Octets 9 to n	**Definition of the vCard data stream** These octets represent the sequence of instructions composing the vCard data stream. Instructions are represented in a text format using the UTF8 encoding (8-bit aligned).

telephone number and a corresponding comment can be grouped together. Note that a vCard data stream is composed of a main vCard object, which can also contain nested vCard objects. Properties for a vCard object are included between the two delimiters BEGIN:VCARD and END:VCARD. A property (name, parameters and value) takes the following form:

<PropertyName> [";" <PropertyParameters>] ":" <PropertyValue >

A property can expand over more than one line of text. Property names in a vCard data stream are case insensitive. The property name, along with an optional grouping label, must appear as the first characters on a line. In a data stream, individual text lines are separated by a line break (ASCII character 13 followed by ASCII character 10). Note that long lines of text can be represented with several shorted text lines using the 'RFC 822 folding technique'. Two types of groups are allowed in the vCard format: vCard object grouping and property grouping. With vCard object grouping, several distinct vCard objects are grouped together whereas in property grouping, a collection of properties within a single vCard object are grouped.

Several generic parameters can be used for the properties of a vCard data stream. The encoding of a property value can be specified with the ENCODING property parameter. Values that can be assigned to this parameter are BASE64, QUOTED-PRINTABLE or 8BIT. Similarly, the character set can be specified with the CHARSET property parameter. Values that can be assigned to this parameters are UTF-8, ASCII and ISO-8859-8. The default language for a vCard data stream is US English (en-US). This default configuration for a property value can be overridden with the LANGUAGE property parameter. Values that can be assigned to this parameter are identified in [RFC-1766]; e.g. en-US or fr-CA. The value to be assigned to a property is usually specified inline (INLINE) as part as the property definition, as shown above. However, the value to be assigned to a property can also be located outside the vCard data stream. The location of the property value is specified with the VALUE parameter. Values that can be assigned to this parameter are INLINE (default) and URL (for an external value accessible via a uniform resource locator).

Figure 5.11 shows a vCard data stream containing one vCard object. The properties given

```
BEGIN:VCARD
VERSION:2.1
UID:6666-601
N:Le Bodic, Gwenael
TEL;HOME:+33-1-23-45-67-89
TEL;CELL:+33-6-23-45-67-89
ORG:Alcatel;Research and Development
AGENT:
BEGIN:VCARD
VERSION:2.1
UID:6666-602
N:Tayot, Marie
TEL;WORK;FAX:+33-1-23-45-67-90
END:VCARD
END:VCARD
```

Figure 5.11   vCard format/example

in Table 5.23 can be used for the definition of a vCard data stream. The only two properties that should always be present in a vCard data stream (version 2.1) [VERSIT-vCard] are the name (N) and the version (VERSION).

5.17 vCalendar Data Stream

The vCalendar format is used to represent items generated by calendaring and scheduling applications [VERSIT-vCalendar]. As for the vCard format, the vCalendar format is widely used with Personal Digital Assistants and is becoming the de facto format for the exchange of calendaring and scheduling information. A vCalendar data stream is composed of one or more objects of types *event* and *todo*. An event is a calendaring and scheduling object representing an item in a calendar. A todo is a calendaring and scheduling object representing an action item or assignment. As for a vCard data stream, the vCalendar data stream is composed of a collection of properties with a main vCalendar object including nested vCalendar objects (event and todo objects). Properties for a vCalendar object are included between the two delimiters BEGIN:VCALENDAR and END:VCALENDAR. Similarly, properties of event and todo objects are included between the pairs of delimiters, BEGIN:VEVENT/END:VEVENT and BEGIN:VTODO/END:VTODO, respectively. The information element enabling the inclusion of a vCalendar data stream in a message is structured as shown in Table 5.24.

The format of a vCalendar property is the same as that of a vCard property. Generic vCard parameters (ENCODING, CHARSET, LANGUAGE and VALUE) can also be used for the configuration of properties in vCalendar data streams.

For a property representing a date, the value to be assigned to the property is structured as follows:

< year >< month >< day > T < hour >< minute >< second >< type designator >

Table 5.23 vCard properties

Property name	Description	Example
FN	Formatted name (include honorific prefixes, suffixes, titles, etc.)	FN:Dr. G. Le Bodic
N	Name (structured name of the person, place or thing). For a person, the property value is a concatenation of family name, given name, etc.	N:Le Bodic, Gwenael
PHOTO	Photograph. The photograph is represented as a picture. Parameters of this property are the picture format such as: • GIF: Graphics Interchange Format • BMP: Bitmap format • JPEG: Jpeg format etc.	
BDAY	Birthday. The value assigned to this property represents the birthday of a person	BDAY:1973-07-06
ADR	Delivery address. Property parameters: • DOM: Domestic address • INTL: International address • POSTAL: Postal address • PARCEL: Parcel delivery address • HOME: Home delivery address • WORK: Work delivery address The structured property value is a concatenation of the post office address, extended address, street, locality, region, postal code and country	ADR;HOME:P.O. Box 6;;16 Wood street;London;UK;SW73NH
LABEL	Delivery label. The property value represents a postal delivery address. Unlike the previous property, the value assigned to this property is not structured. The property is associated with the same parameters as the ones associated with the ADR property	LABEL;INTL;ENCODING = QUOTED-PRINTABLE:P.O. Box 6 = 0D = 0A 16 Wood street = 0D = 0A London = 0D = 0A United Kingdom = 0D = 0A SW73NH
TEL	Telephone. Property parameters: • PREF: Preferred number • WORK: Work number • HOME: Home Number • VOICE: Voice number • FAX: Facsimile number • MSG: Messaging service number • CELL: Cellular number • PAGER: Pager Number • BBS: Bulletin board service number • MODEM: Modem number • CAR: Car-phone number • ISDN: ISDN number • VIDEO: Video-phone number	TEL;PREF;CELL: + 33-6-12-13-14-15

Table 5.23 (*continued*)

Property name	Description	Example
EMAIL	Electronic Mail. Main property parameters are (list is not exhaustive): • INTERNET: Internet address • AOL: America on-line address • CIS: Computer Information Service • TELEX: Telex number • X400: X.400 number	EMAIL;INTERNET:Gwenael@Le-Bodic.net
MAILER	Mailer. The value assigned to this property indicates which mailer is used by the person associated with the vCard object	MAILER:ccMail 2.2
TZ	Time zone. The value assigned to this property indicates the offset from the Coordinated Universal Time (UTC). In the corresponding examples, the time zone is respectively −8 hours (Pacific standard time) and −5 hours (Eastern standard time)	TZ: − 08:00 or TZ: − 0500
GEO	Geographic position. The value assigned to this property indicates a global position in terms of longitude and latitude (in this order)	GEO:37.24, − 17.85
TITLE	Title. The value assigned to this property indicates the job title, functional position or function of the person associated with the vCard object	TITLE:Architect, Research and Development
ROLE	Business category. The value assigned to this property indicates the role, occupation or business category of the person associated with the vCard object	ROLE:Executive
LOGO	Logo (e.g. company logo). The logo is represented as a picture. Parameters of this property are the picture format such as: • GIF: Graphics Interchange Format • BMP: Bitmap format • JPEG: Jpeg format etc.	LOGO;ENCODING = BASE64;TYPE = GIF:R01G0dhfg ...
AGENT	Agent. The value assigned to this property is another vCard object that provides information about the person acting on behalf of the person associated with the main vCard object of the vCard data stream	AGENT: BEGIN:VCARD VERSION:2.1 N:Tayot;Marie TEL;CELL: + 1-612-253-3434 END:VCARD
ORG	Organization name and unit. The structured property value is a concatenation of the company name, company unit and organizational unit	ORG:Alcatel;Marketing

Table 5.23 (*continued*)

Property name	Description	Example
NOTE	Comment.	NOTE;ENCODING = QUOTEDPRINTABLE:Do not use the mobile phone number after 9:00pm
REV	Last revision of the vCard object[a].	REV:2002-12-23T21:12:11Z
SOUND	Sound. The value assigned to this property specifies a sound annotation associated with the vCard object. One property parameter indicates the sound format: • WAVE: WAVE format • PCM: PCM format • AIFF: AIFF format	
URL	Uniform Resource Locator (URL). The value assigned to this property represents a URL which refers to additional information related to the vCard object	URL:http: www.le-bodic.net
UID	Unique identifier. The value assigned to this property indicates a globally unique identifier for the vCard object	UID:6666666-123456-34343
VERSION	Version. The value assigned to this property indicates the version of the vCard format to which complies the vCard writer	VERSION:2.1
KEY	Public key. Property parameters: X509: X.509 public key certificate PGP: PGP key	

[a] See definition of date values in Section 5.17.

Table 5.24   IE/vCalendar

IEI		0x14 (extended object) from release 5 vCalendar data stream (version 1.0)	
IEDL		variable	
IED	Extended Object Header	Octet 1	*Reference number*
		Oct. 2-3	*Object length*
		Octet 4	*Object handling status (UPI+ODI)*
		Octet 5	*Extended object type* ***0x0A (hex)*** ***vCalendar data stream***
		Oct.6-7	*Extended object position*
	Extended Object Definition	Octets 9 to n	**Definition of the vCalendar data stream** These octets represent the sequence of instructions composing the vCalendar data stream. Instructions are represented in a text format using the UTF8 encoding (8-bit aligned).

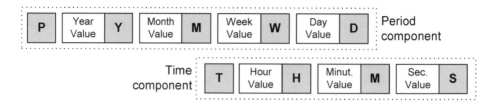

Figure 5.12 vCalendar/period format

For instance, 9:30(:00) AM on April 20, 2002 local time is formatted as

20020420T093000

The same date/time in UTC-based time is formatted as

20020420T093000Z

Values assigned to vCalendar properties can also refer to time periods. Such period values are formatted as shown in Figure 5.12. A value representing a time period is prefixed with the 'P' character. After the P character, optional blocks represent the time period in terms of years, months and days. Optionally, the value can also contain a time component starting with the character 'T' and followed by optional blocks refining the time duration in terms of hours, minutes and seconds. For instance, a period of 1 year and 2 weeks is represented by the following character string:

P1Y2W

A period of 15 minutes is represented by the following character string:

PT15M

A vCalendar object can characterize recurring events. A recurrence rule is a character string with the format shown in Figure 5.13. A recurrence rule starts with a character which identifies the event *frequency* ('D' for daily, 'W' for weekly, 'M' for monthly and 'Y' for yearly). This is followed by an *interval* indicating how often the event repeats (daily, every fourth day, etc.). Optionally, the interval can be followed by one or more *frequency modifiers*, followed by an optional *end date* or *duration*. The following list shows examples of daily and weekly recurrence rules:

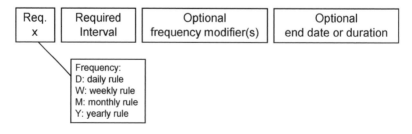

Figure 5.13 vCalendar/recurrence rule

```
D1 #10                      Daily for ten occurrences
D1 20020420T093000Z         Daily until 20th April 2002
W2 #0                       Every other week, forever.
```

Two types of monthly recurrence rules are available in the vCalendar format: the *by-position* rule and the *by-day* rule. The by-position rule allows weekdays in the month to be specified in relation to their occurrence in the month (e.g. 2nd Sunday of the month). The by-day rule allows specific days to be specified (e.g. 6th day or 3rd day). A by-position rule starts with the prefix 'MP' whereas a by-day rule starts with the prefix 'MD'. The following list shows examples of monthly recurrence rules:

```
MP1 1 +  TU #6              Monthly on the first Tuesday for six occurrences
MP2 2 +  TU 1- SU #2        Every other month on the second Tuesday and last Sunday
                            for two occurrences
MD1 1 1- #3                 Monthly on the first and last day for three occurrences.
```

Two types of yearly recurrence rules are available in the vCalendar format: the *by-month* rule and the *by-day* rule. The by-month rule allows specific months to be specified (e.g. yearly in July). The by-day rule allows specific days to be specified (e.g. 206th day of the year). A by-month rule starts with the prefix 'YM' whereas a by-day rule starts with the prefix 'YD'. The following list shows examples of yearly recurrence rules:

```
YM1 6 7 #2                  Yearly in June and July for two occurrences
YD5 206 #3                  Every fifth year on the 206th day for three occurrences
YD3 2 20050601T000000Z      Every third year on the 2nd day until the 1st June 2005.
```

Table 5.25 shows the properties that can be used for the definition of a vCalendar data stream. The only property that should always be present in a vCalendar data stream (version 1.0) is the version (VERSION).

Table 5.26 lists the properties that can be used for the definition of event and todo objects within a vCalendar data stream. According to the vCalendar format v1.0 [VERSIT-vCalendar], properties that should always be supported for an event object are category (CATEGORIES), description (DESCRIPTION), start date/time (DTSTART), end date/time (DTEND), priority (PRIORITY), status (STATUS) and summary (SUMMARY).

According to the vCalendar format v1.0 [VERSIT-vCalendar], properties that should always be supported for a todo object are category (CATEGORIES), date/time completed (COMPLETED), description (DESCRIPTION), due date/time (DUE), priority (PRIORITY), status (STATUS) and summary (SUMMARY).

Figure 5.14 gives an example of a vCalendar data stream. The data stream contains an event object only. Figure 5.15 gives an example of a vCalendar data stream. The data stream contains a todo object only.

5.18 MIDI Melody

In basic and extended EMS, it is possible to insert user-defined melodies in the form of iMelody objects. The iMelody format is a text-based monophonic format that has many limitations and consequently does not allow the definition of sophisticated melodies such as those currently available on most personal computers. Extended EMS allows high-quality

Table 5.25   vCalendar properties

Property name	Description	Example
DAYLIGHT	Daylight saving. The value assigned to the property indicates the daylight saving of the application that generated the vCalendar data stream. The first example indicates that the application does not observe any daylight savings time (value is FALSE). The second example indicates that daylight savings time is observed and that the summer time is 6 hours behind UTC (summer time is from 7 April to 27 November 2002)	`DAYLIGHT:FALSE` or `DAYLIGHT:TRUE;06;` `20020407T025959;` `20021027T010000T010000;` `EST;EDT`
GEO	Geographic position. The value assigned to this property indicates a global position in terms of longitude and latitude (in this order)	`GEO:37.24, − 17.85`
PRODID	Product identifier	`PRODID:Manufacturer PIM`
TZ	Time zone. The value assigned to this property indicates the offset from the Coordinated Universal Time (UTC). In the corresponding examples, the time zone is respectively −8 hours (Pacific standard time) and −5 hours (Eastern standard time)	`TZ:-08:00` or `TZ:-0500`
VERSION	Version. The value assigned to this property indicates the version of the vCalendar format to which complies the vCalendar writer	`VERSION:1.0`

Table 5.26   vCalendar properties/event and todo objects

Property name	Description	Example
ATTACH	Attachment. The attachment can be a document to be reviewed or a list of actions for a todo object	`ATTACH;VALUE =` `URL:file:///` `le-bodic.net/todo.ps`
ATTENDEE	Attendee to a group event or a person associated with a todo object. Optional parameters for this property are: • ROLE (possible values are ATTENDEE, ORGANIZER, OWNER and DELEGATE/ default is ATTENDEE) • STATUS (possible values are ACCEPTED, NEEDS ACTION, SENT, TENTATIVE, CONFIRMED, DECLINED, COMPLETED and DELEGATED/default is NEEDS ACTION)	`ATTENDEE;ROLE =` `OWNER;STATUS =` `COMPLETED:gwenael@le-` `bodic.net`

Table 5.26 (*continued*)

Property name	Description	Example
	• RSVP (possible values are YES or NO/ default is NO) • EXPECT (possible values are FYI, REQUIRE, REQUEST and IMMEDIATE/ default is FYI)	
AALARM	Audio reminder. Sound to be played to notify the corresponding event. An additional property parameter is the sound format (see possible values for the SOUND property in the vCard format). This property value is composed of the run time/start time, snooze time, repeat count and audio content	AALARM;TYPE = WAVE; VALUE = URL:20020706T114500;;; file:///notify.wav
CATEGORIES	Categories. More than one category value may be assigned to this property. Possible values are APPOINTMENT, BUSINESS, EDUCATION, HOLIDAY, MEETING, MISCELLANEOUS, PERSONAL, PHONE CALL, SICK DAY, SPECIAL OCCASION, TRAVEL and VACATION	CATEGORIES:APPOINTMENT; BUSINESS
CLASS	Classification. The value assigned to this property indicates the accessibility of the associated information. Possible values are PUBLIC, PRIVATE and CONFIDENTIAL. Default value is PUBLIC	CLASS:CONFIDENTIAL
DCREATED	Date/time of creation for a todo or event object	DCREATED: 20020421T192500
COMPLETED	Date/time of completion for a todo object	COMPLETED: 20020421T112000
DESCRIPTION	Description	DESCRIPTION:Meeting part of the workshop
DALARM	Display reminder. This property value is composed of the run time/start time, snooze time, repeat count and display string	DALARM:20020421T080000; PT5M;3;Wake up!
DUE	Due date/time for a todo object	DUE:20020421T192500
DTEND	End date/time for an event object	DTEND:20020421T192500
EXDATE	Exception date/times. This property defines a list of exceptions for a recurring event	EXDATE: 20020402T010000Z; 20020403T010000Z
XRULE	Exception rule. This property defines a list of exceptions for a recurring event by specifying an exception rule. In the corresponding example: except yearly in June and July for ever	XRULE:YM1 6 7 #0
LAST-MODIFIED	Last modified	LASTMODIFIED: 20020421T192500Z
LOCATION	Location	LOCATION:Conference room

Table 5.26 (*continued*)

Property name	Description	Example
MALARM	Mail reminder. This property contains an Email address to which an event reminder is to be sent. This property value is composed of the run time/start time, snooze time, repeat count, Email address and note	MALARM:20020421T000000; PT1H;2; gwenael@lebodic.net; Do not forget the meeting
RNUM	Number recurrences. This property defines the number of times a vCalendar object will reoccur	RNUM:6
PRIORITY	Priority. Value 0 is for an undefined priority, value 1 is the highest priority, value 2 is for the second highest priority and so on	PRIORITY:2
PALARM	Procedure reminder. A procedure reminder is a procedure, or application that will be executed as an alarm for the corresponding event. This property value is composed of the run time/start time, snooze time, repeat count and procedure name	PALARM;VALUE = URL: 20020421T100000;PT5M;1; file:///lebodic.net/ notify.exe
RELATED-TO	Related-to. This property allows the definition of relationships between vCalendar entities. This is performed by using globally unique identifiers as specified by the UID parameter	RELATED-TO:6666-87-90
RDATE	Recurrence date/times. This property defines a list of date/times for a recurring vCalendar object	RDATE:20020421T010000Z; 20020422T010000Z
RRULE	Recurrence rule. This property defines a recurring rule for a vCalendar object. The corresponding example indicates that the corresponding event occurs weekly on Tuesday	RRULE:W1 TU
RESOURCES	Resources. Possible values for this property are CATERING, CHAIRS, COMPUTER, PROJECTOR, VCR, VEHICLE, etc.	RESOURCES:VCR;CHAIRS
SEQUENCE	Sequence number. This property defines the instance of the vCalendar object in a sequence of revisions	SEQUENCE:3
DTSTART	Start date/time for an event object. This property defines the date and time when the event will start.	DTSTART:20020421T235959
STATUS	Status. This property defines the status of the related vCalendar object. Possible values for this property are ACCEPTED, NEEDS ACTION, SENT, TENTATIVE, CONFIRMED, DECLINED, COMPLETED and DELEGATED. Default value is NEEDS ACTION	STATUS:ACCEPTED

Table 5.26 (*continued*)

Property name	Description	Example
SUMMARY	Summary. This property defines a short summary (or subject) for the vCalendar object	SUMMARY:Debriefing
TRANSP	Time transparency. This property defines whether the event is transparent to free time searches. Possible values are: 0: object will block time and will be factored into a free time search 1: object will not block time and will not be factored into a free time search other: implementation specific transparency semantics	TRANSP:0
URL	Uniform Resource Locator (URL). The value assigned to this property represents a URL which refers to additional information related to the vCalendar object	
UID	Unique identifier. The value assigned to this property indicates a globally unique identifier for the vCalendar object	UID:66666-32432-4532

```
BEGIN:VCALENDAR
VERSION:1.0
BEGIN:VEVENT
UID:6666-542-2813
CATEGORIES:MEETING
STATUS:TENTATIVE
PRIORITY:1
DTSTART:20020422T030000Z
DTEND:20020422T050000Z
SUMMARY:Debriefing
DESCRIPTION:Analysis of solutions and decision
CLASS:PRIVATE
END:VEVENT
END:VCALENDAR
```

Figure 5.14   vCalendar format (event)/example

melodies to be included in messages. These melodies are represented in the format defined in the Musical Instrument Digital Interface (MIDI). The information element representing a MIDI melody as an extended object is shown in Table 5.27.

This book does not provide a full description of MIDI. Full MIDI specifications can be obtained from [MMA-MIDI-1].

```
BEGIN:VCALENDAR
VERSION:1.0
BEGIN:VTODO
UID:6666-542-2814
CATEGORIES:BUSINESS
STATUS:COMPLETED
COMPLETED:20020422T030000Z
DUE:20020423T050000Z
PRIORITY:1
SUMMARY:Analyse solution no2
DESCRIPTION:Conduct purchasing and software analyses
CLASS:PRIVATE
END:VTODO
END:VCALENDAR
```

Figure 5.15   vCalendar format (todo)/example

Table 5.27   IE/polyphonic MIDI melody

IEI		0x14 (extended object) from release 5	
		Polyphonic MIDI melody	
IEDL		variable	
IED	Extended Object Header	*Octet 1*	*Reference number*
		Oct. 2-3	*Object length*
		Octet 4	*Object handling status (UPI+ODI)*
		Octet 5	*Extended object type*
			0x0C (hex) MIDI melody
		Oct.6-7	*Extended object position*
	Extended Object Definition	Octets 8...n	**MIDI melody** These octets represent the melody in the MIDI format as described in [MMA-MIDI-1] or [MMA-SP-MIDI].

5.18.1 Introduction to MIDI

Since the release of MIDI 1.0 in 1982, MIDI has become the most widely used synthetic music standard among musicians and composers. The MIDI standard encompasses not only the specifications of the connector and cable for interconnecting MIDI-capable devices, but also the format of messages exchanged between these devices. Only the format of messages concerns extended EMS melodies. Melodies in the MIDI format are represented by a sequence of instructions that a *sound synthesizer* can interpret and execute. In an EMS-enabled device, this MIDI sound synthesizer may be implemented either as a software-based synthesizer or as a hardware MIDI chipset. Instructions rendered by the sound synthe-

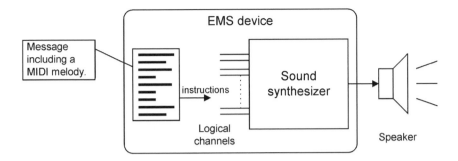

Figure 5.16   Sound synthesizer

sizer are in the form of MIDI messages. For instance, a MIDI message can instruct the synthesizer to use a specific instrument or to play a given note.

5.18.2 MIDI Messages

A MIDI melody is defined as a sequence of MIDI messages. These MIDI messages can be seen as a set of instructions telling the sound synthesizer embedded in the EMS device how to play a melody. For instance, MIDI instructions represent events such as an instrument key being pressed, a foot pedal being released or a slider being moved. MIDI instructions are mainly directed at one of the 16 logical channels of the sound synthesizer, as shown in Figure 5.16. An instrument can be dynamically assigned to a given channel. Only one instrument can be assigned to a channel at a time. After this assignment, each note to be played by the synthesizer on the channel is played using the selected instrument.

The *polyphony* of a sound synthesizer refers to its ability to render several notes or voices at a time. Voices can be regarded as units of resources required by the synthesizer to produce sounds according to received MIDI messages. Complex notes for given instruments may require more than one synthesizer voice to be rendered. Early synthesizers were monophonic, meaning that they were able to render only one note at a time. A sound synthesizer with a polyphony of 16 notes, for instance, is able to render up to 16 notes simultaneously.

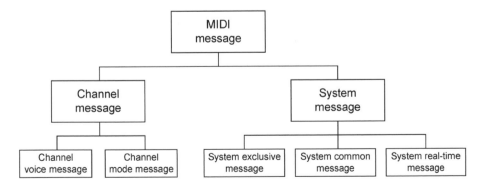

Figure 5.17   MIDI message types

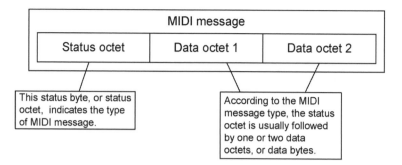

Figure 5.18   Structure of a MIDI message

A sound synthesizer is said to be *multitimbral* if it is able to render notes for different instruments simultaneously. For instance, a sound synthesizer able to render a note for the piano and a note for the saxophone at the same time is a multitimbral synthesizer.

Several types of MIDI messages have been defined and are grouped as shown in Figure 5.17. Any MIDI message is composed of an initial status octet (also known as status byte) along with one or two data octets (also known as data bytes). The structure of a MIDI message is shown in Figure 5.18. At the highest level, MIDI messages are divided into channel messages and system messages.

A *channel message* is an instruction that applies to a selected channel. The identification of the channel, to which a MIDI message is to be applied, is specified as part of the message status octet. Two types of channel messages have been defined: channel mode messages and channel voice messages. *Channel mode messages* affect the way the sound synthesizer generates sounds according to musical information received on one of its logical channels. A *channel voice message* contains the musical information that is to be rendered on one specific logical channel. This includes the instructions shown in Table 5.28.

Table 5.28   MIDI instructions

Channel voice message	Description
Program change	The program change command is used to specify the instrument which is to be used for rendering notes on a specific logical channel
Note on/note off	The rendering of a note is carried out by providing two note-related events to the sound synthesizer: note on and note off. The note on command indicates that a particular note should be rendered whereas the note off command indicates that the note being rendered should be released
Control change	The control change command indicates to the synthesizer that a key (e.g. pedal, wheel, lever, switch, etc.) is activated. The activation of the key affects the way notes are rendered (e.g. modulation, sustain, etc.)
After touch	This message is used to modify the note being played (after-pressure key)
Pitch bend change	This message is used for altering pitch

Unlike channel messages, a *system message* is not an instruction to be applied to a specific channel. A system message is either a system real time message or a system exclusive message. A *system real time message* is used for synchronizing several clock-based MIDI modules together. This message is usually ignored by EMS sound synthesizers. A *system exclusive message* (`sysex`) allows manufacturers to specify proprietary instructions to be interpreted by their sound synthesizers.

5.18.3 General MIDI and MIDI 2.0

After the introduction of early MIDI-capable devices, interoperability problems rapidly occurred while using the MIDI 1.0 format. These problems were due to a lack of a common understanding between manufacturers regarding the identification of notes and instruments. To cope with these issues, the manufacturer Roland proposed an addendum to the MIDI 1.0 format in the form of the General MIDI (GM) format. GM supplements MIDI 1.0 by identifying a set of 128 instruments (also known as patches) and a set of notes. Instruments are classified into 16 families as follows:

- Piano
- Chromatic percussion
- Organ
- Bass
- Ensemble
- Reed
- Synth lead
- Synth effects
- Percussive
- Guitar
- Solo strings
- Brass
- Pipe
- Synth pad
- Ethnic
- Sound effects.

GM has now been adopted as part of MIDI 2.0.

5.18.4 Transport of MIDI Melodies

In theory, MIDI messages can be sent in real-time to the synthesizer. With EMS, the MIDI melody is first conveyed as part of a message and later rendered by the sound synthesizer when requested by the subscriber. For this purpose, additional timing information needs to be associated with MIDI messages in order to tell the synthesizer when to play the melody notes. To achieve this, timing information along with MIDI messages are formatted as Standard MIDI Files (SMF).

A melody formatted as a Standard MIDI File belongs to one of three SMF formats:

- *SMF format 0:* with this format, all MIDI messages are stored in one single track.
- *SMF format 1:* with this format, MIDI messages are stored as a collection of tracks.
- *SMF format 2:* this format is seldom supported by sound synthesizers.

5.18.5 Scalable-polyphony MIDI and 3GPP Profile

Available MIDI synthesizers, have various rendering capabilities. Some synthesizers only support a low level of polyphony while others support a high level of polyphony. With EMS, the synthesizer is embedded in a mobile device which usually has limited processing capabilities. In most cases, device resources available to the MIDI synthesizer for rendering sounds depend on the state of other applications being executed in the device. To cope with devices with various capability levels, the MIDI format has evolved to support the concept of *scalable polyphony*. This evolved MIDI format is known as the Scalable-Polyphony MIDI (SP-MIDI) [MMA-SP-MIDI]. Scalable polyphony consists of indicating, in the melody, what instructions can be dropped without significant quality degradation when the receiving device is running short of resources. For this purpose, the scalable polyphony information is provided as part of a specific system-exclusive message known as a *Maximum Instantaneous Polyphony* (MIP) message. The scalable polyphony information indicates the note usage and priority for each MIDI channel.

For instance, a melody composer can generate a MIDI melody that can be best rendered with a maximum polyphony of 16 but can still be rendered by synthesizers supporting a maximum polyphony of 10 or 8 by dropping low-priority instructions.

Regarding the use of SP-MIDI, the MIDI Manufacturers Association (MMA), in cooperation with the Association of Music Electronics Industries (Japan), has defined an SP-MIDI profile to be supported by devices with limited capabilities (e.g. mobile devices). This profile, known as the *3GPP profile for SP-MIDI* [MMA-SP-MIDI], provides a means of ensuring interoperability between devices supporting a level of note-polyphony ranging from 5 to 24. This profile identifies MIDI messages and SP-MIDI features which shall be supported by mobile devices for the sake of interoperability.

5.18.6 Recommendations for the Creation of MIDI Melodies

Note that EMS-enabled devices may have various levels of support for MIDI. Simple devices can have support for MIDI 1.0 only, whereas more sophisticated devices could support SP-MIDI. If SP-MIDI is supported, then the device is required to support at least the 3GPP profile for SP-MIDI.

It is important to reduce the resources required to transport MIDI melodies as part of messages. The size of melodies can be significantly reduced by applying the following recommendations:

- Use the SMF format which provides the smallest melody size
- Use running status
- Use one instrument per track
- Use one tempo only for the melody
- Use beat, instead of SMPTE, to set the tempo
- Remove all controller messages from the melody.

Avoid the use of the following options:

- Sequence numbers
- Embedded text
- Sequence/track name
- Instrument name
- Lyric
- Synchronization markers
- Cue point
- MIDI channel prefix
- Sequence specific settings.

Due to limitations of EMS devices, content creators should not expect a full support of the following MIDI instructions:

- MIDI message for channel pressure
- MIDI message for pitch bend
- Individual stereophonic (pan)
- MIDI message master volume.

Content creators can expect any EMS device supporting MIDI melodies to support a level of polyphony of at least six notes. Consequently, the MIP message of a SP-MIDI melody should specify at least how to render a melody with a device supporting a polyphony of six notes only.

5.19 Vector Graphics

In basic and extended EMS, dimensions of bitmap pictures are somewhat limited because of the significant amount of resources required to convey them as part of messages. The bitmap format is not always the most resource efficient way of representing images. A better way of representing images, composed of simple geometrical elements (lines, circles, etc.), is to use a vector graphic format. Unlike bitmap formats, a vector graphic format represents an image by identifying geometrical elements composing the image.

For instance, a circle in an image is represented in a bitmap format by a matrix of states for bitmap pixels. The pixel state is white or black, a greyscale level or a colour for respectively black-and-white, greyscale and colour bitmap pictures (e.g. a pixel is represented with a 6-bit state for 64-colour bitmap pictures). The same image represented with a vector graphic format consists in indicating which geometrical elements are composing the image. Characteristics of these graphical elements are provided such as the radius, colour and line width for a circle.

It may happen that an image needs to be scaled up or down from its original size to be rendered correctly on the receiving device. For such scaling operations, a clear advantage of the vector graphic format is that the representation of the image on the screen of the receiving device can be recalculated dynamically. Performing such a scaling operation with a bitmap image usually leads to a problem with pixelization where the quality of the image presentation is significantly degraded.

The vector graphic format is well suited for line drawings such as hand-written maps and Asian characters.[5] For this purpose, it is possible, in extended EMS, to insert vector graphic images in messages by using a format called *Wireless Vector Graphics* (WVG) [3GPP-23.040] (from release 5). An additional interesting feature of WVG is that the format can also be used for the representation of animated images. Three different methods have been defined for inserting a WVG image in an EMS message:

- A *character-size WVG image* is an image with the same height as that of text characters in which it is placed. It is represented with a dedicated information element (not represented

---

[5] Asian characters can also be encoded with the UCS2 alphabet. However, it was noted that predictive text input mechanisms for complex languages, such as Asian languages, are difficult to implement. It is therefore expected that vector graphics will provide new opportunities for developing more user-friendly built-in features or accessories (touch-screen displays, etc.) for entering text in complex languages with EMS devices.

as an extended object). This is an appropriate way of inserting basic line drawings as part of the text.

- A *configurable-size WVG image as an independent information element* is usually used for representing, in the message, an image for which the dimensions can be configured by the message originator (e.g. a geographical map).
- A *configurable-size WVG image as an extended object* is also used for representing an image for which the dimensions can be configured. The advantage of representing the image as an extended object is that the image representation can expand over several message segments and compression may be applied to reduce the amount of message space required to transport/store the image.

5.19.1 Character-size WVG Image

A character-size WVG image is used for representing an image for which the height is equal to the height of text characters in which it is placed. For this purpose, a dedicated information element is used for inserting the image in a message. The information element is structured as given in Table 5.29.

The width of a character-size WVG image may be shorter than, equal to or greater than the width of text characters. The width of the image is determined according to the aspect ratio specified in the image definition and according to the text character height.

5.19.2 Configurable-size WVG Image with Independent Information Element

A WVG image can also be included in a message in the form of a configurable-size image. For this purpose, the information element shown in Table 5.30 can be used. With this method, the image dimensions are specified as part of the image definition.

5.19.3 Configurable-size WVG Image as an Extended Object

A method for inserting a WVG image in a message consists of incorporating the image in the form of an extended object. The information element representing a WVG image as an extended object is shown in Table 5.31.

Table 5.29   IE/character-size WVG image

IEI	0x19 from release 5 Character-size WVG image	
IEDL	variable	
IED	Octet 1	**WVG image position** This octet represents the position in the text where the WVG image shall be displayed.
	Octets 2...n	**WVG image definition** These octets represent the definition of the image in the WVG format.

Table 5.30   IE/configurable-size WVG image

IEI	0x18 from release 5 Configurable-size WVG image	
IEDL	variable	
IED	Octet 1	**WVG image position** This octet represents the position in the text where the WVG image shall be displayed.
	Octets 2...n	**WVG image definition** These octets represent the definition of the image in the WVG format.

Table 5.31   IE/configurable-size WVG image (extended object)

IEI		0x14 (extended object) from release 5 Configurable-size WVG image	
IEDL		variable	
IED	Extended Object Header	Octet 1	*Reference number*
		Oct. 2-3	*Object length*
		Octet 4	*Object handling status (UPI+ODI)*
		Octet 5	*Extended object type* **0x0B (hex) WVG image**
		Oct.6-7	*Extended object position*
	Extended Object Definition	Octets 8...n	**WVG image definition** These octets represent the definition of the image in the WVG format.

5.19.4 WVG Format Definition

The WVG format is a compact vector graphic format. In comparison with the bitmap representation, images can be represented in the WVG format by a limited amount of message space. However, additional processing capabilities are usually required from the EMS device to be able to render vector graphics (integer and floating point calculations). The following WVG features enable very compact image representations:

- *Linear and non-linear coordinate systems:* graphical elements forming a WVG image can be represented using two types of coordinate systems: linear and non-linear systems. Both coordinate systems can be used in the same image and the selection of one against the other is carried out with the objective of minimizing the overall image size.
- *Bit packing:* the representation of numerical values is efficiently compressed by representing these values over a varying number of bits (from 1 to 16 bits).

- *Global and local envelopes:* local envelopes inserted in a global envelope are used for providing a finer coordinate system for a local zone of the image only.
- *Variable resolution:* different resolutions can be used for representing graphical element coordinates, sizes, angles, etc. The benefit of using a variable resolution resides in the reduction of the number of bits necessary for the representation of numeric values (dimensions, lengths, etc.).
- *Colour palettes:* a colour palette is provided for an image. This reduces the number of bits necessary to identify colours associated with graphical elements.
- *Default values:* many values characterizing graphical elements and animations can be omitted. If these values are omitted in the definition of the image, then default values are used instead.

An image represented in the WVG format is composed of an image header and an image body. The *WVG image header* contains elements such as general information (version, author name, date), colour palette and default colours, codec parameters (coordinate systems, etc.) and animation settings (looping mode and frame timing). The *WVG image body* contains graphical elements such as:

- Ellipses
- Rectangles
- Animations elements
- Text elements
- Polylines including:

  - Simple line polylines
  - Circular polylines
  - Bezier lines

- Grouping elements
- Reuse elements
- Frame elements
- Local elements
- Arbitrary polygons

  - Regular polygons
  - Star shaped polygons
  - Regular grid elements.

Table 5.32 show three images. For each image, the size is given for the image represented in the WVG format and compared with the size of the picture represented in the corresponding bitmap format. Note that the two first images are black-and-white images whereas the third image is a colour picture (no animation). The size of an image represented in the WVG format is independent of the original image dimensions.

5.20 Support of Colour for Text Formatting

In the previous chapter, a feature for formatting text was presented for basic EMS. This feature allows the text of a message to be formatted on the following aspects: text alignment (left, right and centre), font style (bold, italic, underlined, strikethrough) and font size (small,

Table 5.32 Examples of WVG images

Image	WVG size	Bitmap size

	186 octets	Black-and-white bitmap picture
WVG – Bear (reproduced by permission of Bijitec)		32 × 32 pixels 128 octets 64 × 64 pixels 512 octets
	97 octets	
WVG – Bike (reproduced by permission of Bijitec)		
	208 octets	Colour bitmap picture 32 × 32 192 octets 64 × 64 3072 octets
WVG – House, tree and bridge (reproduced by permission of Bijitec)		

normal and large). In extended EMS, this feature is extended for supporting text foreground and background colours. For this purpose, one octet has been added to the structure of the information element dedicated to text formatting. The structure, which originally had a payload size (IED length) of three octets, now has an additional fourth octet which indicates text foreground and background colours. The evolved information element has the structure shown in Table 5.33.

Table 5.33   IE/text formatting with colour

IEI	0x0A from release 4 updated in release 5 Text formatting (with support for colour)	
IEDL	0x04 (4 octets)	
IED	Octet 1	**Start position** as defined in Section 4.3
	Octet 2	**Text formatting length** as defined in Section 4.3
	Octet 3	**Formatting mode** as defined in Section 4.3
	Octet 4	**Text and background colours** This octet represents the text and background colours to be applied to the identified text section. If this octet is omitted (can be determined by checking the value assigned to the IEDL field), then the information element does not specify any colour for text or background. This octet is structured as follows: Bit 7-4 Text foreground colour Bit 3-0 Text background colour The list of possible colours is the following: 0000 (black) 1000 (grey) 0001 (dark grey) 1001 (white) 0010 (dark red) 1010 (light red) 0011 (dark yellow) 1011 (light yellow) 0100 (dark green) 1100 (light green) 0101 (dark cyan) 1101 (light cyan) 0110 (dark blue) 1110 (light blue) 0111 (dark magenta) 1111 (magenta)

Note that a receiving device supporting the evolved information element should also support the basic version of this information element. A receiving device supporting only the basic information element for text formatting should also be able to interpret partly the extended version of this information element (usually by ignoring the value assigned to the fourth octet of the information element payload).

5.21 Hyperlink

With extended EMS, an originator SME can include a hyperlink in a message. A hyperlink is composed of a hyperlink title and a URI. In its simplest form, a hyperlink title is just a short text.

Table 5.34   IE/hyperlink

IEI	0x21 hyperlink	from release 5	
IEDL	variable		
IED	Octets 1 - 2	**Absolute hyperlink position** These two octets indicate the absolute position of the hyperlink in the entire text of the message.	
	Octet 2	**Hyperlink title length** This octet represents the length of the hyperlink title expressed in number of characters (integer representation).	
	Octet 3	**URL length** This octet represents the length of the URI expressed in number of characters (integer representation).	

It can also be a combination of several graphical elements such as text, images, animations, etc. The URI points to a location where additional information, associated with the hyperlink title, can be retrieved. A dedicated information element allows the inclusion of a hyperlink in a message. The dedicated information element is structured as shown in Table 5.34.

Note that this information element only indicates the position of the hyperlink in the message and the length of the hyperlink title and URI. Elements representing the hyperlink title and URI are conveyed in the text part of the message and are not conveyed as part of the information element itself. This allows earlier SMEs (release 99/4), that do not support the hyperlink dedicated information element, to still be able to present separately the hyperlink title and the hyperlink URI to the subscriber. In this situation, no graphical association is made between the hyperlink title and the hyperlink URI. If the hyperlink information element is supported by the SME, the SME graphically associates the hyperlink title with the URI. For instance, the hyperlink URI may be hidden and the hyperlink title underlined. In this case, the SME can provide a means for the subscriber to select one of the hyperlinks contained in the message. If the subscriber selects and activates the hyperlink title, then the SME may offer the possibility of launching the microbrowser with the hidden hyperlink URI or alternatively the SME may offer the possibility of saving the hyperlink as a local bookmark.

The hyperlink in the text part of the message is a character string composed of the concatenation of the hyperlink title, a space character and the hyperlink URL. All elements (text, image, animation, etc.) for which the position is in the following range are part of the hyperlink title:

[Absolute hyperlink position ... Absolute hyperlink position + hyperlink title length]

Additionally, all characters for which the position is in the following range are part of the hyperlink URL:

[Absolute hyperlink position + hyperlink title length + 1... Absolute hyperlink position+ hyperlink title length + 1 + hyperlink title length]

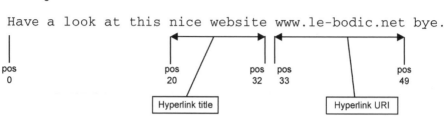

Absolute hyperlink position=20
Hyperlink title="nice website"
Hyperlink title length=12
Hyperlink URI="www.le-bodic.net"
URI length=16

Figure 5.19   Hyperlink information element/example

Figure 5.19 shows an example of a hyperlink contained in a message.

5.22 Exchange of Capability Information

With extended EMS, devices can exchange capability information. The availability of capability information allows an originator SME to format a message according to whatever the receiving SME is capable of rendering. The capability information indicates which extended object formats are supported by the associated SME.

The method for exchanging capability information consists of four successive steps:

1. *Capability request in the message submission*: the originator SME initiates the exchange of capability information by inserting a capability request information element in a message submitted to the SMSC. The message may also contain other information elements, text and EMS elements. The originator SME shall mark the message for automatic deletion (see Section 3.7.6).
2. *Capability request in the message delivery*: the SMSC delivers the submitted message (with the capability request information element) to the receiving SME in the normal way.
3. *Capability return in the delivery report*: the receiving SME analyses the content of the delivered message and identifies the request for returning its capability information. In response, the receiving SME inserts a specific extended object information element in the corresponding delivery report. This specific extended object information element informs on the capabilities of the receiving device. After analysis, the initial message may be discarded by the receiving SME without presentation to the subscriber.
4. *Capability return in the status report*: the SMSC receives the delivery report. The extended object contained in the report is copied to the associated status report and sent to the originator SME which initiated the exchange of capability information.

Figure 5.20 shows the four steps leading to the exchange of capability information between a mobile device and an application server. The originator SME initiates the exchange of capability information by inserting a dedicated information element in a submitted message (step 1). The information element is conveyed to the recipient SME as part of the delivered message (step 2). In this book, this dedicated information is known as the capability request

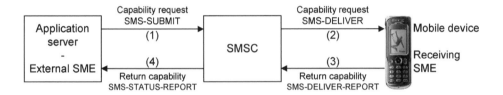

Figure 5.20   Exchange of capability information

Table 5.35   IE/capability request

IEI	0x1A from release 5
	Capability request
	(a.k.a. Extended object data request command)
IEDL	0x00 (0 octet - no IED)

information element (also known as the extended object data request command in 3GPP technical specifications). This information element is structured as shown in Table 5.35.

In response to the message requesting the provision of capability information, the receiving SME inserts a dedicated information element in the corresponding delivery report (step 1). This dedicated extended object is conveyed to the message originator as part of a status report (step 4). This dedicated capability return extended object is structured as shown in Table 5.36. The value 0 shall be assigned to all unused or reserved bits.

5.23 Guidelines for the Creation of Extended Objects

In some situations, an element (sound/melody, image or animation) may be conveyed in a message as a basic EMS element or as an extended EMS element. Making the assumption that extended EMS devices also support basic EMS features, the following considerations should be taken into account when creating EMS messages:

- If *compression is not supported* by the device generating the message, then it is always better to convey the element as a basic EMS element. In this situation, the element represented in the basic EMS format can be interpreted by the largest audience (basic and extended EMS-enabled devices).
- If *compression is supported* by the device generating the message, then the two following solutions are possible:

  - The ability to reach the largest audience is considered more important than the mini-mization of the number of segments composing the message. In this situation, convey-ing the element as a basic EMS element is preferred (compression will not be used).
  - The minimization of the number of segments composing the message is considered more important than the ability to reach the largest audience. In this situation, convey-ing the element as an extended EMS element is preferred (compression will be used).

Table 5.36 IE/capability return

IEI		0x14 (extended object) from release 5 Capability return a.k.a. Data format delivery request	
IEDL		variable	
IED	Extended Object Header	*Octet 1*	*Reference number*
		Oct. 2-3	*Object length*
		Octet 4	*Object handling status (UPI+ODI)*
		Octet 5	*Extended object type* ***0xFF (hex)*** ***Capability return***
		Oct.6-7	*Extended object position*
	Extended Object Definition	Octet 8	This octet indicates whether or not extended objects numbered 0 to 7 are supported by the SME. Bit 0: predefined sounds Bit 1: iMelody sound/melody Bit 2: black-and-white bitmap image Bit 3: greyscale bitmap image Bit 4: colour bitmap image Bit 5: predefined animation Bit 6: black-and-white animation Bit 7: greyscale animation Value 1 assigned to a bit means that the associated extended object is supported, otherwise the associated extended object is not supported by the SME.
		Octet 9	This octet indicates whether extended objects numbered 8 to 12 are supported by the SME. Bit 8: colour animation Bit 9: vCard data stream Bit 10: vCalendar data stream Bit 11: vector graphic (WVG) Bit 12: polyphonic melody (MIDI) Bit 13 to 15: reserved for future use Value 1 assigned to a bit means that the associated extended object is supported, otherwise the associated extended object is not supported by the SME.
		Octets 10 ...n	Reserved for future use

5.24 Pros and Cons of Extended EMS

This chapter has presented extended EMS, an application-level extension of SMS. Extended EMS supersedes SMS and basic EMS features by allowing the inclusion of large objects in messages. It also enables objects to be compressed and supports a substantial list of object formats.

Note that, so far, operators do not differentiate SMS from EMS traffic for billing purpose (segment-based billing). Consequently, the cost of sending EMS messages may become high if large objects are included in messages. Operators commonly see three-segment messages as typical messages subscribers are willing to pay for. Subscribers are probably not ready to pay for the exchange of much larger EMS messages. In any case, eight-segment messages are considered as the largest messages to be exchanged by mobile subscribers in the person-to-person scenario. In the machine-to-person scenario, billing may be more complex and therefore larger messages may be exchanged.

In the GSM network configuration, the transport of EMS messages (basic and extended) is still performed over signalling channels of mobile networks. These signalling channels are bandwidth-limited. Due to these bandwidth limitations and to difficulties evolving from the legacy SMS billing model, limits of the SMS system have probably been reached with extended EMS. It becomes very difficult to develop further extensions for this service.

When writing this book, there was no commercial products on the market, supporting extended EMS. Basic and extended EMS are intended to offer, to mobile subscribers, a foretaste of what mobile messaging could become in the coming years. GPRS and UMTS networks are being deployed and they support high-bandwidth connections. Mobile messaging applications will, of course, benefit from the deployment of these networks. For this purpose, standardization development organizations have designed a framework for the development of a new messaging service: the Multimedia Messaging Service (MMS). Unlike EMS, MMS is not an application-level extension of SMS. MMS benefits from a new service architecture, relies upon high-bandwidth transport connections and is intended to offer a truly multimedia experience to subscribers. In this context, the window of introduction for extended EMS is very narrow. A rapid and successful introduction of MMS in the market would probably prevent extended EMS devices gaining wide acceptance by the mass market. However, if the introduction of MMS faces technical problems or billing issues, then extended EMS certainly has a chance to meet the needs for rich-media messaging until the successful introduction of MMS.

The next chapter provides an in-depth description of MMS.

Further Reading

Tutorial on MIDI and Music Synthesis, the Manufacturer MIDI Association, available from http://www.midi.org/about-midi/tutorial/tutor.htm.

6

Multimedia Messaging Service

As described in previous chapters, SMS has been a very successful service with 2G mobile networks such as GSM. Access to higher bandwidth, with 2.5G and 3G networks, enables the development of alternative and new services. Sophisticated capabilities are integrated into these services to meet the ever-increasing user demands for multimedia features. In addition, what is now expected by the mobile communications market is a better convergence with existing services available to Internet users. In order to meet this requirement, the 3GPP and the WAP Forum have defined a new messaging service to be supported by 2G, 2.5G and 3G networks. This new messaging service is known as the Multimedia Messaging Service (MMS). MMS allows the exchange of multimedia messages in the context of person-to-person and machine-to-person scenarios. In comparison with SMS and EMS messages, a message in the MMS environment is composed of truly multimedia elements. This includes the possibility of composing multimedia messages as 'slideshow' presentations (i.e. combination of text, audio and pictures, all choreographed over time). Interoperability with the Internet electronic mail service is a primary consideration for MMS. With MMS, a subscriber is able to compose a message and send it to Internet mail recipients. In addition, the MMS subscriber can retrieve message originating from Internet users.

The wheel has not been reinvented for the development of the Multimedia Messaging Service. MMS capitalizes on the best features of existing fixed and mobile messaging systems such as SMS, EMS and the Internet electronic mail. Design objectives for MMS include the use of existing transport protocols and content formats widely used in the Internet world.

As shown in previous chapters, the Short and Enhanced Messaging services have been specified almost entirely in the scope of the ETSI and 3GPP standardization activities. The setting of the Multimedia Messaging Service is more complex and the definition of MMS has required a tremendous workload from several standardization development organizations. Mainly, the work has been carried out in the scope of the 3GPP and WAP Forum standardization processes. The 3GPP deals with the high-level service requirements, architectural aspects of MMS, message structure and content formats. It also defines the technical realizations for selected interfaces allowing interactions between communicating network elements. On the other hand, the WAP Forum has focused on the technical realizations of MMS on the basis of WAP and Internet transport protocols. Consequently, specifications defining the service are scattered over a large number of documents maintained by different organizations. Practically, this does not facilitate the work of implementers. At the time of writing this book, the 3GPP has provided three releases of the service definition and the WAP Forum has finished the completion of two of them. The first technical realization is based on WAP transport protocols and is named MMS 1.0. The second technical realization is based on either WAP or Internet transport protocols and is named MMS 1.1 (codenamed MMS Voyager).

This chapter first introduces the service, its context of development and associated use cases. A synopsis of available technical specifications is presented and dependencies between specifications are explained. This chapter also describes the structure of a multimedia message and shows how such a message can be transported between network elements. All transactions, including message/notification management and reports handling are presented along with illustrative examples.

6.1 Service Description

The Multimedia Messaging Service (MMS) is the result of a series of evolutionary steps. It all started with the introduction of SMS in 2G networks. The next milestone in mobile messaging was the development of EMS with the support of limited rich-media content for the composition of messages. Now, while the MMS is emerging, the key enabler for this technology is the availability of 2.5G and 3G networks offering access to high-bandwidth connections. Initially, the preferred transport technology for MMS is the packet-switched connections of GPRS networks. However, the transport of messages over Circuit Switched Data (CSD) and High Speed CSD (HSCSD) are also alternative transport technologies for MMS. Eventually, it is likely that network operators will migrate to 3G networks. These mobile networks are ideally scaled for the support of MMS.

MMS is intended to meet ever-increasing demands for multimedia communications. It also aims to break interoperability barriers between mobile communications and the Internet that prevent the development of an ubiquitous messaging service. This is performed by adopting transport protocols and message formats already widely used in the Internet world.

Mobile subscriber needs have evolved from voice-centric applications in favour of messaging and visual content. The cocktail of messaging and visual content is captured in the Multimedia Messaging Service. With MMS, subscribers are able to compose and receive messages ranging from simple plain text messages, as found in SMS and EMS, to complex multimedia messages, as found over the Internet. With MMS, a message can be structured as a multimedia slideshow, similar to a Microsoft PowerPoint presentation. Each slide is composed of elements such as text, audio, video and images organized over a graphical layout. The period of time for the presentation of each slide can be configured by the message originator or, alternatively, the recipient can press a key to switch from one slide to the following one. Features available with existing Internet messaging systems have not been forgotten in MMS. Indeed, MMS supports group sending and the management of delivery and read-reply reports. It allows messages to have different priorities and classes. The concept of message notifications has been introduced in MMS to allow immediate or deferred retrieval of messages. This retrieval mechanism allows an efficient and flexible method for transporting messages over scarce radio resources. In order to cope with the storage capacity limitations of mobile devices, MMS supports the concept of persistent network-based storage. With this concept, messages can be stored persistently in the network and manipulated remotely via mobile devices. The fact that mobile devices have heterogeneous message handling capabilities (e.g. support for video, maximum image dimensions, etc.) does not help content providers to offer homogeneous services to a large audience of mobile subscribers. In order to cope with this issue, MMS supports content adaptation allowing message content to be tailored to the capabilities of the receiving device.

The commercial introduction of MMS started in March 2002 for a limited number of market segments. It is expected that MMS will later become accessible to all market segments and will eventually become the de facto messaging service for the mass market. The initial availability of the service to a limited number of market segments is explained by the fact that the deployment of MMS requires the use of multimedia devices which are still costly to produce. Consequently, it is difficult for mobile manufacturers to offer MMS features for low-end devices. Note that compared with EMS, network operators need to upgrade network infrastructures if they wish to offer MMS to their subscribers. Business analysts have already identified the high potential of MMS and this has led to significant investments in the service from network operators, device manufacturers and service providers in various parts of the world.

Box 6.1 Commercial Availability of the Multimedia Messaging Service

Telenor was the first operator to launch MMS in Europe in March 2002. This initiative was followed by D2 Vodafone (April 2002), T-Mobile UK (May 2002), Telecom Italia Mobile (May 2002), Swisscom (June 2002) and T-Mobile Germany/Austria (summer 2002). Other major operators such as Telia, Telefonica Moviles, mm02, Vodafone UK are expected to launch the service by the end of 2002. Industry analysts expect a significant market impact from MMS around 2003 with the support of MMS in the mid-to-lower price range. Furthermore, usage of MMS should overtake SMS usage in the mass market around 2005.

6.2 MMS Use Cases

This section describes a selection of possible use cases for the Multimedia Messaging Service. Such use cases represent the basis for the identification of high-level requirements for MMS by standardization organizations. Unlike SMS, MMS is a very recent technology and still has to be accepted by the mass market.

- *Picture/video messaging:* the use of the Multimedia Messaging Service in the person-to-person scenario is tightly associated with the availability of multimedia devices such as the camera or the camcorder. These multimedia devices may be provided built into the mobile handset or as independent accessories. These devices allow subscribers to capture still images and video sequences to be inserted in multimedia messages. In this category, the picture messaging refers to the typical scenario where the subscriber shoots a snapshot of a scene while on the move and sends it as part of a multimedia message to one or more recipients.

- *Next generation voicemail:* in GSM, subscribers can leave voice messages in voicemail servers. It is common for GSM operators to deliver short messages to subscribers indicating that they have messages waiting in the voicemail server. Upon receipt of these voice message waiting notifications, subscribers establish voice calls to the voicemail system in order to manipulate voice messages (listen, delete, etc.). With MMS, operators can extend their voicemail service by sending multimedia messages containing the voice message (as a speech attachment to the message) along with some graphical and textual elements indicating the priority of the message, the date and time the voicemail message has been created. With MMS, the voicemail message can be forwarded by the subscriber to the recipients who may be interested in its content.

- *Building block for other services:* the Multimedia Messaging Service can be considered as a building block for developing more complex services. For instance, immediate messaging is often seen as the combination of two concepts: presence and messaging. The presence refers to the availability of some information related to the state of a subscriber. This includes a network status such as 'disconnected' or 'attached', the subscriber location or an application-level status that represents the subscriber's mood for communicating such as 'having a meeting', 'having a coffee break', etc. Technologies for managing presence information in mobile networks have been defined by organizations such as the 3GPP, the Wireless Village, the WAP Forum or the IETF (see Chapter 7). Additionally, several technologies are considered as suitable candidates for the messaging aspect of the immediate messaging service. The Multimedia Messaging Service is one of them.
- *Value added services:* a VAS provider is an organization that offers an added-value service based on MMS. A VAS application may provide weather notifications, news updates, location-based information, etc. For this purpose, the provider sets up a VAS application which generates multimedia messages and sends them to one or multiple recipients. In many cases, the user needs to first subscribe to the value added service in order to receive corresponding messages. This subscription can be performed by sending a message to the VAS application. Mass distribution of information can be achieved with a value added service. In order to operate a value added service, the VAS provider has to establish a service agreement with the MMS provider. In particular, such an agreement specifies how the revenue generated by the value added service is shared between the MMS provider and the VAS provider. This use case is also referred to as the machine-to-person scenario.

6.3 The MMS Architecture

The Multimedia Messaging Service (MMS) is organized over a set of several network elements. One of the objectives of MMS is to deliver a messaging service interoperable with existing mobile messaging systems such as SMS and EMS but which is also interoperable with existing fixed messaging systems such as the Email service.

The MMS architecture includes the software messaging application in the MMS-capable mobile device necessary for the composition, sending and receipt of messages. In addition, other elements in the network infrastructure are required to route messages, to adapt the content of messages to the capabilities of receiving devices, etc. Figure 6.1 shows the general architecture of a typical configuration of the MMS environment.

The *MMS user agent* is the software application shipped with the mobile handset which allows the composition, viewing, sending, and retrieval of multimedia messages. For the exchange of a multimedia message, the MMS user agent which generates and sends the multimedia message is known as the *originator MMS user agent* whereas the MMS user agent which receives the multimedia message is known as the *recipient MMS user agent*. The *MMS Environment* (MMSE) refers to the set of MMS elements, under the control of a single

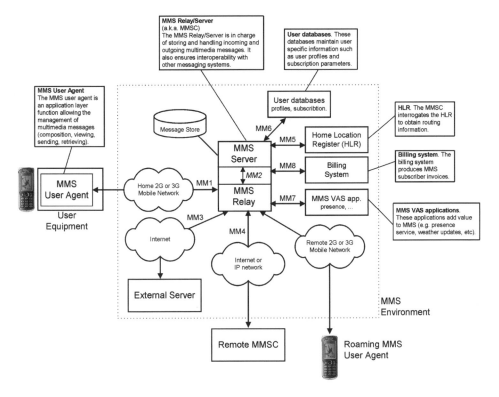

Figure 6.1   MMS architecture

administration (MMS provider), in charge of providing the service to MMS subscribers. Recipient and originator MMS user agents are attached respectively to the recipient and originator MMSE. A key element in the MMS architecture is the MMS Relay/Server. The MMS Relay is responsible for routing messages within the MMSE but also outside the MMSE whereas the MMS Server is in charge of storing messages that are awaiting retrieval. The MMS Relay and MMS Server may be provided separated or combined. In the latter configuration, the name usually given to the combined MMS Relay/Server is the MMS Centre (MMSC).

Box 6.2 Commercial Availability of MMSCs

Several manufacturers are already offering MMSC solutions. Identified MMSC manufacturers are Alcatel, CMG, Comverse, Ericsson, Logica, Materna, Motorola, Nokia and Openwave.

Box 6.3 Commercial Availability of MMS-enabled Mobile Devices

Several manufacturers also plan to provide MMS-enabled mobile handsets from 2002. Ericsson and Nokia have already released MMS-capable phone handsets. Other major device manufacturers such as Alcatel, Motorola and Siemens are likely to release products by the end of 2002 or during 2003. Section 6.38 describes the features of several MMS-capable handsets already available on the market. In order to produce MMS-capable handsets, the manufacturer has to choose between developing the MMS user agent in-house or alternatively to ship the handset with an off-the-shelve MMS user agent developed by a third party organization. Organizations that currently offer off-the-shelve MMS user agents include Magic 4, Nokia, Openwave (also a provider of WAP microbrowsers) and Teleca (AU-systems).

6.4 MMS Centre – MMS Relay/Server

The MMS Relay/Server, also known as the MMSC, is a key network element in the MMS architecture. This element is responsible for storing and managing messages, associated reports and notifications.

The MMSC interconnects with other messaging systems (SMSC, Email servers, etc.). This ensures that MMS subscribers can exchange messages with Internet users and subscribers who do not have MMS-capable handsets (SMS or EMS only handsets).

An MMSC usually provides two different transport modes for the delivery of multimedia messages to recipient MMS user agents. An MMSC always supports at least the *batch mode* which consists of delivering the entire message with all its attachments prior to being presented to the recipient subscriber. A MMSC may optionally support the *streaming mode* which allows parts of a message to be rendered on-the-fly by the recipient MMS user agent while the message content is being transferred. The major drawback of the batch mode resides in the fact that the entire multimedia message has to be transferred and stored first prior to being presented to the recipient subscriber. This method requires substantial storage capacities from the receiving mobile device. Unlike the batch mode, the streaming delivery allows large amount of message content to be rendered by the receiving device even if the device has very limited storage capacities.

MMS user agents developed for different market segments have various message handling capabilities (support of black-and-white images only, support of streaming, etc.). One responsibility of the recipient MMSE is to identify the capabilities of a receiving MMS user agent before message retrieval. This allows the message content to be adapted to whatever the receiving MMS user agent is capable of rendering. This process is known as the *content adaptation* process. Content adaptation in the WAP environment was presented in Chapter 1.

One of the MMSC's responsibilities is to generate Charging Data Records (CDRs) when handling multimedia messages and associated reports/notifications. These CDRs are used by network operators to bill MMS subscribers for network resource usage. This method offers flexibility for network operators to design various billing systems fulfilling the requirements of different business models.

Additionally, MMSC responsibilities include the possibility to forbid access to MMS subscribers (e.g. blocking the access to prepaid subscribers running out of credits), the management of a blacklist for avoiding spam messages to be delivered to MMS subscribers, etc.

6.5 The MMS User Agent

The MMS User Agent (UA) is the software application that resides in User Equipment (3G network), in a Mobile Station (2G and 2.5G network) or in any other device that may exchange multimedia messages. The MMS user agent can offer the following features:

- Viewing and composition of multimedia messages
- Submission (sending) and retrieval (reception) of multimedia messages
- Presentation of message notifications (indicating that a message is awaiting retrieval)
- Presentation of read-reply and delivery reports
- Handling of locally and remotely stored messages (delete, forward, etc.)
- Management of MMS settings (MMSC address, request for delivery reports, etc.).

6.6 User Databases

User databases hold information on the MMS subscriber. This may include subscription details, user profiles and even subscriber locations.

6.7 MMS Interfaces

In an MMS environment, network elements communicate via a set of interfaces. Each interface supports a number of operations such as message submission, message retrieval, message forwarding. Each operation is associated with a set of parameters (also known as information elements). Several interfaces have been standardized in order to ensure interoperability between devices produced by various manufacturers. Other interfaces have not been standardized and are therefore the subject of proprietary implementations.

The *MM1 interface* is a key interface in the MMS environment. It allows interactions between the MMS user agent, hosted in the mobile device, and the MMSC. Operations such as message submission, message retrieval can be invoked over this interface. The 3GPP has defined the functional requirements of this interface. Based on these requirements, the WAP Forum has designed associated MM1 technical realizations for the WAP environment.

The *MM2 interface* is the interface between the MMS Relay and the MMS Server. Most commercial solutions offer a combined MMS Relay and MMS Server in the form of an MMSC. In this case, the interface between the two elements is developed in a proprietary fashion. At the time of writing this book, no technical realization of this interface has been standardized.

The *MM3 interface* is the interface between an MMSC and external servers. Operations over this interface allow the exchange of messages between MMSCs and external servers such as Email servers and SMSCs.

The *MM4 interface* is the interface between two MMSCs. This interface is necessary for exchanging multimedia messages between distinct MMS environments.

The *MM5 interface* is needed to allow interactions between the MMSC and network elements such as the HLR. Through the MM5 interface, an MMSC can request information maintained by the HLR. This includes the retrieval of routing information for forwarding a message to another messaging domain or some information about a particular subscriber.

The *MM6 interface* allows interactions between the MMSC and user databases (e.g. presence server). Unfortunately, the MM6 interface has yet to be standardized. Consequently, this interface is not covered in this book.

The *MM7 interface* fits between the MMSC and external Value Added Service (VAS) applications. This interface allows a VAS application to request services from the MMSC (message submission, etc.) and to obtain messages from remote MMS user agents.

The *MM8 interface* is needed for allowing interactions between the MMSC and a billing system. Unfortunately, such an interface has not been standardized yet. However, charging and billing aspects in the MMS environment are considered in Section 6.25.

6.8 WAP Forum Technical Realizations of MM1

So far, the WAP Forum has published stable specifications for two technical realizations of the MM1 interface, for different network configurations of the WAP framework. The specification of a third technical realization is under development in the scope of the WAP Forum standardization process. The two sets of published specifications are named MMS 1.0 and MMS 1.1 (codenamed Voyager) and correspond to two different levels of features. MMS 1.0 features represent a subset of features supported in MMS 1.1. MMS 1.0 maps MMS features in a WAP 1.x network configuration. In addition, the WAP 2.0 network configurations are also supported in MMS 1.1. Possible network configurations defined in WAP 1.x and WAP 2.0 specification suites are described in Chapter 1.

The advantage of using WAP as the transport technology for the realization of MMS resides in the fact that WAP hides the specifics of underlying transport bearers from MMS-based applications. Transport bearers that can be used in the WAP environment include SMS, GSM CSD, GPRS, EDGE and W-CDMA. Furthermore, WAP has become widely adopted by major network operators and therefore ensures interoperability between different MMS solutions, at the transport level, and facilitates the rapid deployment of MMS solutions.

In this book, the 3GPP terminology for MMS is used. The mapping between 3GPP and WAP Forum terms is presented in Box 6.4.

6.8.1 MM1 Technical Realization – WAP MMS 1.0

With WAP MMS 1.0, the configuration of elements interacting over the MM1 interface is based on recommendations of the WAP 1.x specification suite. In this configuration, a network element, called the WAP Gateway, bridges the mobile network and the Internet/ Intranet domain. The network configuration and associated protocol stacks are depicted in Figure 6.2.

Interactions between the MMS user agent and the MMSC are carried out over a *wireless section* and a *wired section*. The transport of data over the wired section, between the MMSC and the WAP gateway, is performed over the HTTP protocol. On the wireless section, between the MMS agent and the WAP gateway, the transport is performed over the WSP protocol. In this configuration, the WAP gateway converts HTTP requests/responses into equivalent WSP requests/responses. The WAP gateway also converts all incoming WSP

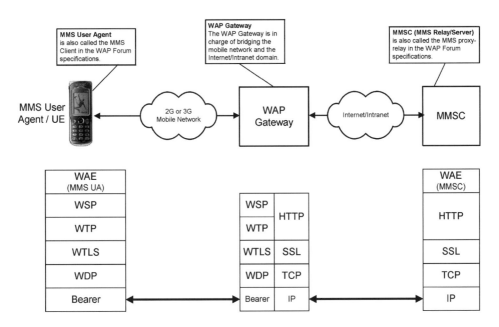

Figure 6.2   MMS architecture with WAP MMS 1.0

requests/responses from the MMS user agent into equivalent HTTP requests/responses to be delivered to the MMSC. The objective of this network configuration is to optimize the transport of data over the wireless section by transcoding requests/responses into a compact binary format and by performing other transport-level optimizations.

Box 6.4 WAP Forum Terminology

In technical specifications published by the WAP Forum, the MMS user agent is called the *MMS Client* and the MMSC (known as the MMS Relay/Server in 3GPP specifications) is called the *MMS Proxy-Relay*. Interfaces also bear different names as shown below:

3GPP terminology	*WAP Forum terminology*
MM1	MMS_M
MM2	MMS_S
MM3	E (Email server) and L (Legacy mobile messaging servers)
MM4	MMS_R
MM5	not referred to by the WAP Forum
MM6	not referred to by the WAP Forum
MM7	not referred to by the WAP Forum
MM8	not referred to by the WAP Forum

At the low level of the protocol stack, several bearers can be used on the wireless section. The bearer choice depends on the user subscription, terminal capabilities/state and network conditions. The most common bearers are SMS, CSD and GPRS for 2G and 2.5G networks and W-CDMA for 3G networks.

Box 6.5 Dedicated or shared WAP gateway for MMS traffic?

A WAP gateway could be shared for the transport of WAP browsing and MMS traffics. However, for efficiency reason and because the envisaged billing model for MMS is diverging from the one for the WAP browsing (see section 6.25), it is likely that operators will acquire additional WAP gateways dedicated to the management of MMS traffic.

6.8.2 MM1 Technical Realization – WAP MMS 1.1

The MMS 1.0 network configuration described in the previous section is also supported in MMS 1.1. In addition, MMS 1.1 also supports network configurations presented in the WAP specification suite 2.0 and introduced in Chapter 1. An additional network configuration has a wireless IP router bridging the wireless section and the wired section. Another additional network configuration has a router augmented with a WAP proxy supporting wireless-profiled TCP and wireless-profiled HTTP. The first additional network configuration is depicted in Figure 6.3.

With the two additional network configurations introduced with MMS 1.1, the MMS user agent and the MMSC can interact directly using a common protocol: HTTP. The possible low-level bearers for the wireless section are the same as those for MMS 1.0.

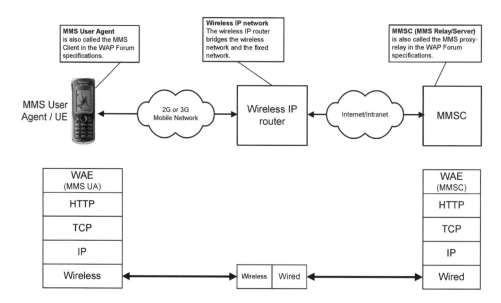

Figure 6.3   MMS architecture with WAP MMS 1.1

Table 6.1   MMS technical specification synopsis

3GPP MMS specifications	WAP Forum recommendations	MMS features
3GPP MMS release 99	WAP MMS 1.0 in WAP 2.0 spec. suite	Basic MMS features such as: • message composition • message sending/reception • capability negotiation • address hiding • read/reply reports • WAP configuration: WSP
3GPP MMS release 4	WAP MMS 1.1 (codenamed MMS Voyager)	• reply charging • message forwarding • support of MMS in USIM • WAP configurations: WSP or HTTP
3GPP MMS release 5	Likely to be published as MMS 2.0 (codenamed MMS Cubed)	• persistent network-based storage • security enhancements • address resolution • MM1 updates • MM7 definition • MM4 updates

6.9 Technical Specification Synopsis

In comparison with SMS and EMS, MMS is a far more sophisticated messaging service. The definition of this service required a tremendous standardization workload. In order to achieve a timely market introduction for the service, the standardization work was appropriately divided between the 3GPP and the WAP Forum.

The 3GPP is responsible for the definition of high-level service requirements, overall MMS architecture and transactions flows. Additionally, the 3GPP provides the technical specifications for the realization of several interfaces. On the other hand, the WAP Forum concentrated on the technical realizations of the MM1 interface in the WAP environment. The 3GPP has defined three levels of technical specifications referred to as 'releases' (see the 3GPP standardization process in Chapter 2). To each release corresponds a level of features and releases are developed in a backward compatible manner. The three 3GPP releases are respectively known as MMS release 99, MMS release 4 and MMS release 5. So far, the WAP Forum has produced stable technical realizations corresponding to the two first 3GPP releases (referred to as MMS 1.0 and MMS 1.1, respectively) and is currently working on a third technical realization (codenamed MMS Cubed) expected to fulfil the requirements of 3GPP release 5. Table 6.1 gives the level of features of each MMS release.

Figure 6.4 shows the relationships between 3GPP technical specifications and the WAP Forum recommendations for 3GPP release 99, release 4 and release 5.

6.9.1 3GPP MMS Specifications

Table 6.2 lists 3GPP technical specifications related to MMS along with their availability in

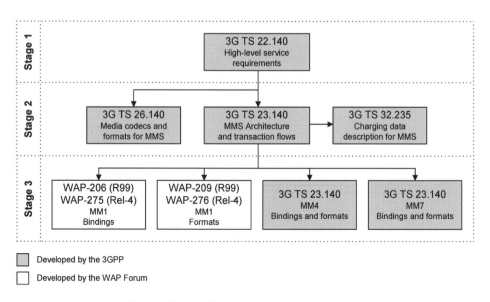

Developed by the 3GPP

Developed by the WAP Forum

Figure 6.4 MMS technical specification synopsis

terms of 3GPP releases. The technical specification [3GPP-22.140] (stage 1) identifies the high-level service requirements for MMS. The technical specification [3GPP-23.140] defines the overall MMS architecture and transaction flows (stage 2), but also technical realizations for MM4 and MM7 interfaces (stages 2 and 3). The technical specification [3GPP-26.140] identifies content formats and defines format profiles for the representation of objects in multimedia messages. Finally, the technical specification [3GPP-32.235] defines the procedures for the management of charging data records used by billing systems for the production of invoices for MMS subscribers.

6.9.2 WAP Forum MMS Recommendations

Technical specifications for the realization of the MM1 interface, published by the WAP Forum, are available in three sets. Two sets of specifications have already been publicly released as MMS 1.0 and MMS 1.1, respectively. Another set of specifications is under development and should be published by the beginning of 2003 (likely to be published as

Table 6.2 3GPP technical specifications

Specification number	Specification title	Availability		
		R-99	Rel-4	Rel-5
3GPP TS 22.140	Service aspects (MMS Stage 1)	Yes	Yes	Yes
3GPP TS 23.140	Functional description (MMS Stage 2)	Yes	Yes	Yes
3GPP TS 26.140	Media formats and codecs	No	No	Yes
3GPP TS 32.235	MMS charging management	No	Yes	Yes

Table 6.3   WAP Forum specifications/first set MMS 1.0 corresponding to 3GPP release 99

Specification number	Specification title
WAP-205-MMSArchOverview-20010425-a	MMS architecture overview
WAP-206-MMSCTR-20010612-a	MMS client transactions
WAP-209-MMSEncapsulation-20010601-a	MMS encapsulation

Table 6.4   WAP Forum specifications/second set MMS 1.1 corresponding to 3GPP release 4

Specification number	Specification title
WAP-274-MMSArchOverview-20020409-d	MMS architecture overview
WAP-275-MMSCTR-20020410-d	MMS client transactions
WAP-276-MMSEncapsulation-20020409-d	MMS encapsulation

MMS 2.0). The three sets of WAP Forum technical specifications fulfil the functional requirements defined in MMS technical specifications of 3GPP release 99, release 4 and release 5, respectively. The composition of the two published sets of WAP Forum technical specifications is shown in Tables 6.3 and 6.4.

Documents, titled *MMS architecture overview* [WAP-205][WAP-274], outline how MMS can be realized in the WAP framework. Documents, titled *MMS client transactions* [WAP-206][WAP-275], present the transactions between the MMS user agent and the MMSC over the MM1 interface whereas documents, titled *MMS encapsulation* [WAP-209][WAP-276], present the structure of protocol data units exchanged over the MM1 interface.

6.9.3 W3C Multimedia Standards

As indicated in previous sections, the 3GPP and the WAP Forum have produced a set of specifications allowing implementers to build the necessary components for the realization of the Multimedia Messaging Service. Once these components are in place and appropriately configured, MMS subscribers are able to exchange multimedia messages composed of various elements such as text, images, audio and video sequences. To allow this, several multimedia formats for building messages have been identified by the 3GPP, manufacturers and operators. The identification process was carried out with the objective of allowing a good convergence with the Internet. It is achieved by selecting formats already widely used in the Internet world. Formats/languages derived from XML such as SMIL, XHTML and SVG have been identified as being suitable for representing parts of multimedia messages in the MMS environment. The definition of these XML-derived formats/languages has been published by the World Wide Web Consortium (W3C).

The development of XML started in 1996 and has been a W3C recommendation since 1998. XML is based on a mature technology, called SGML, developed in the 1980s. SGML is a complex mark-up language for structuring any kind of data and has obtained the status of

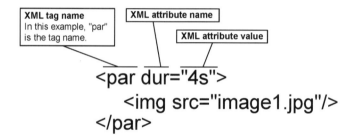

Figure 6.5   Structure of an XML tag

ISO standard since 1986. The best parts of SGML constituted the basis for the design of XML. XML also benefited from experiences gained from the use of another widely known mark-up language: HTML. As a result, XML became a 'best-of-breed' mark-up language almost as powerful as SGML for structuring data but far simpler to manipulate. XML is particularly well suited for representing objects such as technical drawings, multimedia scene descriptions, financial transactions, technical specifications, etc. Key characteristics of XML are its extensibility, its platform independence and the fact that XML is becoming the de facto standard for representing many types of information. Like HTML, XML makes use of tags and attributes as shown in Figure 6.5.

6.10 Structure of a Multimedia Message

Multimedia messages transported in MMS environments include basic short messages with text only but also truly multimedia messages with video and audio attachments synchronized over time and appropriately laid out over a graphical region. The message format for MMS is based on existing content formats used by Internet systems augmented by presentation formats (also known as scene description formats) used in the Web.

On Internet and MMS domains, a message is formatted according to RFC 822 specifications [RFC-822][RFC-2822]. According to RFC 822, each message consists of an envelope and message contents. The envelope contains the instructions required to deliver and interpret message contents. As an extension to RFC 822, the inclusion of non-textual elements such as image, audio and video in multimedia messages is made possible by formatting message contents in the form of a *multipart message* according to Multipurpose Internet Mail Extensions (MIME).

Instead of presenting non-textual message elements as a list of message attachments to the subscriber, the multimedia message may optionally contain a description of the way textual and non-textual elements should be presented to the subscriber (graphical layout and temporal synchronization). Such a description is known as scene description and is defined according to format/languages such as SMIL, XHTML or WML.

For first implementations of MMS, the size of a multimedia message is likely to vary between several Koctets and 100 Koctets.

Table 6.5   RFC 822 header fields

RFC822 header field name	Meaning
Transport related	
To	Address of primary message recipient(s)
Cc	Carbon copy
	Address of secondary message recipient(s)
Bcc	Blind carbon copy
	Address of secondary message recipient(s)
From	Address of the message originator
Sender	Address of the user who sent the message
Received	Added by a transfer agent during message transfer
Return-path	Used to identify a path back to the message sender
Message related	
Date	Date and time when the message was sent
Reply-to	Address to which a reply should be sent
Message-ID	Unique message identifier
In-reply-to	Message to which this message is a reply to
References	Any relevant message references
Keywords	Message keywords characterizing the message content
Subject	Message subject

6.10.1 Message Basic Format

In 1982, the IETF published a standard defining message formats to be used by the Internet community. The standard, known as RFC 822 [RFC-822],[1] defines a message representation for messages composed of a header (message envelope) and a body (message content) in the form of US-ASCII text.

An RFC 822 message is composed of a primitive envelope with several header fields, a blank line and the message body. Each header field is a single line of plain text (US-ASCII) containing the header field name and the assigned value(s) separated by a colon (:).

The main message header fields defined in RFC 822 are listed in Table 6.5. An RFC 822 message contains a set of header fields related to the transport of the message. The To field identifies the primary recipient(s) of the message. Multiple recipients may be identified with this field. The Cc and Bcc fields identify the secondary recipient(s) of the message. Both fields accept multiple secondary recipients. Whether a message recipient is a primary or a secondary recipient does not affect the way the message is transported and delivered. With these fields, the message originator only indicates the level of importance recipients should attach to the message. As a rule of thumb, the message content is of direct interest to primary recipients and is copied to secondary recipients for 'information only'. A message originator can hide his/her intention to send the message to one or more secondary recipients by listing the recipients in the Bcc field rather than in the Cc field. The fields From and Sender indicate who originated the message (message originator) and who sent the message

---

[1] Note that, in 2001, an extended version of the Internet message format was published by the IETF in [RFC-2822]. Consequently, [RFC-2822] replaces [RFC-822].

```
From: Gwenael@le-bodic.net
To: Armel@amorepro.com
Sent: 12/06/01 17:14
Subject: Lunch on Sunday

Do not forget! We are having lunch on Sunday.
```

Figure 6.6   Example of RFC 822 message

(message sender), respectively. For instance, a company director may write a message (message origination) but may delegate the sending to his/her secretary (message sending). The Received field can be added by transfer agents during the transfer of the message from the message originator user agent to the message recipient user agent(s). This field indicates the name of intermediary transfer agents along with the transfer date and time. The analysis of the different Received fields available in a message allows the determination of the complete message transfer route. The Return-path field indicates how the message recipient can reply back to the message originator. This field usually contains the message originator address.

Additionally, an RFC 822 message contains a set of header fields related to the message itself. The Date field indicates when the message was sent. The Reply-to indicates the address to which the message recipient(s) should reply. The Message-ID uniquely identifies the message. The In-reply-to field represents the identification of the original message to which the message is a reply. The field References identifies other related messages that may be related to the content of a message. The keyword and subject characterize the content of the message. Figure 6.6 shows a very short Email message in the RFC 822 format.

6.10.2 Multipurpose Internet Mail Extensions/RFC 204x

The RFC 822 message representation has been widely adopted well beyond the Internet. While it was a very successful approach for basic messaging systems, it did not fulfil the needs of more advanced messaging systems allowing the exchange of messages composed of multiple parts (sounds, images, etc.) and for which the text can be richer than the US-ASCII (Asian languages, some specific European languages, etc.). To cope with these limitations, the IETF introduced the Multipurpose Internet Mail Extensions (MIME), a set of extensions that complements the RFC 822 format by allowing the representation of non-textual message contents. This set of extensions is completely defined in the five documents listed below:

[RFC-2045]	MIME Part 1: format of Internet message bodies
[RFC-2046]	MIME Part 2: media types
[RFC-2047]	MIME Part 3: message header extensions for non-ASCII text
[RFC-2048]	MIME Part 4: registration procedures
[RFC-2049]	MIME Part 5: conformance criteria and examples.

In addition, the WAP Forum has defined and published in [WAP-203] a binary representation of multipart messages. This binary representation is used for the transfer of messages over the

Table 6.6   MIME header fields

MIME header field	Meaning
MIME-version	Indicates the MIME version
Content-type	Nature of the message
Content-transfer-encoding	Message encoding for transfer
Content-ID	Unique identifier
Content-description	Textual description of the message content

MM1 interface in the WAP framework. The binary representation is explained in Section 6.30.

MIME defines the five additional message header fields listed in Table 6.6 in order to complement existing RFC 822 header fields. The MIME-Version field allows processing entities to distinguish MIME messages from older messages (which lack such a field):

MIME-Version: 1.0

6.10.2.1 Content Type and Subtype

The content type identifies the nature of the message content. The content type is composed of a media type, a media subtype and optional parameters. Parameters are specific to the media being referred to. The first example below shows the definition of a media type for an image in the JPEG format (media type indicates that the corresponding object is of type image and the subtype indicates the image format – JPEG). The second example shows the identification of text formatted with the US-ASCII character set.

Example 1:   Content-type: image/jpeg
Example 2:   Content-type: text/plain; charset = "us-ascii"

In MIME, there are seven high level content types. Five of them are known as *discrete media* types, the other two are known as *composite media* types. Table 6.7 shows the five discrete media types. Examples of subtypes are given but the list of subtypes is not exhaustive in this table.

For the text content type, an optional subtype parameter can indicate the character set that is used for representing characters. The character set is used in MIME to refer to a method for converting a sequence of bits (septets, octets, etc.) into a sequence of characters. If omitted, the default character set is US-ASCII (see Appendix C).

Table 6.8 shows the two composite media types. The multipart content type is of key importance for formatting multimedia messages composed of multiple elements. Examples of subtypes are given in the table but the list of subtypes is not exhaustive.

Another multipart subtype was later introduced for the representation of compound objects. This content type is known as the Multipart/Related and is defined in [RFC-2387]. This content type is used for aggregating multiple inter-related body parts in a message (e.g. various elements composing a scene description). The optional Start parameter, for the Multipart/Related content type, indicates the starting body part

Table 6.7 Content type for discrete media types

Content type	Subtype	Description
Text	Plain	Plain text containing no formatting commands. Plain text is intended to be displayed, as received by the message recipient(s), without any further processing
	Rich text	Rich text is text along with formatting instructions such as italic, bold, justified, left aligned, etc.
Image	Gif	Image in the GIF format
	Jpeg	JPEG format using JFIF encoding
Audio	Basic	Sound
Video	Mpeg	Video in the MPEG format
Application	Octet-stream	A sequence of octets for which the structure is unknown
	Postscript	A printable document in the PostScript format

in the aggregate (e.g. the file describing the scene description structure). If omitted, then the first body part is considered as the starting body part. Another optional parameter for this content type, called Content-disposition, indicates the style to be used for displaying body parts contained in the aggregate. The two possible styles, defined in [RFC-1806], are

Table 6.8 Content type for composite media types

Content type	Subtype	Description
Message	RFC 822	An RFC 822 message encapsulated in a MIME message
	Partial	Large elements can be broken into several parts. This subtype identifies such a part belonging to a large element
	External-body	This subtype indicates that the element is not included as part of the message. The element location is provided as a parameter of this subtype
Multipart	Mixed	This is the most basic multipart subtype. A multipart/ mixed message is composed of one or more parts. There is no relation or order specified between the parts
	Alternative	A multipart/alternative message has two or more parts representing the same content. For instance, such a message can contain several scene descriptions such as one formatted in SMIL and another one formatted in XHTML. Both scene descriptions represent the same content. The recipient user agent selects the scene description which is the most appropriate according to the receiving device capabilities
	Parallel	A multipart/parallel message has two or more parts that have to be rendered simultaneously. This subtype is seldom used in the scope of MMS
	Digest	A multipart/digest message is composed of an ordered sequence of RFC 822 messages. This subtype is seldom used in the scope of MMS

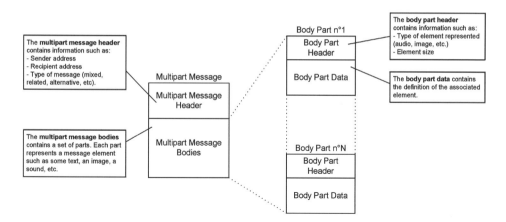

Figure 6.7   Structure of a multipart message

'inline' for presenting the list of body parts in a single document where all body parts are expanded inline (value INLINE assigned to Content-disposition) and 'attachment' for presenting the list of body parts as a list of attachments (value ATTACHMENT assigned to Content-disposition).

Multiple parts of a message are differentiated with boundary delimiters. The name of a boundary delimiter is assigned to the subtype parameter called boundary. The name of a boundary delimiter has a size ranging from 1 to 70 characters. In the multipart message, each part begins with two hyphens (- -) followed by the boundary delimiter name. The multipart message is terminated by a carriage return/line feed followed by two hyphens, the boundary delimiter name and two additional hyphens. Figure 6.7 shows the structure of a multipart message.

Registration of MIME Content Types and Subtypes   This section has shown a number of content types and subtypes commonly used for the generation of multipart messages. New content types and subtypes can always be registered. The organization, which deals with the registration of such types, is the Internet Assigned Numbers Authority (IANA at http://www.iana.org). The registration process for MIME content types and subtypes is described in [RFC-2048]. MIME types are classified into four main trees:

- *IETF tree:* gathers types of general interest for the Internet community; formats in this tree are defined in the form of an RFC.
- *Vendor tree:* used for media types associated with commercially available products.
- *Personal or vanity tree:* media types created, experimentally or as part of products that are not distributed commercially may be registered in the personal or vanity tree.
- *Special'X-' tree:* these types are unregistered, experimental, and should be used only with the active agreement of the parties exchanging them.

The registration process usually takes a minimum of 2 weeks to complete from the day the proposed registration has been submitted.

Figure 6.8 Base64 and quoted-printable encoding/decoding methods

6.10.2.3 Content Transfer Encoding

The transfer encoding field indicates which encoding mechanism has been used to encode the original message content for transport purposes. The encoding is usually necessary for allowing data to be transported over 7-bit transfer protocols such as SMTP. The indication of which encoding has been applied to the original message is necessary for the message recipient to be able to reconstruct the message in its original form after transfer. The example below indicates that the original 8-bit encoded message has been encoded in 7-bit data for transfer purpose according the base64 encoding method:

```
Content-Transfer-Encoding: "base64": "8bit"
```

Three encoding/decoding methods are commonly used in MIME for the following domains: binary, 8-bit data and 7-bit data. These methods are known as the identity, quoted-printable and base64. The identity method means that no encoding has been applied to the original message. The encoding/decoding processes for quoted-printable and base64 methods are depicted in Figure 6.8.

The quoted-printable method encodes an original message in a form that is more or less readable by a subscriber. On the other hand, the base64 method encodes the message in a form which is not user-readable but which presents the advantage of being a more compact data representation. Such encoding methods are particularly relevant for transfers over the MM3 and MM4 interfaces. Note that transfers over the MM1 interface are carried out in a binary domain. Consequently, there is no need to use one of the three encoding methods, as identified in this section, for the MM1 interface.

6.10.2.4 Content identification and description

The `content-id` field provides a unique identification for one given body part of a multipart message. This is to be compared with the `Message-ID` of the RFC 822 format which provides a unique identifier for the message itself.

The optional `content-description` field provides a textual description of the associated body part. The example below shows a textual description for an image:

```
Content-Description: "This image represents a group of small houses"
```

6.10.2.5 Example of a Multipart Message

Figure 6.9 shows a multipart message composed of a rich-text body part.

```
From: gwenael@le-bodic.net
To: armel@armorepro.com
MIME-Version: 1.0
Message-ID: <66645456.645@le-bodic.net>
Content-Type: multipart/mixed; boundary=msgbdry
Sent:      12/06/01 17:14
Subject: Lunch on Sunday

--msgbdry
Content-Type: text/richtext

<bold> Do not forget! </bold>
We are having lunch on Sunday.

--msgbdry--
```

Figure 6.9   Example of a multipart message

6.11 Elements of a Multimedia Message

Previous sections have shown how simple messages can be structured with the RFC 822 format and how more sophisticated multipart messages can be organized with MIME. RFC 822 and MIME are the two most common formats used for structuring messages on the Internet. Consequently, it was decided to use them for MMS in order to achieve a better convergence with existing Internet technologies.

This section specifies which types of elements can be included in multipart messages. Note that the trend is to favour XML-based formats since these formats facilitate the interoperability of MMS with existing services available over the Internet.

6.11.1 Text and SMS/EMS Encapsulation

In MMS, text is supported as part of a simple message or as part of a scene description (XHTML or SMIL) in a multipart message. With SMIL, a text section is included as an independent body part in the message and is referred to in the SMIL scene description. Supported characters sets are US-ASCII for basic text and UTF8 or UTF16 for representing extended character sets.

Is it possible to encapsulate a message segment (SMS, basic or extended EMS) in a multimedia message. For this purpose, the content type `application/vnd.3gpp.sms` is used and the message segment is encoded as defined in [3GPP-24.011] (binary encoding/ RP-DATA RPDU). Note that initial MMS-enabled devices may be using the `application/x-sms` for this purpose. However, because, the MMSC is usually aware of content types supported by MMS-enabled devices (via the content adaptation mechanism), the MMSC can transcode the MIME type to ensure interoperability. In any case, the content type `application/vnd.3gpp.sms` should be preferred for any new development. Note, however, that there is no clear use cases for the encapsulation of short messages in multimedia messages.

6.11.2 Images

It was shown in Chapter 5 that, in extended EMS, images can be represented with different formats. In extended EMS, vector graphics are suitable representations for simple or animated line drawings. Alternatively, images can be represented as bitmap pictures. Similarly, in MMS, several formats can be used for representing images in multimedia messages. *Bitmap images* are expected to become widely used in MMS because of the large availability of such image formats over the Internet. Bitmap images are exact pixel-by-pixel mappings of the represented image (except if a lossy compression is used to reduce the bitmap size). Common formats for bitmap images are JPEG, GIF, Portable Network Graphic (PNG) and WBMP [WAP-190]. The PNG is a file format whose definition is published in the form of a W3C recommendation [W3C-PNG]. PNG is used for representing bitmap images in a lossless fashion. PNG ensures a good compression ratio and supports image transparency. PNG represents a patent-free substitute for the GIF format.

On the other hand, other types of images can be included in multimedia messages. These images are known as *scalable vector graphics*. Scalable vector graphics are based on descriptions of graphical elements composing the represented image. These descriptions are usually made using instructions similar to those of a programming language. These instructions are processed by a graphics processor to reconstruct graphical elements contained in the represented image. *Metafile* and the emerging *Scalable Vector Graphics* (SVG) are two well known vector graphic formats used for representing images. SVG is an open standard derived from XML and published as a W3C recommendation [W3C-SVG]. Advantages of scalable vector graphics include the possibility of dynamically scaling the represented image according to the capabilities of the receiving device. Furthermore, the size of scalable vector graphics representing synthetic images can be very low compared with the size of equivalent bitmap representations. However, scalable graphic formats are not appropriate for representing all types of images. For instance, photographs are usually not well represented with vector graphics. Representing photographs with a scalable vector graphic format may lead to very large representations, larger than equivalent bitmap representations. In addition, note that additional processing capabilities are usually required to render vector graphics (integer and floating point calculations).

Figure 6.10 shows a very simple image represented in the SVG format. Note that, in the SVG format, the size of the image representation is 200 octets against 1600 octets for an equivalent representation (128 × 128 pixels) in the GIF format.

The sequence of SVG instructions necessary for representing the image in Figure 6.10 is provided in Figure 6.11. The SVG representation instructs the graphic processor to draw a

SVG example

Figure 6.10   SVG example/image representation

```
<svg width="128" height="128" >

<circle cx="45" cy="45" r="40" fill="yellow" />
<rect x="45" y="45" width="80" heigth="35" fill="blue" />
<text x="15" y="120">SVG example</text>

</svg>
```

Figure 6.11   SVG example/SVG instructions

plain yellow circle (centre coordinates are 45,45 and radius is 40 pixels), a plain blue rectangle (width 80 pixels and height 35 pixels) and some text ('SVG example').

Box 6.6 Web Resources for Images

SVG at the W3C: http://www.w3c.org/Graphics/SVG/
PNG at the W3C: http://www.w3c.org/Graphics/PNG/

6.11.3 Audio

Several audio formats can be supported in MMS. Audio formats are usually grouped into two categories:

- *Natural audio* formats are used to represent recorded materials such as speech.
- *Synthetic audio* formats consist of specifying commands instructing a synthesizer on how to render sounds.

MIDI and iMelody formats, presented in Sections 5.18 and 4.5.2, respectively, are both synthetic audio formats. Both formats can be supported in MMS. Note, however, that the iMelody format is not a format suggested for MMS by the standardization organizations. Nevertheless, at least one MMS-capable device, already available on the market, supports iMelody objects in multimedia messages (see Section 6.38). In the natural audio category, formats that can be supported are AMR and MP3.

The Adaptive Multirate (AMR) codec is commonly used to represent speech in mobile networks. This codec can adapt its rate according to available resources. Possible rates range from 4.75 to 12.2 kbps. Note that AMR, when configured at given rates, becomes directly compatible with technical characteristics of other codecs specified by several standardization organizations. For instance, AMR, at the 12.2 kbps rate, is compatible with the Enhanced Full Rate (EFR) speech codec defined by ETSI for GSM. Furthermore, AMR, at the 7.4 kbps rate, becomes compatible with the IS-136 codec. Finally, AMR, configured at the 6.7 kbps rate, becomes compatible with the speech codec used in the Japanese PDC. AMR is an Algebraic Code Linear Predictive codec which achieves excellent performance for the representation of recorded audio samples.

MP3 stands for MPEG Layer-3 and is an audio compressed format. MP3 offers an excellent sound quality and a high compression ratio. MP3 is based on perceptual coding techniques addressing the perception of sound waves by the human ear.

6.11.4 Video

Initial MMS devices are not expected to support the rendering of video. However, as technology improves, the support of video could become a popular feature for MMS-enabled devices. The 3GPP suggests the use of the H.263, defined in [ITU-H.263], and MPEG-4 for MMS devices compliant with [3GPP 23.140] release 99. Furthermore, the 3GPP mandates the use of the H.263 format for MMSCs and MMS user agents supporting video from release 4.

MPEG stands for Moving Pictures Expert Group and is an organization that develops technical specifications for representing and transporting video. The first specification from this organization was MPEG-1, published in 1992. MPEG-1 allows video players to render video in streaming mode. MPEG-2, introduced in 1995, supersedes MPEG-1 features and is mainly used for compression and transmission of digital television signals. In December 1999, the group released the specification for MPEG-4 (ISO/IEC 14496). MPEG-4 is based on an object-oriented paradigm where objects are organized in order to compose a synthetic scene. This is a major evolution from previous MPEG formats. The compact MPEG-4 language used for describing and dynamically changing scenes is known as the Binary Format for Scenes (BIFS). BIFS commands instruct the MPEG-4 player to add, remove and dynamically change scene objects in order to form a complete video presentation. Such a technique allows the representation of video scenes in a very compact manner. This compactness makes MPEG-4 a very suitable video format for MMS. MPEG-4 also supports streaming in a low bit rate environment. MPEG-4 has proved to provide acceptable streaming services over 10 kbps channels. Mobile networks often provide very variable and unpredictable levels of resources for services. To cope with these network characteristics, MPEG-4 can prioritize objects and only transmits the most important objects when the system is running short of resources.

Box 6.7 Web Resources for Video

Moving Pictures Expert Group: http://mpeg.telecomitalialab.com
MPEG-4 Industry Forum: http://www.m4if.org

6.12 Scene Description with SMIL or XHTML

A *scene description* (also known as a message presentation), contained in a multimedia message, organizes the way elements should appear over the regions of a graphical layout and defines how elements should be rendered over a common timeline. A scene description allows message objects (sounds, images, etc.) to be rendered by the receiving device in a meaningful order. Message scene descriptions are typically used for news updates, advertisements, etc. The scene description of a multimedia message can be adapted to the receiving device capabilities by means of content adaptation. The 3GPP recommends the use of three formats/languages for presentation/scene description: WML, SMIL and XHTML. Note, however, that SMIL is becoming the de facto format for scene descriptions in MMS and its support is mandatory for release 5 devices accepting scene descriptions.

6.12.1 Introduction to SMIL

The *Synchronized Multimedia Integration Language* (SMIL), pronounced 'smile', is an XML-based language published by the W3C. A major version of this language, SMIL 2.0 [W3C-SMIL], is organized around a set of modules defining the semantics and syntax of multimedia presentations (for instance, modules are available for the timing and synchronization, layout and animation, etc.). SMIL is not a codec or a media format but rather a technology that allows media integration. With SMIL, the rendering of a set of media objects can be synchronized over time and organized dynamically over a predefined graphical layout to form a complete multimedia presentation. SMIL is already supported by a number of commercial tools available for personal computers including RealPlayer G2, Quicktime 4.1 and Internet Explorer (from version 5.5). Because of small device limitations, a subset of SMIL 2.0 features has been identified by the W3C to be supported by devices such as PDAs. This subset, called *SMIL basic profile*, allows mobile devices to implement some of the most useful SMIL features without having to support the whole set of SMIL 2.0 features. Unfortunately, the SMIL basic profile appeared to be still difficult to implement in first MMS-capable mobile devices. To cope with this difficulty, a group of manufacturers designed an even more limited SMIL profile, known as the *MMS SMIL*, to be supported by early MMS-capable devices. In the meantime, the 3GPP is producing specifications for an extended SMIL profile, known as the packet-switched streaming SMIL profile (*PSS SMIL profile*), that is to become the future standard profile for all MMS-capable devices. The MMS SMIL is an interim de facto profile until devices can efficiently support the PSS SMIL profile. The PSS SMIL profile is still a subset of SMIL 2.0 features, but a superset of the SMIL basic profile, and is published in [3GPP-26.234].

Designers of SMIL multimedia presentations can:

- Describe the temporal behaviour of the presentation
- Describe the layout of the presentation on a screen
- Associate hyperlinks with media objects
- Define conditional content inclusion/exclusion based on system/network properties.

Box 6.8 Resources for SMIL

A good two-part tutorial on SMIL is identified in the further reading section (Bulterman, 2001, 2002).
SMIL at the W3C: http://www.w3c.org/AudioVideo

6.12.1 SMIL 2.0 Organization

A major version of the language, SMIL version 2.0, has been publicly released by the W3C in August 2001. The 500-page SMIL 2.0 specifications define a collection of XML tags and

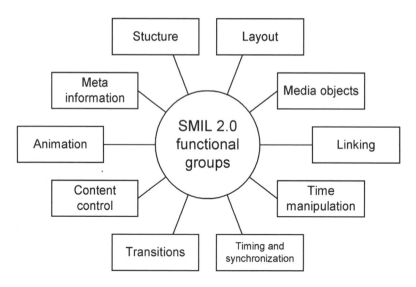

Figure 6.12   SMIL 2.0 functional groups

attributes that are used to describe temporal and spatial coordination of one or more media objects forming a multimedia presentation. This collection is organized into ten major *functional groups* as shown in Figure 6.12.

Each functional group is composed of several *modules* (from 2 to 20). The aim of this SMIL organization is to ease the integration of SMIL features into other XML-derived languages. A number of profiles have been defined on the basis of this organization. A SMIL profile is a collection of modules. So far, several profiles have been introduced such as the SMIL 2.0 language profile, XHTML + SMIL profile and the SMIL 2.0 basic profile (as introduced earlier).

6.12.1.2 Spatial Description

SMIL 2.0 content designers are able to define complex spatial layouts. The presentation rendering space is organized with regions. Each region in the layout can accommodate a graphical element such as an image or a video file. Regions can be nested in each other in order to define sophisticated presentations. The tag `root-layout` is used to define the main region of the presentation. Sub-regions to be positioned within the main region are defined with the `region` tag. The SMIL example in Figure 6.13 shows how two sub-regions, one accommodating an image and the other one some text, can be defined within the main region.

6.12.1.3 Temporal Description

Objects in a SMIL presentation can be synchronized over a common timeline. For this purpose, a set of time containers can be used such as the sequential and the parallel time containers:

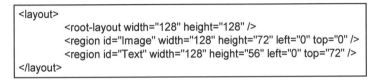

```
<layout>
        <root-layout width="128" height="128" />
        <region id="Image" width="128" height="72" left="0" top="0" />
        <region id="Text" width="128" height="56" left="0" top="72" />
</layout>
```

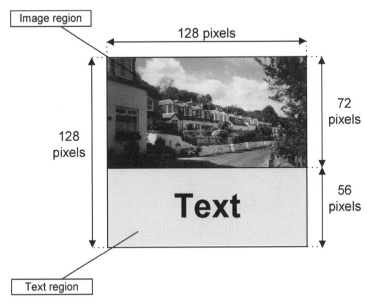

Figure 6.13 SMIL/layout container. The picture in this example was shot in the quaint village of Corrie, located on the wild coast of the island of Arran, Scotland.

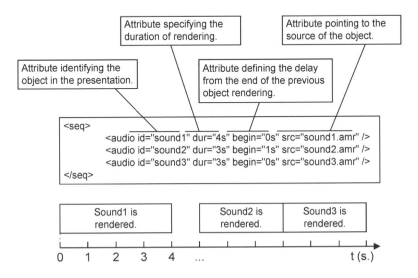

Figure 6.14 SMIL/sequential time container

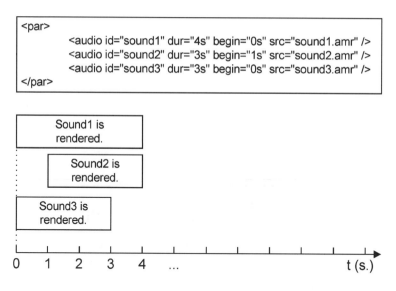

```
<par>
        <audio id="sound1" dur="4s" begin="0s" src="sound1.amr" />
        <audio id="sound2" dur="3s" begin="1s" src="sound2.amr" />
        <audio id="sound3" dur="3s" begin="0s" src="sound3.amr" />
</par>
```

Figure 6.15   SMIL/parallel time container

- *Sequential time container:* this container identified by the `seq` tag enables the sequencing of an ordered list of objects. Each object is rendered in turn and the rendering of each object starts when the rendering of the previous object has terminated. An absolute time duration for the rendering of each object may also be specified as shown in Figure 6.14.
- *Parallel time container:* this container identified by the `par` tag enables the rendering of several objects in parallel as shown in Figure 6.15.

In a scene description, containers can be nested in order to create a whole hierarchy of time containers for the definition of complex multimedia presentations.

6.12.2 SMIL Basic Profile

A indicated in previous sections, the W3C has defined a *SMIL basic profile* for SMIL 2.0. The SMIL basic profile is a subset of the full set of SMIL 2.0 features which is appropriate for small devices such as PDAs. In the short term, this profile does not appear to be suitable for mobile devices which still have very limited capabilities. In order to cope with such limitations, a group of mobile manufacturers, called the MMS interoperability group, designed a new profile for MMS called the MMS SMIL.

6.12.3 MMS SMIL and MMS Conformance Document

For first implementations of MMS-capable devices, manufacturers lacked an appropriate message scene description language. A language rich enough to allow the definition of basic multimedia presentations but simple enough to be supported by devices with very limited capabilities. To cope with this situation, the MMS interoperability group[2] designed

---

[2] At the time of writing, the MMS interoperability group is composed of CMG, Comverse, Ericsson, Nokia, Logica, Motorola, Siemens and Sony-Ericsson.

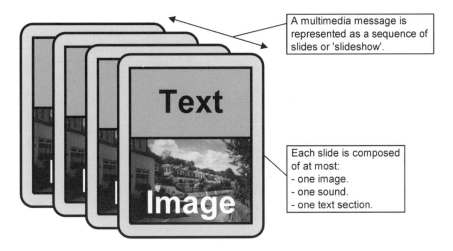

A multimedia message is represented as a sequence of slides or 'slideshow'.

Text

Each slide is composed of at most:
- one image.
- one sound.
- one text section.

Image

Figure 6.16   Message presentation with MMS SMIL

a profile for SMIL that fulfils the need of first MMS-capable devices. The profile is known as the MMS SMIL and has been defined outside standardization processes in a document known as the *MMS conformance document* [MMS-Conformance]. In addition to the definition of MMS SMIL, the conformance document also provides a set of recommendations to ensure interoperability between MMS-capable devices produced by different manufacturers.

In MMS SMIL, a presentation is composed of a sequence of slides (a slideshow) as shown in Figure 6.16. All slides from the slideshow have the same region configuration. A region configuration can contain at most one text region, at most one image region and each slide can be associated with at most one sound. For slideshows containing both, a text region and an image region, the term layout refers to the way regions are spatially organized: vertically (portrait) or horizontally (landscape). Mobile devices have various screen sizes, it may therefore happen that a particular layout for a message scene description does not fit in the device display. In this situation, the layout may be overridden by the receiving device (from portrait to landscape and vice versa).

A message presentation may contain timing settings allowing an automatic slideshow presentation, during which the switching from one slide to the following one is made automatically after a time period defined as part of the message scene description. However, MMS-capable devices may allow an interactive control by ignoring the timing settings and by allowing the user to switch from one slide to the next one by pressing a key.

The MMS conformance document identifies SMIL 2.0 features that can be used for constructing the slideshow. A slideshow should have a sequence of one or more parallel time containers (instructions within the two tags $<$ par $>$ ... $<$ /par $>$). Each parallel time container represents the definition of one slide. Time containers should not be nested. Each media object identified inside the parallel time container represents one component of a slide. Supported elements in MMS SMIL are listed in Table 6.9.

In addition to the definition of MMS SMIL, the MMS conformance document also contains the following rules:

• Dimensions of regions inside the root region should be expressed in absolute terms (i.e.

Table 6.9 MMS SMIL module elements

Modules	Elements	Attributes	Content model
Layout	`Layout`		`region, layout`
	`Region`	`left, top, height, width, fit, id`	
	`Root-layout`	`width, height`	
Media	`Text`	`src, region, alt, begin, end`	
	`Img`	`src, region, alt, begin, end`	
	`Audio`	`src, alt, begin, end`	
	`Ref`	`src, region, alt, begin, end`	
Structure	`Smil`		`head, body`
	`Head`		`layout`
	`Body`		`par`
Timing and synchro.	`Par`	`dur`	`text, img, audio, ref`
Meta information	`Meta`	`name, content`	

pixels) or in percentage relative to the dimension of the root region. It is not recommended to mix absolute and relative dimensions in the same message presentation.

- The `src` attribute must refer to a media type compliant with the associated element. For instance, the value associated with the `src` attribute of an `Img` element must refer to an image, and nothing else.
- Timing information should be expressed in integer milliseconds.
- Maximum image dimensions for which interoperability is ensured is 160×120 pixels.
- First MMS-capable devices are able to interpret all messages whose size is less than 30 Koctets. More advanced devices may also support larger messages.

The MMS conformance document specifies that a slide may be composed of a text region only, an image region only or of two regions, one for the text and another one for the image. Considering the two types of layouts for messages containing both an image region and a text region, a slideshow can be formatted according to the six configurations shown in Figures 6.17–6.22. In its simplest form, a slideshow is configured for representing only text or only an image on each slide. Such configurations are shown in Figures 6.17 and 6.18. With the portrait layout, the image region may be positioned at the bottom or at the top of the screen. Examples of portrait layouts are depicted in Figures 6.19 and 6.20. With the landscape layout, the image may be positioned at the left or at the right of the screen. Examples of landscape layouts are depicted in Figures 6.21 and 6.22.

Figure 6.23 shows an example of MMS SMIL scene description corresponding to a two-slide configuration organized in the portrait layout. In this example, the message is composed of two slides. The two slides are structured according to the portrait layout with an upper image and a lower section for text. The first slide is displayed during 4 seconds followed by the second slide for 10 seconds. The entire slide presentation time is therefore 14 seconds.

According to the MMS conformance document, if the multimedia message contains a

Figure 6.17 Image only

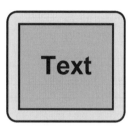

Figure 6.18 Text only

Figure 6.19 Portrait/text on top

Figure 6.20 Portrait/image on top

Figure 6.21 Landscape/text on left side

Figure 6.22 Landscape/image on left side

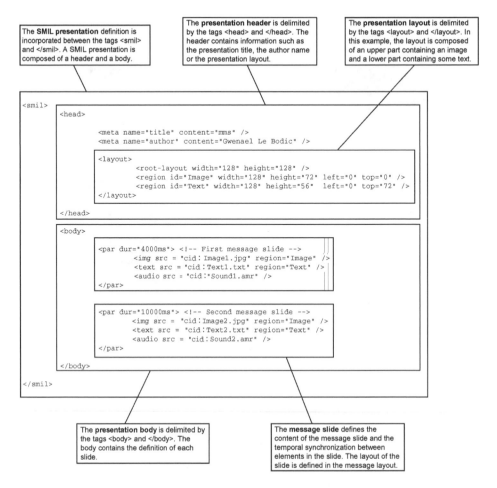

Figure 6.23   SMIL presentation/example

scene description, then the `Content-type` header field of the multipart message header is set to `application/vnd.wap.multipart.related`, otherwise it is set to `application/vnd.wap.multipart.mixed`. Furthermore, if a presentation is available, the `start` parameter of the `Content-type` header field refers to the presentation part in the message and the type parameter indicates that the presentation is a SMIL presentation as shown below:

```
Content-type : application/vnd.wap.multipart.related;
          start =< 12345@le-bodic.net >;
          type = "application/smil"
```

In a MMS SMIL scene description, the `src` attribute of `img`, `text` and `audio` tags refer to `content-ID` values assigned to body parts in the multipart message as shown below:

```
<img src = cid:Image1.jpg>
```

Optionally, a `content-location` parameter [RFC-2577] can be associated with each body part. The value assigned to this parameter is used as default file name if the recipient wishes to extract the corresponding body part as an independent file.

6.12.4 3GPP PSS SMIL Language Profile

The Packet-switched Streaming Service (PSS) SMIL language profile, defined by the 3GPP in [3GPP-26.234], is a subset of the SMIL 2.0 language profile but a superset of the SMIL 2.0 basic profile. The PSS SMIL language profile contains the following SMIL 2.0 modules:

- Content control (BasicContentControl, SkipContentControl)
- Layout (BasicLayout)
- Linking (BasicLinking)
- Media object (BasicMedia, MediaClipping, MediaAccessibility, MediaDescription)
- Meta-information(Metainformation)
- Structure (Structure)
- Timing and synchronization (BasicInlineTiming, MinMaxTiming, BasicTimeContainers, RepeatTiming and EventTiming)
- Transition effects (BasicTransitions)

3GPP technical specification [3GPP-26.140], from release 5, mandates the support of the PSS SMIL language profile for all MMS user agents and MMSCs that do accept scene descriptions.

6.12.5 XHTML

As an alternative to SMIL, XHTML is a language that can also be used for representing message scene descriptions. In particular, XHTML Mobile Profile (XHTML MP) extends HTML Basic Profile published by the W3C. HTML MP is a subset of HTML 1.1 but a superset of HTML Basic Profile. XHTML MP has been specifically tailored for resource-constrained devices. However, HTML MP remains a suitable language for the definition of rich MMS scene descriptions.

3GPP technical specifications [3GPP-26.140], from release 5, suggest the support of XHTML Mobile Profile by all MMS user agents and MMSCs able to interpret scene descriptions.

6.13 Summary of Supported Formats

Table 6.10 indicates which content formats are supported by MMS user agents according to the associated 3GPP releases. 3GPP requirements regarding the support of formats in MMS are captured in [3GPP-23.140] for release 99 and release 4. They are captured in [3GPP-26.140] for release 5 onwards. Formats are classified into high level format types (character sets, audio, image, video, PIM, inter-media synchronization and scene description).

6.14 Addressing Modes

For exchanging messages within a single MMS environment, users are identified according to one of the two addressing modes: *Email addressing* [RFC-2822] and *MSISDN addressing*

Table 6.10 MMS format support table

Media types	R99	Rel-4	Rel-5	Conformance document 2.0
Character sets	Mandatory	Mandatory	Mandatory	Mandatory
US-ASCII	–	–	–	Mandatory
UTF 8	–	–	–	Mandatory
UTF 16	–	–	–	Mandatory
Natural and synthetic audio	Optional	Optional	Optional	Optional
(natural audio) AMR	Suggested	Mandatory	Mandatory	Suggested
(natural audio) MP3	Suggested	Suggested	–	–
(natural audio) MPEG-4 AAC	Suggested	Suggested	Suggested	–
(synthetic audio) MIDI or SP-MIDI	Suggested MIDI	Suggested MIDI	Suggested SP-MIDI	–
(synthetic audio) iMelody	–	–	–	–
Image	Optional	Optional	Optional	Optional
(bitmap image) JPEG	Suggested	Mandatory	Mandatory	Mandatory
(bitmap graphic) GIF	Suggested	Suggested GIF89a	Suggested GIF89a	Suggested GIF87a and GIF89a
(bitmap graphic) WBMP	–	–	–	Suggested
(vector graphics) SVG	–	–	–	–
Video	Optional	Optional	Optional	–
H.263	Suggested	Mandatory baseline	Mandatory baseline	–
MPEG 4	Suggested	Suggested Visual simple profile level 0	Suggested Visual simple profile level 0	–
PIM elements	Optional	Optional	Optional	Optional
vCard	–	–	–	Suggested
vCalendar	–	–	–	Mandatory[a]
Inter-media synchronization	Optional	Optional	Optional	–
MP4	–	Mandatory	Mandatory	–
Scene description/message presentation	Optional	Optional	Optional	Optional
WML	–	–	Suggested	–
SMIL	–	–	Mandatory	Mandatory MMS SMIL
XHTML	–	–	Suggested Mobile profile	–

[a] The support of vCalendar objects is mandatory only if the MMS-capable device has agenda capabilities.

> **Box 6.9 How to Read the 'Format Support Table'**
>
> The left column of the table identifies seven formats types: character sets, audio, image, video, PIM, inter–media synchronization and scene description. Each format type is associated with a status for each 3GPP release. The status can be optional or mandatory. At the time of writing, only the support of text is mandatory. This means that the minimum requirement for an MMS-capable handset is to support at least one character set. For each optional format type, one particular format may be mandatory. This is the case for video support in 3GPP releases 4 and 5 where the support of H.263 is mandatory. This means that the support of video is optional but, if the MMS-capable device supports the video format type, then the support of the video format H.263 becomes mandatory. This ensures that all MMS-capable handsets supporting the video format type can interoperate well by using the video format H.263. However, nothing prevents an MMS-capable handset from supporting more than one video format. A format is suggested in a 3GPP release when its use is recommended but not mandatory. No status is given for formats which are neither mandatory nor suggested.

[ITU-E.164]. With Email addressing, a subscriber address is expressed in the form of an email address in the following format: `user@system`.[3] With MSISDN addressing, users are identified with MSISDN numbers. For the exchange of messages between MMS user agents attached to different MMS environments, the addressing is based on the Email addressing. For this purpose, each MMSE has a unique domain name in the form: `mms.serviceprovider.net` or `mms.network.net`.

Several parameters in messages refer to MMS subscribers (message originator, message recipient, etc.). This is the case for parameters such as `To`, `Cc`, `Bcc` and `From`. Examples of values assigned to the addressing parameters are provided below:

Email addressing:

`To: User Name <name@mms.network.net>`

or

`To: User Name <name@mms.network.net>/TYPE = rfc822`

MSISDN addressing:

`To: + 33647287345/TYPE = PLMN`

Address hiding refers to the possibility for message originators to hide their address from message recipients. In this case, the anonymous multimedia message is delivered without providing the originator details to the message recipients. Note that the originator MMSC always knows the originator address even if address hiding is requested by the message originator.

Note that one significant limitation of SMS and EMS is the lack of support for a standard mechanism for group sending at the transport level. With MMS, group sending is supported. This means that multiple recipient addresses can be provided for one single message submission.

---

[3] In the Email address, `system` can be a domain name (for instance operator domain) or a fully qualified host address.

6.15 Message Submission and Reports

In the MMS environment, transactions between entities are usually performed in an asynchronous fashion. In this environment, a transaction is initiated by a request and to each request usually corresponds a response containing the transaction results. In this book, the convention for naming requests and responses is the one used by the 3GPP. This convention consists of suffixing the request name by .REQ and suffixing the response name by .RES. In addition, the request and the response names are prefixed by the associated interface name. For instance, the request for submitting a message over the MM1 interface is named MM1_submit.REQ and the corresponding response is named MM1_submit.RES. Each request or response is associated with a set of features such as message addressing, provision of message qualifiers (class, priority, subject, etc.), definition of reply charging parameters, etc.

In the person-to-person scenario, the message submission refers to the sending of a message from an originator MMS user agent to its associated MMSC over the MM1 interface. The support for the submission of a message is optional for the MMS user agent whereas it is mandatory for an MMSC. Upon reception of a submitted message, the originator MMSC consults routing tables and transfers the message to the recipient MMSC (or to other message servers such as SMSCs or Email servers). The recipient MMSC stores the message and builds a notification. The notification, indicating that there is a new message waiting, is delivered to the recipient MMS user agent. The MMS user agent can use this notification to retrieve the message immediately or later. The transaction flow describing the delivery of the notification and the retrieval of a message is presented in Section 6.16.

For a message submission request, the message originator can request a *delivery report* to be generated upon delivery of the message to the message recipient(s). If the message is sent to more than one recipient, then several delivery reports may be received by the message originator. Additionally, the message originator can request a *read-reply report* to be generated once the message has been read by the message recipient.

The message originator can indicate, as part of a message submission request, that he/she is willing to pay for a message reply. This feature is called *reply-charging* (defined from 3GPP release 4). With reply-charging, the message originator can define several conditions to be honoured by the message recipient(s) as a requirement for the message reply to be paid for. These conditions are:

- *Reply deadline:* the message originator can indicate that the message reply will not be paid for if the reply is sent after a given reply deadline.
- *Reply size:* the message originator can specify a maximum size for the reply message. If the reply is bigger than the specified size, then the reply can be delivered to the message originator but the message originator does not pay for it.

If the reply charging conditions are fulfilled, then the reply is paid for by the message originator and the message reply is therefore 'free of charge' for the message recipient. This applies to the first successfully delivered reply for each recipient of the corresponding original message. If the message is forwarded to another MMS user agent, then the reply charging no longer applies to the forwarded message.

At the time of writing, technical specifications (3GPP releases 4 and 5) restrict the support of reply charging to MMS user agents and VAS applications belonging to the same MMS environment. Additionally, the reply to a message, when reply charging has been enabled, is

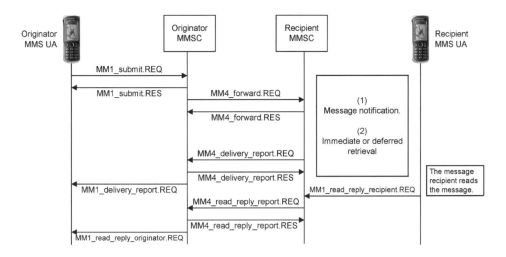

Figure 6.24   Message submission, transfer and reports

limited to text only. These limitations may be elaborated in the future by standardization organizations.

Figure 6.24 shows the transaction flow between various MMS elements involved in the submission and transfer of a multimedia message. The transaction flow also shows the handling of delivery and read-reply reports.

6.15.1 Message Submission

The originator MMS user agent can request the submission of a multimedia message to the originator MMSC over the MM1 interface (`MM1_submit.REQ`). The submission request is associated with the following features:

- *Message identification:* the message is always identified uniquely with a reference number. In the case of reply charging, a message reply contains the reference number of the corresponding original message.
- *Addressing:* addresses for one or more message recipients are specified for a submission request. The address of the originator is also specified as part of the message but the message originator may request the address to be hidden from the message recipient(s).
- *Message subject, priority and class:* the message subject is a short text provided by the message originator (usually summarizing the message content). The priority indicates the message importance (low, normal or high) and the class indicates the message content category (personal, advertisement, informational, etc.).
- *Time stamping:* the originator MMS user agent may provide the time and date of the submission request.
- *Time constraints:* the message originator may request an earliest time of delivery for the submitted message. Additionally, the message originator may also specify a validity period for the message.
- *Reporting:* the message originator can request delivery and/or read-reply reports to be generated upon message delivery and/or message reading, respectively.

- *Reply charging:* the message originator can indicate whether or not he/she is willing to pay for a message reply. If the originator is willing to pay for a reply, then an optional reply deadline indicates the date until which the message recipient can send a prepaid reply. Furthermore, the message originator may indicate that the reply will not be paid for if the size of the reply message is larger than the one specified by a given size. If the message being submitted is in response to a message for which reply charging was enabled, then the identification of the message, for which the submitted message is a reply, is provided as part of the reply submission request.
- *Content type and content:* the submission request indicates the type of message content. It also contains the message content itself.

After the submission request, the originator MMSC responds to the request (`MM1_submit.RES`). The response is associated with the following features:

- *Message identification:* the identification of the message to which the request status refers. The message identification can be reused later for delivering reports back to the originator MMS user agent.
- *Request status:* the status of the submission request. The submission may have been successful or may have failed (no subscription, corrupt message structure, service not available, etc.).

Upon receipt of the submission response, the originator MMS user agent analyses the request status and usually indicates to the message originator whether or not the message has been successfully submitted. If the submission request acknowledgement is not received after a predefined period of time, then the originator MMS user agent can conclude that the request for message submission has failed.

6.15.2 Message Transfer

After successful message submission, the originator MMSC identifies the messaging server(s) which manages messages for the message recipient(s). Three alternatives are possible:

1. *Message originator and message recipient belong to the same MMS environment.* In this situation, the originator MMSC is also the recipient MMSC.
2. *Message originator and message recipient belong to two distinct MMS environments.* In this situation, the identified messaging server is also an MMSC, known as the recipient MMSC. The transfer of the message from the originator MMSC to the recipient MMSC is performed over the MM4 interface as shown in Figure 6.24.
3. *The message recipient is not an MMS subscriber.* This is the case when the message is addressed to an Email user or an SMS subscriber. In this situation, the message is transferred from the originator MMSC to the recipient messaging server (Email server or SMSC) over the MM3 interface.

In the second alternative, the originator MMSC forwards the message to the recipient MMSC over the MM4 interface. Over this interface, the forward request is associated with all submission request features (addressing, time stamping, time constraints, message qualifiers, message identification, reporting, identification, content and content type). The forward request (`MM4_forward.REQ`) is also associated with the following additional features:

- *Version:* the protocol used for forwarding a message over the MM4 interface provides a means of uniquely identifying the version of the MMS environment (usually the 3GPP MMS version).
- *Message type:* the message type indicates which operation is being performed over the MM4 interface. In this context, the message type refers to the message forward request (`MM4_forward.REQ`).
- *Transaction identification:* the transaction identification allows MMSCs to identify the pairs of forward request/response (`MM4_forward.RES/REQ`).
- *Acknowledgement request:* the recipient MMSC can indicate whether or not a forward response/acknowledgement is requested by the originator MMSC. If this is requested, then the recipient MMSC responds back to the forward request (`MM4_forward.RES`) as shown in the transaction flow in Figure 6.24.
- *Forwarding:* a forwarding counter indicates the number of times the message has been forwarded. In addition, the address of the user who requested the message to be forwarded is provided. If the message has been forwarded several times, then a sequential list of all forwarding user addresses can be specified as part of the forward request.

If the message originator requested his/her address to be hidden, then the message originator address is not specified as part of the forward request.

If the originator MMSC requested a forward response from the recipient MMSC (see parameter acknowledgement request of the forward request), then the recipient MMSC acknowledges the forward request with the `MM4_forward.RES` response. This response is associated with several of the corresponding forward request features (version, transaction identification and message identification). In addition, the response is associated with the following features:

- *Message type:* the message type indicates which operation is being performed over the MM4 interface. In this context, the message type refers to the message forward response (`MM4_forward.RES`).
- *Request status:* the request status informs on the status of the request (forward successful, no subscription, bad address, network not reachable, etc.).

6.15.3 Delivery Report

As part of the submission request, the message originator can request the generation of a delivery report (see `MM1_submit.REQ`). In this situation, the recipient MMSC generates a delivery report after confirmation of message delivery or upon message deletion. It has to be noted that the message recipient can deny the generation of delivery reports. Once the delivery report has been generated by the recipient MMSC, the delivery report is forwarded to the originator MMSC over the MM4 interface (`MM4_delivery_report.REQ`). In this section, the original message refers to the message with which the delivery report is associated. The request for forwarding a delivery report over the MM4 interface is associated with the following features:

- *Addressing:* both the address of the originator and the address of the recipient of the original message are provided as part of the forward request.
- *Version:* the protocol used for forwarding a message over the MM4 interface provides a

means of uniquely identifying the version of the MMS environment (usually the 3GPP MMS version).

- *Message type:* the message type indicates which operation is being performed over the MM4 interface. In this context, the message type refers to the request for the forwarding of a delivery report (MM4_delivery_report.REQ).
- *Transaction identification:* the transaction identification allows MMSCs to identify pairs of request/response (MM4_delivery_report.RES/REQ).
- *Original message identification:* the identification of the original message with which the delivery report is associated is provided as part of the forward request.
- *Original message time stamping:* the date and time when the original message was handled (rejection, deletion, retrieval, etc.) is included as part of the forward request.
- *Acknowledgement request:* the recipient MMSC can indicate whether or not an acknowledgement is requested for the forwarding of the delivery report over the MM4 interface. If this is requested, then the originator MMSC acknowledges the forward request with the MM4_delivery_report.RES response as shown in the transaction flow in Figure 6.24.
- *Delivery status:* the recipient MMSC indicates the status of the delivery such as 'the message has been retrieved', 'the message has been deleted' (expiry), 'the message has been rejected', etc.

If this was requested by the recipient MMSC, the originator MMSC confirms the receipt of the delivery report with a delivery report forward response (MM4_delivery_report. REQ). This response is associated with several of the corresponding request features (version, message type, transaction identification and message identification). The response is also associated with the following additional feature:

- *Request status:* the request status indicates the status of the request for forwarding a delivery report.

Once the delivery report has been received, the MMSC delivers it over the MM1 interface to the originator MMS user agent. For this purpose, the delivery report is transferred over the MM1 interface as part of a delivery report request (MM1_delivery_report.REQ). The request is associated with several features of the request for forwarding a delivery report over the MM4 interface (addressing, original message identification, time stamping and delivery status). The request is also associated with the following feature:

- *Addressing:* the address of the recipient of the original message is specified as part of the request.

Upon receipt of the delivery report, the originator MMS user agent presents the status of the delivery to the MMS subscriber. The presentation of a delivery report could take the form shown in Figure 6.25.

6.15.4 Read-reply Report

During message submission, the message originator can request to be notified when the message has been read by the message recipient. For this purpose, the recipient MMS user agent generates a read-reply report when the message has been read by the message recipient.

From: Armel@armorepro.com
To: Gwenael@le-bodic.net
Sent: 12/06/01 17:14
Subject: Delivery: First Message

**Your message has been successfully delivered
on 12/06/01 17:11.**

Figure 6.25   Example of delivery report presentation

Note that the message recipient can deny the generation of read-reply reports. Once the read-reply report has been generated by the recipient MMS user agent, the read-reply report is submitted over the MM1 interface to the recipient MMSC (MM1_read_reply_ recipient.REQ). In this section, the original message refers to the message with which the read-reply report is associated. The request for submitting a read-reply report over the MM1 interface in the recipient MMS environment is associated with the following features:

- *Original message identification:* the identification of the original message to which the read-reply report corresponds is provided as part of the request.
- *Addressing:* the address of the recipient of the original message is specified as part of the request.
- *Original message time stamping:* the date and time indicate when the original message was handled (read or deleted without being read).
- *Read-reply status:* the read-reply status indicates the status of the corresponding message such as 'the message has been read' or 'the message has been deleted without being read'.

Once the read-reply report has been received by the recipient MMSC, the recipient MMSC forwards the delivery report via the MM4 interface to the originator MMSC (MM4_read_reply_report.REQ). The recipient MMSC can request the originator MMSC to provide a response for the delivery report forward (MM4_read_reply_ report.RES). Features associated with the response/request pair for the forwarding of a read-reply report are similar to the ones for the forwarding of a delivery report over the same interface. Once the read-reply report has been received by the originator MMSC, the originator MMSC delivers the read-reply report to the originator MMS user agent over the MM1 interface (MM1_read_reply_originator.REQ). The request for the delivery of a read-reply report over the MM1 interface in the originator MMS environment is associated with the same features as the ones associated with the submission request of a read-reply report in the recipient MMS environment.

Upon receipt of the read-reply report, the recipient MMS user agent indicates when the message recipient read the message (or when the message was deleted without being read). The presentation of a read-reply report could take the form shown in Figure 6.26.

This section has described the submission of a multimedia message and its transfer between distinct MMS environments. The section has also shown how delivery and read-reply reports for submitted messages are managed. The next section shows how the recipient MMS user agent is notified that a multimedia message awaits retrieval. The next section also presents two message retrieval methods: immediate and deferred retrievals.

From: Armel@armorepro.com
To: Gwenael@le-bodic.net
Sent: 12/06/01 17:14
Subject: Read: First Message

Your message was read on 12/06/01 17:11.

Figure 6.26   Example of read-reply report presentation

6.16 Message Notification, Immediate and Deferred Retrieval

Once a multimedia message has been received by an MMSC, the message is first stored. The MMSC extracts some information from the multimedia message to form a compact notification. The notification is sent to the recipient MMS user agent to indicate that a multimedia message is awaiting retrieval in the MMSC message store. The MMSC keeps the message in the message store at least until the message validity period expires or until the message is successfully retrieved.

Upon receipt of the notification, message retrieval can be initiated by the recipient MMS user agent according to two distinct methods. The first method consists of hiding the message notification from the message recipient and requesting the immediate delivery of the corresponding multimedia message to the recipient MMS user agent. The message recipient perceives the message retrieval as if the message had been pushed to the MMS-capable device. This method is known as the *immediate retrieval* process and has close similarities with the way SMS and EMS messages are delivered to mobile devices. Alternatively, the notification may be stored locally in the receiving device and used later, by the message recipient, to retrieve the corresponding message at his/her own convenience. This method is known as the *deferred retrieval* process. The transaction flow for notification and retrieval processes is shown in Figure 6.27.

6.16.1 Message Notification

The notification of a multimedia message is an indication sent by the recipient MMSC to the MMS user agent in order to indicate that a multimedia message is awaiting retrieval (MM1_notification.REQ). The request for delivering a notification over the MM1 interface is associated with several features already associated with the request for submitting a message (message qualifiers, time constraints and reply charging). In addition, the request for notification delivery is associated with the following features:

- *Addressing:* the address of the message originator may be provided as part of the notification.
- *Message size:* the size of the message awaiting retrieval.
- *Time constraints:* the time of expiry of the message awaiting retrieval.
- *Reporting:* indication whether or not the message originator requested a delivery report to be generated. The recipient MMS user agent may disable the generation of a delivery report by indicating it in the notification response (MM1_notification.RES).

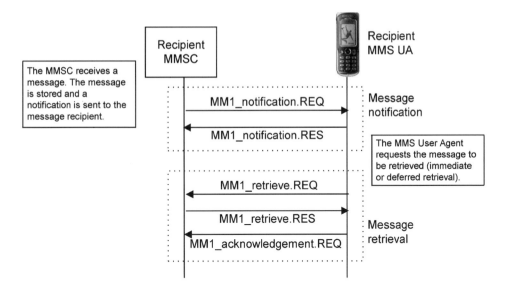

Figure 6.27   Message notification and retrieval

- *Message reference:* the notification always provides a reference to the message awaiting retrieval. This reference is necessary for the retrieval process. This reference is usually provided in the form of a URI and uniquely identifies the corresponding message in the MMSC store.
- *Element descriptor:* an element descriptor is a parameter which contains complementary information qualifying one of the message elements. For instance, element descriptors can indicate the name, size and format for elements included in a message. A notification may contain more than one element descriptor.

The recipient MMS user agent acknowledges the receipt of a notification by responding back to the recipient MMSC (`MM1_notification.RES`). This response is associated with the following features:

- *Message status:* the recipient MMS user agent can indicate how the message is intended to be handled (immediate rejection, etc.).
- *Reporting:* the recipient MMS user agent can indicate as part of the notification response that the delivery report (requested by the message originator) shall not be generated.

6.16.2 Message Retrieval

The message reference specified as part of a notification is necessary for retrieving the corresponding message. For retrieving a message, the recipient MMS user agent requests the message retrieval over the MM1 interface (`MM1_retrieve.REQ`). The request only contains the message reference.

Once the recipient MMSC receives the request for message retrieval, the MMSC prepares the message for retrieval (content adaptation, etc.). After message preparation, the message is delivered to the recipient MMS user agent as part of the retrieval response (`MM1_retrieve.RES`). The response is associated with several features associated with the following requests:

- Request for submission of a message/`MM1_submit.REQ` (time constraints, message qualifiers, reporting, reply charging, message identification, content and content type).
- Request for message forward/`MM4_forward.REQ` (forwarding).
- Request for notification/`MM1_notification.REQ` (message reference, addressing).

In addition, the response is also associated with the following features:

- *Time stamping:* the time and date when the message was submitted or forwarded.
- *Request status:* the recipient MMSC indicates what the status of the corresponding request is (e.g. 'the media format could not be converted', 'insufficient credit for retrieval', etc.).

Upon receipt of the message included in the retrieval response, the originator MMS user/agent acknowledges the receipt with an acknowledgement request (`MM1_acknowledgement.REQ`). The request is associated with the following feature:

- *Report allowed:* the recipient MMSC indicates whether or not a delivery report can be generated for the retrieved message.

6.17 Message Forward

In MMS, as in any messaging system, a message may be forwarded to other recipients. In the context of MMS, two methods are available for forwarding a message. The first method consists of submitting again a previously retrieved message over the MM1 interface (`MM1_submit.REQ`). Alternatively, MMS allows the forwarding of messages which have not been retrieved yet by the MMS user agent. This can be performed by requesting the forwarding of a message identified by the message reference included in the corresponding message notification. The MMSC, which handles the forwarding request, is known as the forwarding MMSC.

The transaction flow, for the message forwarding without prior download of the message, is depicted in Figure 6.28. The request for forwarding a message (`MM1_forward.REQ`) prior to retrieval is associated with the following features:

- *Addressing:* the request for forwarding a message identifies the recipient(s) to whom the message is to be forwarded. Additionally, the address of the forwarding MMS user agent may be specified as part of the request.
- *Time stamping:* the date and time when the message has been forwarded can be specified in the request.
- *Time constraints:* the forwarding MMS user agent can specify an earliest time of delivery for a message. It can also specify a time of expiry for the forwarded message.
- *Reporting:* the forwarding MMS user agent can request delivery and/or read-reply reports to be generated upon message delivery and/or message reading, respectively.
- *Message reference:* the message reference, specified as part of the associated notification, needs to be included in the forward request. The message reference is required so that the forwarding MMSC can identify, in the message store, which message is to be forwarded.

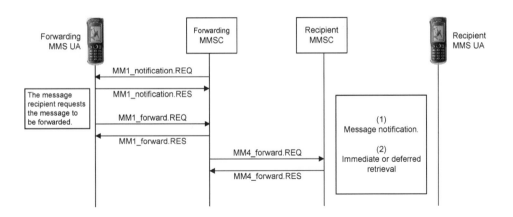

Figure 6.28   Message forward and transfer

Upon receipt of the forward request, the MMSC responds with a `MM1_forward.RES` response. The response is associated with the following features:

- *Forward status:* the forwarding MMSC indicates the status of the forward request.
- *Message identification:* the forwarding MMSC provides a message identification for the message being forwarded. This identification allows the forwarding MMS user/agent to identify messages with which incoming reports may be later associated.

The forwarding MMSC forwards the message to the recipient MMSC(s) over the MM4 interface or over another interface to a non-MMS messaging environment (see the three alternatives identified for message forwarding over the MM4 interface). The forwarding MMSC updates the forwarding parameters of the message (forwarding counter and list of addresses of MMS user agents that requested the message to be forwarded).

6.18 MMS Value Added Services – Submission, Replacement and Reports

In the MMS environment, a VAS application has the ability to interact indirectly with MMS user agents via the MM7 interface. In this machine-to-person scenario, the message submission refers to the sending of a message from the VAS application to the associated MMSC. Upon receipt of this message, the MMSC delivers the message to the recipient(s) in the normal way over the MM1 interface. As part of the submission request, the VAS application can request the generation of delivery and/or read-reply reports. Note that if the message was directed to multiple recipients, then several reports may be received by the VAS application. Once a message has been submitted, the MMSC can cancel the delivery of the message or replace the submitted message by a new message. The message cancellation can only be applied to a message whose associated notification has not yet been sent to the recipient. Additionally, the replacement can only be applied to a message which has not yet been retrieved or forwarded. Figure 6.29 shows the transaction flow between various MMS elements involved in the submission, replacement and retrieval of a message. The transaction flow also shows the handling of delivery and read-reply reports.

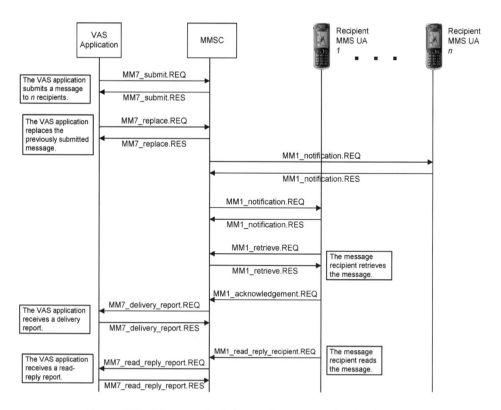

Figure 6.29   Message submission, replacement, retrieval and reports

6.18.1 Message Submission

A VAS application can request the submission of a multimedia message to the MMSC over the MM7 interface. The message can be directed to a single recipient, to multiple recipients or to a distribution list managed by the MMSC. The submission request (`MM7_submit.REQ`) is associated with the following features:

- *Authorization:* the VAS application supplies the identification of the VAS provider along with the service identification.
- *Addressing:* one or more recipients are specified for a submission request. The address of the message originator may also be indicated as part of the submission request.
- *Version, message type and transaction identification:* the submission request is always associated with a specific version of the MM7 interface, with the type of transaction over the interface (in this case message submission) and with an identification of the transaction so that a request can be easily correlated with the corresponding response.
- *Linked message identification:* the submitted message can be associated with a previously received message. In this situation, the identification of the previously received message is provided as part of the submission request.
- *Message subject, priority and class:* the message subject is a short text provided by the message originator (usually summarizing the message content). The priority indicates the

importance of the message and the class indicates the message content category (advertisement or informational).

- *Service code:* the message submission request can be associated with a service code. This service code may be used by the MMS provider for billing purposes.
- *Time stamping:* the VAS application may provide the time and date of the submission request.
- *Time constraints:* the VAS application may request an earliest time of delivery for the submitted message. Additionally, the message originator may also specify a validity period for the message.
- *Reply charging:* the VAS application can indicate whether or not a reply to the submitted message will be paid for by the VAS provider. If the VAS provider is willing to pay for a reply, then an optional reply deadline indicates the date until which the message recipient can send a paid reply. Furthermore, the VAS application may indicate that the reply will not be paid for if the size of the reply message is larger than the one specified by a given size. If the message being submitted is in response to a message for which reply charging was enabled, then the identification of the message for which the submitted message is a reply, is provided as part of the submission request.
- *Reporting:* the VAS application can request delivery and/or read-reply reports to be generated upon message delivery and/or message reading, respectively.
- *Content adaptation restrictions:* the VAS application can request that content adaptation not be applied to the submitted multimedia message.
- *Content type and content:* the submission request indicates the type of message content. It also contains the message content itself.

After submission request, the MMSC acknowledges the request with the `MM7_submit.RES` response. The response is associated with the following features:

- *Version, message type and transaction identification:* as defined for the corresponding request.
- *Message identification:* if the MMSC accepts the submission request, then it provides a message identification to the VAS application.
- *Request status:* the status of the submission request. The submission may have been successful or may have failed (service error, server error, etc.).

Upon receipt of the submission response, the VAS application analyses the request status and determines whether or not the request was successful.

6.18.2 Message Replacement and Cancellation

After successful message submission, the VAS application can cancel the message delivery. This can only be performed if the notification of the message to be cancelled has not yet been sent to the message recipient(s). The cancellation request (`MM7_cancel.REQ`) is associated with the following features:

- *Version, message type and transaction identification:* the replacement/cancellation request is always associated with a specific version of the MM7 interface, to the type of transaction over the interface (in this case, message replacement/cancellation) and to an identification of the transaction so that a request can be easily correlated with the corresponding response.

- *Addressing:* the address of the originator VAS application is provided as part of the request.
- *Authorization:* the VAS application supplies the identification of the VAS provider along with the service identification.
- *Message identification:* the identification of the message to which the request applies. The identification is provided by the MMSC as part of the response of the corresponding message submission request.

Alternatively, a previously submitted message can also be replaced by a new message. A message can be replaced only if it has not yet been forwarded or retrieved by the recipient. The replacement request (MM7_replace.REQ) is associated with the features of the cancellation request. It is also associated with the following additional features:

- *Content adaptation restrictions:* the VAS application can request that content adaptation not be applied to the new multimedia message.
- *Service code:* the message replacement request can be associated with a service code. This service code may be used by the MMS provider for billing purposes.
- *Reporting:* the VAS application can request a read-reply report to be generated upon message reading.
- *Time stamping:* the VAS application may provide the time and date of the request.
- *Time constraints:* the VAS application may request an earliest time of delivery for the new message.
- *Content adaptation restrictions:* the VAS application can request that content adaptation not be applied to the new multimedia message.
- *Content type and content:* the replacement request indicates the type of message content. It also contains the message content for the new message.

The MMSC acknowledges the acceptance of cancellation and replacement requests with MM7_cancel.RES and MM7_replace.RES responses, respectively. These responses are associated with the following features:

- *Version, message type and transaction identification:* as defined for the corresponding request.
- *Request status:* the status of the cancellation/replacement request. The submission may have been successful or may have failed (service error, server error, etc.).

6.18.3 Delivery and Read-reply Reports

As part of the submission request, the VAS application can request the generation of a delivery report and/or read-reply report. In this situation, the MMSC provides the delivery and reply reports to the VAS application with MM7_delivery_report.REQ and MM7_read_reply_report.REQ requests, respectively. These requests are associated with the following features:

- *Version, message type and transaction identification:* the report request is always associated with a specific version of the MM7 interface, with the type of transaction over the interface (in this case, report request) and with an identification of the transaction so that a request can be easily correlated with the corresponding response.

- *Addressing:* addresses of the VAS application and the originator of the message, to which the delivery report relates, are provided as part of the delivery/read-reply report.
- *Message identification:* the identification of the message to which the report relates. The identification is provided by the MMSC as part of the response of the message submission request.
- *Time stamping:* the report carries the date and time at which the associated message was handled (retrieved, deleted, read, etc.).
- *Message status:* the report informs on the status of the associated message (retrieved, deleted, read, etc.).

Upon receipt of the report request, the VAS application acknowledges the request with `MM7_delivery_report.RES` and `MM7_read_reply_report.RES` responses. These responses are associated with the following features:

- *Version, message type and transaction identification:* as defined for the corresponding request.
- *Request status:* the status of the report request. The request may have been successful or may have failed (service error, client error, etc.).

6.19 MMS Value Added Services – Message Delivery and Errors

An MMS user agent can also submit a message addressed to a VAS application. The VAS application is typically identified by a VAS short code for the submission request over the MM1 interface. The MMSC checks that the VAS short code is valid before attempting any delivery of the message over the MM7 interface. Upon receipt of the message, the VAS application may reply to the message. In order to maintain the relation between the reply and the original message, the reply can contain the *linked identification* information element, which refers to the message with which the reply is associated. The transaction flow for the delivery of a message to a VAS application is shown in Figure 6.30.

The MMSC can request the delivery of a message to the VAS application with the `MM7_delivery.REQ` request. This request is associated with the following features:

- *Addressing:* the address of the originator MMS application and addresses of intended recipients are provided as part of the request.
- *Authentication:* the MMSC can supply its own identification as part of the delivery request.
- *Version, message type and transaction identification:* the delivery request is always associated with a specific version of the MM7 interface, with the type of transaction over the interface (in this case, message delivery) and with an identification of the transaction so that a request can be easily correlated with the corresponding response.
- *Linked message identification:* the MMSC provides a message identification as part of the delivery request. This identification can be used by the VAS application for a subsequent submission of a message related to a previously delivered message.
- *Message subject, priority and class:* the message subject is a short text provided by the message originator (usually summarizing the message content). The priority indicates the message importance and the class indicates the message content category (advertisement or informational).

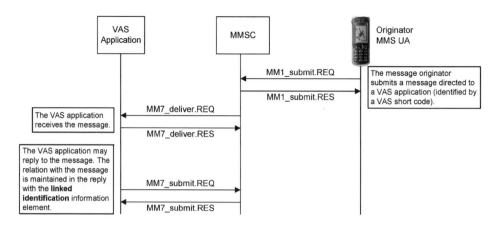

Figure 6.30   Message delivery to a VAS application

- *Time stamping:* the delivery request can contain the time and date of message submission.
- *Reply charging:* the message originator can indicate whether or not he/she is willing to pay for a message reply. If the originator is willing to pay for a reply, then an optional reply deadline indicates the date until which the message recipient can send a paid reply. Furthermore, the originator may indicate that the reply will not be paid for if the size of the reply message is larger than the one specified by a given size. If the message being submitted is in response to a message for which reply charging was enabled, then the identification of the message, for which the submitted message is a reply, is provided as part of the submission request.
- *Content type and content:* the submission request indicates the type of message content. It also contains the message content itself.

Upon receipt of the delivery request, the VAS application acknowledges the request with the MM7_delivery.RES response. This response is associated with the following features:

- *Version, message type and transaction identification:* as defined for the corresponding request.
- *Service code:* the VAS application can indicate a service code as part of the delivery response. This service code may be used by the MMS provider for billing purposes.
- *Request status:* the status of the report request. The request may have been successful or may have failed (service error, client error, etc.).

If the MMSC or the VAS application is unable to identify and process a request, then two generic error notifications can be used. The MM7_RS_error.RES notification is used by the MMSC to inform the VAS application that an error occurred during a request processing. The MM7_VAS_error.RES notification is also used by the VAS application to inform the MMSC that an error occurred during a request processing. These error notifications are associated with the following features:

- *Version, message type and transaction identification:* the error notification is always associated with a specific version of the MM7 interface, with the type of transaction over the interface (in this case, error notification) and with an identification of the transaction so that a request can be easily correlated with the corresponding response.

- *Error status:* the MMSC or the VAS application indicates which type of error has occurred during the processing of the corresponding request.

The VAS application or MMSC, which is notified with a generic error, can relate the error to a previous request with the transaction identification of the request contained in the error notification.

6.20 Capability Negotiation and Content Adaptation

In a mobile environment, mobile handsets and PDAs have very diverse capabilities. Devices have various screen dimensions, with colour or greyscale capabilities, with or without the support of video. In this context, it is important to have mechanisms to adapt the content of messages to whatever the device is capable of accepting and rendering.

Capability negotiation refers to the methods in place for exchanging capabilities of the receiving devices with the network element which performs content adaptation. Content adaptation refers to the ability to adapt the content of a message to the capabilities of the network/handset taking into consideration possible user preferences. For this purpose, content adaptation is performed according to capabilities and preferences managed in two profile types:

- *User agent profiles:* the user agent profile contains information for content formatting purposes. For instance, this profile might indicate the handset screen dimensions, supported character sets, etc.
- *User preference profiles:* this profile contains application specific information about user preferences indicating what type of content the user is interested in. For instance, the user might indicate that all advertisement messages should be discarded without retrieval.

Various capability negotiation and content adaptation techniques have been developed by several organizations. The most notable are the W3C with CC/PP and the Resource Description Framework (RDF). Another available technique is the one defined by the WAP Forum. The WAP technique is known as the User Agent Profile (UAProf) and was introduced in Chapter 1.

In an MMS environment, the recipient network may perform content adaptation just before message retrieval by the recipient MMS user agent. The generic UAProf specifications includes a schema that contains attributes describing the capabilities of the client hardware, micro-browser, etc. These generic attributes can be exploited by the network for performing content adaptation (attributes named `ScreenSize`, `CPUType` and `PushMessageSize` for instance). In addition, MMS specific attributes describe the capabilities of the MMS user agent. These attributes are specified as part of the `MMSCharacteristics` component. These MMS specific attributes are listed in Table 6.11.

6.21 Persistent Network-based Storage

With initial versions of MMS devices, multimedia messages are stored locally in the mobile device after retrieval. The drawback of a message local store is that messages, if not kept in the MMSC, can only be accessed via the MMS user agent which retrieved them. Furthermore, storage capacities of mobile devices is limited in comparison to storage capacities of devices

Table 6.11 Content adaptation/MMS attributes

Attribute name	Description	Examples
MMSMaxMessageSize	Maximum message size expressed in octets	2048
MMSMaxImageResolution	Maximum image dimensions expressed in pixels	80x60
MMSCCPPAccept	List of supported content types	image/jpeg, audio/wav
MMSCCPPAcceptCharSet	List of supported character sets	US-ASCII or ISO-8859-1
MMSCCPPAcceptLanguage	List of supported languages	en, fr for English and French, respectively
MMSCCPPAcceptEncoding	List of supported transfer encoding methods	base64, quoted-printable
MMSVersion	List of supported MMS version	1.0, 1.1
MMSCCPPStreamingCapable	Whether or not the MMS user agent supports streaming	Yes or No

available in fixed networks. To cope with this limitation, the concept of persistent network-based storage was introduced. Persistent network-based storage is based on an extension of the MM1 interface for allowing an MMS user agent to interact with a remote mailbox, called MMBox (stands for Multimedia Message Box), managed by the network. Network-based persistent storage offers the following features:

- Storing persistently submitted and retrieved messages in an MMBox (if configured or requested).
- Requesting the persistent storage of a multimedia message which is referenced in a notification.
- Storing persistently a message which is forwarded to other recipients.
- Uploading and retrieving messages to/from an MMBox.
- Forwarding a message stored in an MMBox.
- Deleting messages stored in an MMBox.
- Viewing messages stored in an MMBox and consulting attributes associated with each message.
- Updating states and flags associated with messages stored in an MMBox.

Note that only functional specifications have been developed for persistent storage in [3GPP-23.140] release 5. Consequently, it is too early to find implementations supporting this feature. The WAP Forum is working to define the technical realization for persistent storage. Specifications resulting from this work should be published in the forthcoming technical realization of MMS in the WAP environment (planned publication date is 2003).

A message, stored in an MMBox, can be associated with a state and to a set of flags. A message can be associated with one of the five mutually exclusive states:

- *Draft:* a message is in the draft state when it has been uploaded and stored but not yet submitted.
- *Sent:* a message is in the sent state when it has been stored after submission.

- *New:* a message is in the new state when it has been received by the MMSC and persistently stored in an MMBox without having been retrieved yet.
- *Retrieved:* a message is in the retrieved state when it has been retrieved by the MMS user agent.
- *Forwarded:* a message is in the forwarded state when it has been forwarded by the MMS user agent.

In addition to the state, a message may be associated with a set of keywords. Such keywords enable the MMS user agent to request the list of messages complying with given keyword search criteria.

6.22 Settings for MMS-capable Devices

An MMS-capable mobile device needs to be configured in order to operate properly in an MMS environment. The MMS configuration includes MMS connectivity settings and user preferences.

6.22.1 Connectivity Settings

MMS connectivity settings gather all parameters required to access the network infrastructure for sending and retrieving messages, reports and notifications. This includes the following parameters:

- *MMSC address:* this address is usually provided in the form of a URI.
- *WAP gateway profile:* this profile gathers all parameters required to access the WAP gateway. This includes the type of WAP gateway address (e.g. IPv4, IPv6), the WAP gateway address, the access port, service type (e.g. connection-less, secured), the authentication type, authentication identification and the identification secret.
- *Bearer access parameter (e.g. GGSN):* these parameters are required to establish a bearer-level connection with the core network. This includes the bearer type (e.g. CSD, GPRS), the type of address for the network access point (e.g. MSISDN for CSD and access point name for GPRS), bearer transfer rate, call type (e.g. analogue for CSD), the authentication type, authentication identification and the identification secret.

6.22.2 User Preferences

User preferences are default values that are used for the creation of new messages. Subscribers can update these preferences at their own convenience. Most common parameters are listed below:

- *Request for a delivery report:* whether or not the subscriber wishes delivery reports to be requested upon message submission.
- *Request for read-reply report:* whether or not the subscriber wishes read-reply reports to be requested upon message submission.
- *Sender visibility:* whether or not the subscriber wishes his/her address to be hidden from message recipients.
- *Priority of the message:* the default level of message priority (low, medium or high).

- *Time of expiry:* the time at which the message expires (e.g. ten days after sending, one month after sending, etc.)
- *Earliest time of delivery:* the earliest time at which the message should be delivered to message recipients (e.g. one day after message sending).

Other configuration parameters could include the message retrieval mode (deferred or immediate retrieval), the possibility to automatically reject anonymous messages, etc.

6.22.3 Storing and Provisioning MMS Settings

MMS settings can either be stored in the USIM or in the memory of the user equipment (e.g. flash memory). Both methods allow the persistent storage of settings. For the provisioning of MMS settings, three solutions can be identified:

- *User configuration:* the subscriber can usually configure MMS settings via the user equipment interface. In this scenario, settings may either be stored in the user equipment memory or in the USIM. Alternatively, settings may be scattered over both USIM and user equipment memory. If stored in the user equipment memory, the mobile operator can request the assignment of default values to MMS settings during the device manufacturing process.
- *Configuration of USIM-stored settings by the USIM issuer:* the USIM issuer (usually the network operator), can update USIM-stored settings and provide the USIM to the user upon subscription to a set of services (including MMS). Note that at the time of writing, not many devices support the USIM. Most devices have support for the SIM only.
- *Over-the-air (OTA) provisioning of settings:* OTA provisioning refers to the possibility of sending parameters dynamically via the network in order to configure a mobile device remotely. This can be performed, for instance, by sending an SMS message containing MMS parameters to the mobile device. Upon receipt, the user equipment updates USIM-stored or equipment-stored settings. Note that there is no available standardized solution for OTA provisioning of MMS parameters. The WAP Forum is working on the design of a generic mechanism for OTA provisioning of application settings. In the meantime, Nokia and Ericsson have developed proprietary OTA provisioning of various service settings (including MMS settings) for their handsets.[4]

6.23 USIM Storage of MMS Settings

As shown in the previous section, the USIM may accept the storage of a set of files containing MMS settings. The following categories of information can be stored in USIM files:

- *User preferences* provide default parameter values to be assigned to new messages created by the subscriber.
- *Message notifications* can also be stored in the USIM along with the status of the associated multimedia message (retrieved or not, rejected, forwarded, etc.).
- *Connectivity settings* are necessary for establishing a connection with the network access point and for interacting with the WAP gateway and the MMSC.

[4] Specifications for this OTA provisioning mechanism are available from the Ericsson developer website at http://www.ericsson.com/mobilityworld (section T68/Documentation).

Table 6.12   MMS USIM files

Elementary file	Description	Supported by USIMs from

EF_{MMSN}	Storage of a notification including:
	• notification/message status such as:
	  (a) notification: read or not read
	  (b) corresponding message: retrieved, not retrieved,
	      rejected or forwarded
	• list of supported implementations (WAP MMS, etc.)
	• content of the notification
	• pointer to a notification extension
		Release 4

EF_{EXT8}	Storage of a notification extension	Release 4
EF_{MMSICP}	Storage of MMS issuer connectivity parameters including: • list of supported implementations • MMSC address • Interface to core network and bearer (bearer, address, etc.) • Gateway parameters (address, type of address, etc.) This elementary file may contain a prioritized list of connectivity parameter sets	Release 4
EF_{MMSUP}	Storage of MMS user preferences including: • list of supported implementations • user preference profile name • user preference profile parameters (sender visibility, delivery report, read-reply report, priority, time of expiry, earliest delivery time)	Release 4
EF_{MMSUCP}	Storage of MMS user connectivity parameters (same format as EF_{MMSICP})	Release 4

For this purpose, the elementary files listed in Table 6.12 have been introduced [3GPP-31.102]. Elementary files are stored in the ADF$_{USIM}$ application dedicated file in the USIM.

6.24 Streaming in MMS

In the person-to-person scenario, the message delivery process consists of first notifying the recipient MMS user agent and second retrieving the complete multimedia message (immediate or deferred retrieval). Once the message has been retrieved, then the message content can be rendered on the recipient device upon user request. This process is suitable for most use cases. However, it is not always possible (message too large) or efficient to retrieve the complete message prior to rendering it. In these situations, it is sometimes possible to split the message content into small data chunks and to deliver the message content, chunk per chunk, to the recipient device. With this process, data chunks are directly rendered on the recipient device without waiting for the complete message to be retrieved. Once data chunks have been rendered by the recipient device, they are usually discarded. This means that the whole process of transmitting the message data chunks has to be carried out again each time the MMS subscriber wishes to read the message. This enhanced process for performing message delivery and rendering is known as the *streaming process* [3GPP-26.233][3GPP-26.234].

6.24.1 Example of MMS Architecture for the Support of Streaming

In the MMS environment, it is possible to use the streaming process for the delivery of multimedia messages composed of streamable content. For this purpose, the recipient MMS user agent and recipient MMSE must have streaming capabilities. Note that the support of streaming is not mandatory for the MMS user agent, or for the MMSC. The delivery of streamable multimedia messages involves an additional element in the MMS environment. This element is known as the *media server* and is depicted in Figure 6.31. The media server may be integrated with the MMSC or might be a separate physical entity in the network.

The delivery of a multimedia message with the streaming process consists of five consecutive steps:

1. Upon receipt of the message, the recipient MMSC notifies the recipient MMS user agent that a multimedia message is awaiting retrieval. The recipient MMS user agent requests message retrieval (deferred or immediate). Note that this step does not defer from the normal message delivery process (without support for streaming).
2. After receipt of the delivery request, the recipient MMSC decides whether the message

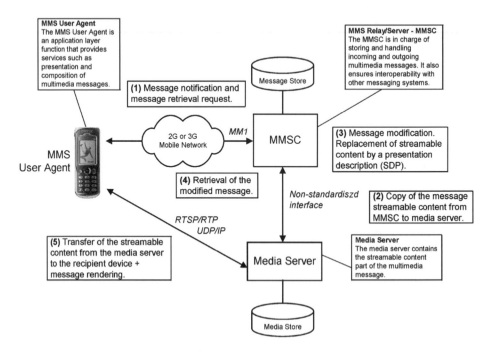

Figure 6.31   MMS environment for support of streaming

should be delivered in the normal way (also known as batch delivery) or in streaming mode. The decision is taken according to the message content, the capability negotiation and/or the user settings and preferences. If the message is to be delivered in a streaming mode, then the recipient MMSC copies the message's streamable content to the media server.

3. After copy of the message's streamable content to the media server, the recipient MMSC modifies the multimedia message prior to its delivery to the recipient MMS user agent. The modification consists of removing the streamable content from the multimedia message and replacing it by a *presentation description*. The presentation description indicates the streaming transport protocols to be used, the network address of the media server, the characteristics of the content to be streamed, etc. The presentation description is formatted according to the Session Description Protocol (SDP) [RFC-2327].

4. The modified message is delivered to the recipient MMS user agent in the normal way and locally stored.

5. At a later stage, the message recipient may decide to read the message for which a modified version is locally stored. In this situation, the recipient mobile device detects that the message contains a presentation description referring to a remote streamable content. In order to render the streamable content, the mobile device establishes an RTP/RTSP session with the media server and requests the SGSN (in a GPRS environment) to activate a secondary PDP context with a level of quality of service fulfilling the streaming mode requirements. The recipient MMS user agent instructs the media server to start the delivery of the streamable content with the RTSP Play command. Once the streamable content has been rendered by the recipient device, the RTSP/RTP session is released and the secondary PDP context is deactivated.

In the person-to-person scenario, the way the multimedia message is delivered is independent of the way it was initially submitted by the message originator. In other words, the message originator does not control the way (batch or stream) a multimedia message is delivered to the message recipients.

The transaction flow, corresponding to the five steps of the message delivery in streaming mode, is shown in Figure 6.32.

This section has presented the support of streaming delivery in the person-to-person scenario. Note that value-added service providers may also send messages, containing streamable content, to MMS subscribers. Two alternatives are possible in the machine-to-person scenario. The first alternative consists of submitting the message, in the normal way, over the MM7 interface. In this situation, the MMSC decides if the message is to be delivered in streaming or batch mode (as in the person-to-person scenario). The second alternative for the value-added service provider consists of submitting to the MMSC, a message which contains a presentation description pointing to a media server. In this situation, the value-added service provider decides that the message is to be delivered in streaming mode.

6.24.2 Streaming Protocols: RTP and RTSP

Two protocols can be used for the transport of streamable content in MMS: the Real-Time Transport Protocol (RTP) [RFC-1889] and the Real-Time Streaming Protocol (RTSP) [RFC-2326].

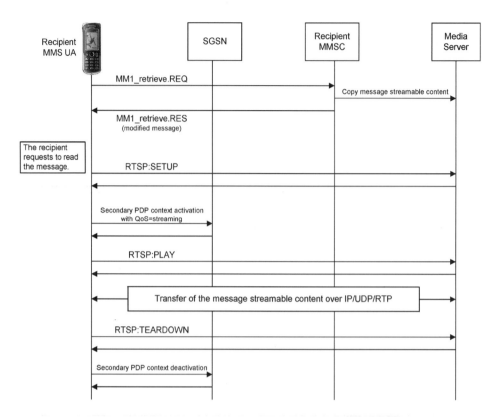

Figure 6.32   Transaction flow for support of streaming in MMS

RTP is a generic transport protocol allowing the transfer of real-time data from a server. In the MMS environment, RTP enables a one-way communication from the media server to the MMS user agent. This communication enables the delivery of streamable content, stored in the media server, to the MMS user agent in streaming mode. RTP usually relies on the connection-less UDP protocol (itself relying on the IP protocol). Compared with the connection-oriented TCP, UDP allows a faster and more resource-efficient transfer of data. However, UDP lacks a mechanism for reporting loss of data chunks. Consequently, with RTP, lost data chunks are not retransmitted (note that for the transfer of real-time data, transport reliability is not as important as timely delivery). RTP relies on three basic mechanisms for the transport of real-time data:

- *Time-stamping:* the media server time-stamps all data chunks prior to their delivery to the MMS user agent. After receiving data chunks, the MMS user agent reconstructs the original timing in order to render the streamable content at the appropriate rate.
- *Sequence numbering:* UDP does not always deliver data chunks in the order they were sent by the media server. To cope with this, the media server tags data chunks with an incremental sequence number. This sequence number allows the MMS user agent to reorganize data chunks in the correct order. The sequence number is also used by the MMS user agent for detecting any loss of data chunks.

Table 6.13 RTSP instructions

Command	Description
SETUP	The MMS user agent requests the media server to allocate resources and to start an RTSP session
OPTIONS	The MMS user agent indicates to the media server what options are supported
DESCRIBE	The MMS user agent retrieves the description of some streamable content from the media server
PLAY	The MMS user agent requests the media server to play (retrieve and render) the streamable content
PAUSE	The MMS user agent temporarily halts the delivery of the message content (without releasing media server resources allocated to the RTSP session)
TEARDOWN	The MMS user agent requests the media server to stop the delivery of the message content. In this situation, the media server can release resources allocated by the media server
REDIRECT	The media server requests the MMS user agent to connect to another media server

- *Payload type identifier:* this identifier indicates the encoding/compression schemes used by the media server. According to the payload type identifier, the MMS user agent determines how to render the message content.

With RTSP, the MMS user agent can control the way the streamable content is delivered from the media server. For this purpose, the MMS user agent instructs the media server to start, pause, play the message content during the content delivery and rendering. In other words, with RTSP, the MMS user agent can control the content delivery from the media server in the same way as a person controls a VCR with a remote controller. This is why RTSP is sometimes known as a 'network remote control' for multimedia servers. RTSP is used for establishing and controlling streaming sessions but is not used for the transport of streamable content. The transport of the streamable content is handled by transport protocols such as RTP. While most streaming content delivery use RTP as a transport protocol, RTSP is not tied to RTP. According, to RTSP instructions, the media server adapts the way the streamable content is delivered. The set of RTSP instructions includes those listed in Table 6.13 (the list is not exhaustive).

6.24.3 Session Description Protocol

In the context of MMS, the Session Description Protocol (SDP) is used for describing streaming sessions as part of modified multimedia messages. SDP is published by the IETF in [RFC-2327]. With a session description, the MMS user agent is able to identify the media server where the streamable content is stored and to initiate the streaming session for content retrieval. SDP is used for session description only and does not include a transport protocol. SDP is intended to use existing transport protocols such as RTP. A session description in the SDP format includes:

- *Session and media information* such as the type of media (audio, video, etc.), the transport protocol (RTP/UDP/IP, etc.), the media format (MPEG4, H.263, etc.). In addition, the

Table 6.14 SDP commands

Type	Description	Status RFC 2327	Status 3GPP
Session			
v	Version identifier for the session	●	●
o	Owner/creator and session identifier	●	●
s	Session name	●	●
i	Session information	○	○
u	URI of description	○	○
e	Email address	○	○
p	Phone number	○	○
c	Connection information	●	●
b	Bandwidth information	○	●
z	Time zone adjustments	○	○
k	Encryption key	○	○
a	Session attributes. Pair (control, range)	○	●
Time			
t	Time the session is active	●	●
r	Zero or more repeat times	○	○
Media			
m	Media name	●	●
i	Media title	○	○
c	Connection information	●	●
b	Bandwidth information	○	●
	Pair (bandwidth information, AS)		
k	Encryption key	○	○
a	Zero or more attribute lines	○	●
1	4-uple (control, range, fmtp, rtpmap)		

media server location is usually provided along with corresponding transport ports for content retrieval.

- *Timing information* such as start and stop times.

A session description in the SDP format consists of a series of text lines in the form <type> = <value> where <type> is a one-character type as shown in Table 6.14) and <value> is a structured text string whose format depends on its associated type. A session description consists of a session-level description and optionally several media-level descriptions. Note that the presence of the SDP text line starting with type 'a = ' indicates that there is a need to open an RTSP session. Figure 6.33 shows the session description included as part of a multipart multimedia message for the streaming of a video content.

6.25 Charging and Billing

The design of charging methods is of key importance for enabling MMS providers to develop billing solutions that meet the requirements of various business models. The 3GPP published

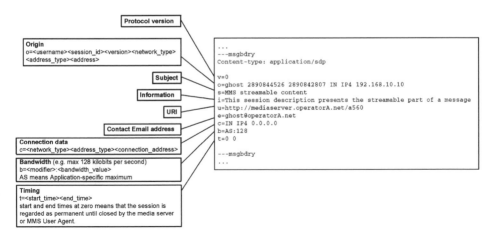

Figure 6.33   SDP example

a set of specifications describing generic charging principles in [3GPP-32.200] and charging principles for MMS in [3GPP-22.140]. For billing purposes, the MMSC collects charging information, when handling messages (submission, delivery, forward and deletion) and reports, and generates Charging Data Records (CDRs) accordingly. CDRs are then provided to the MMS provider billing system in order to produce subscriber invoices.

The 3GPP, or any other standardization organization, does not define the rules for MMS billing. Billing aspects are out of the scope of 3GPP technical specifications. That is to say, the MMSC generates CDRs for specific events but whether or not the subscriber (sender, recipient or both) is invoiced according to the generated CDR(s) is an MMS provider independent choice. However, it is expected that the following billing models will be put in place by MMS providers:

- A flat rate per multimedia message (as for SMS and EMS billing).
- Rate per byte of information transported to reflect network resource usage.
- Specific rate for events (financial information, weather forecasts, etc.).
- A subscription for a number of MMS units per month.

In addition, the following configurations are considered:

- The message sender pays.
- Both sender and recipient pay their respective charges for message handling.
- The recipient pays for retrieving messages from a VAS provider. In this situation, a commercial agreement between the recipient and the VAS provider needs to be set up.
- The sender pays for a reply message on a per message basis (reply charging).
- Billing models should support post-paid MMS for which the subscriber pays the bill after usage (e.g. once a month) and pre-paid MMS where the subscriber pays for a credit of units which can later be turned into MMS usage.

The MM8 interface between the MMSC and a billing system is intended to ensure interoperability between MMSCs and billing systems developed by different manufacturers. In particular, an important feature of this interface consists of enabling the transfer of CDRs from the

Table 6.15 MMS charging data records

Event	Interface	MMSC	CDR name
Submission	MM1	Originator	O1S-CDR
Forward request	MM4	Originator	O4FRq-CDR
Forward response	MM4	Originator	OFRs-CDR
Delivery report	MM4	Originator	O4D-CDR
Delivery report	MM1	Originator	O1D-CDR
Read-reply report	MM4	Originator	O4R-CDR
Read-reply originator	MM1	Originator	O1R-CDR
Originator message deletion		Originator	OMD-CDR
Message forward	MM4	Recipient	R4F-CDR
Notification request	MM1	Recipient	R1NRq-CDR
Notification response	MM1	Recipient	R1NRs-CDR
Retrieve request	MM1	Recipient	R1RtRq-CDR
Retrieve response	MM1	Recipient	R1RtRs-CDR
Acknowledgement	MM1	Recipient	R1A-CDR
Delivery report request	MM4	Recipient	R4DRq-CDR
Delivery report response	MM4	Recipient	R4DRs-CDR
Read-reply recipient	MM1	Recipient	R1RR-CDR
Read-reply report request	MM4	Recipient	R4RRq-CDR
Read-reply report response	MM4	Recipient	R4RRs-CDR
Recipient message deletion		Recipient	RMD-CDR
Forwarding		Forwarding	F-CDR

MMSC to the billing system. Unfortunately, at the time of writing, there is no standardized technical realization for the MM8 interface. Only CDRs have been defined but there is no agreed mechanism for transporting them in an MMS environment.

The identification of events, which trigger the generation of CDRs, and the definition of associated CDRs are provided in [3GPP-32.235]. Twenty-one types of CDRs have been defined, for MM1 and MM4 interfaces, as shown in Table 6.15.

These CDRs include information such as the duration of message transmission, charging information (post-paid, prepaid), message content type, message class, message size, message priority, reply charging instructions, recipient addresses, etc.

6.26 Message Size Measurement

The notion of message size is mentioned in many places in MMS-related technical specifications (information element/parameter of the message notification, charging data records, etc.). In an MMS environment, a multimedia message can take various forms. It is represented in a compact binary form for transfer over the MM1 interface in the WAP environment. It is transferred as a text-based MIME multipart message over MM3 (to an Email server only) and MM4 interfaces. Furthermore, contents of a multimedia message may be adapted, by the recipient MMSC, to the receiving device capabilities and this may also affect the size of the message transferred over the MM1 interface. In this context, it is difficult to derive a generic definition for the size of a message.

However, in order to ensure a common understanding of the notion of message size, the 3GPP provided, in [3GPP-23.140] (release 5), the following definition: the message size represents the number of octets for the entire message as transferred over the MM1 interface. In the WAP environment, the message size includes the entire MMS protocol data unit (binary encapsulated) composed of message headers (addressing, qualifiers, etc.) and message contents (text, scene description, images, etc.).

6.27 Security Considerations

Several secured transport protocols can be used in order to secure communications between the different MMS elements. For instance, WAP WTLS can be used to secure the communications between the MMS user agent and the WAP Gateway. The WAP Identity Module (WIM) and Public Key Infrastructures (PKI) are mechanisms for identifying and authenticating users. Secure MIME, as defined in [RFC-2633], can also be used to secure a message above the transport protocol.

6.28 Digital Right Management in MMS

Digital Right Management (DRM) is an unavoidable consideration for the development of services involving the exchange of content. Any DRM solution is usually based on the following components:

- *Expression of user rights:* this component refers to the ability to express what the subscriber can do with some content. These rights include permissions (play, modify, redistribute, etc.), constraints (four times, for a period of two weeks, etc.), and so on. Usage rights are typically defined by the content provider and the receiving device shall ensure that the received content is used only in the scope of these usage rights.
- *Content protection:* this component relates to means of encrypting content.
- *Trust relationships:* this component defines mechanisms allowing the realization of trust relationships between the content provider and the subscriber.

Unfortunately, due to a lack of time, no DRM solution has been standardized for MMS (release 99, release 4 and release 5). However, 3GPP standardization work is going on for the specification of a generic DRM solution that may become applicable to a number of service capabilities including MMS. Such a DRM solution is expected to become available in the 3GPP release 6 timeframe (2003). In the meantime, the WAP Forum is also working on the definition of a generic DRM solution which should be available before the 3GPP DRM solution is live.

In the meantime, a simple solution will be put in place to limit the distribution of messages containing elements that should not be redistributed. This solution applies only to messages generated by value-added service providers and consists of flagging a message with a Boolean *Message Distribution Indicator* (MDI). With this indicator, the value-added service provider indicates whether the message can be redistributed freely or not. Looking at the 3GPP technical specifications, there is no functional requirement for the MMSC or the MMS user agent to prevent the redistribution of a message for which the value-added service provider indicated that it should not be redistributed. The functional description of such a message distribution has been provided in 3GPP technical specifications release 5. However, the technical realization of such an indicator has not yet been defined. The WAP Forum should incorporate such a technical realization in the version of the specification which is to follow WAP MMS 1.1.

6.29 Technical Realization of Interfaces

In the MMS environment, entities (e.g. MMSC, MMS user agents) interact over eight distinct interfaces. Each interface specifies which operations can be invoked between two interacting entities and provides the list of information elements/parameters associated with each operation.

For the stage 2 definition of an interface, a list of generic information elements is designed for each operation (independent from the technical realization – stage 3). For the stage 3 definition, each stage 2 information element is mapped onto one or more low level parameters.

Several interfaces have been standardized to guarantee interoperability between devices produced by different manufacturers. Other non-standardized interfaces are the subject of proprietary implementations. Table 6.16 gives the status of availability for the definition of each interface (stages 2 and 3).

The following sections provide an overview of interfaces which are not yet standardized and in-depth descriptions of already standardized interfaces. In these descriptions, the 3GPP naming convention for designating transaction requests and responses is used (see Section 6.15).

6.30 MM1 Interface MMSC – MMS User Agent

In the MMS environment, interactions between the MMS user agent and the MMSC are carried out over the MM1 interface. This includes message submission, message notification, message retrieval, message forwarding, handling of associated reports and management of persistent storage. At the time of writing, the only standardized technical realizations (stage

Table 6.16 Availability of interface definitions

Interface	Between the MMSC and	Release 99		Release 4		Release 5	
		Stage 2	Stage 3	Stage 2	Stage 3	Stage 2	Stage 3
MM1	MMS user agent	3GPP	WAP F	3GPP	WAP F	3GPP	WAP F
MM2	Internal to the MMSC	–	–	–	–	–	–
MM3	External servers	–	–	3GPP	–	3GPP	–
MM4	Another MMSC	–	–	3GPP	3GPP	3GPP	3GPP
MM5	HLR	–	–	–	–	–	–
MM6	MMS u ser databases	–	–	–	–	–	–
MM7	MMS VAS Applications	–	–	–	–	3GPP	3GPP
MM8	Billing system	–	–	–	–	–	–

3) defined for the MM1 interface are those defined by the WAP Forum. Two versions of the MM1 interface have been designed and are defined as part of two sets of specifications, respectively known as WAP MMS 1.0 and WAP MMS 1.1 (corresponding to functional requirements defined in 3GPP release 99 and 3GPP release 4, respectively). A third technical realization of the MM1 interface is under development in the WAP Forum. This new version corresponds to the MM1 interface fulfilling the 3GPP functional requirements defined in release 5.

For the MM1 interface, the WAP Forum has defined new names for responses and requests. These names are shown in transaction flow diagrams in this book but are represented in italic. The WAP Forum naming convention consists of suffixing the request name by `.REQ` and suffixing the corresponding response by `.CONF`. A transaction consisting of an indication only (a request in the 3GPP naming convention) has its name suffixed with `.IND`. In addition, an operation name is prefixed by the interface name suffix ('M' for the MM1 interface known as the MMS_M interface in the WAP Forum specifications). For instance, the response for a message submission is known as `MM1_submit.RES` according to the 3GPP naming convention and `M-Send.CONF` according to the WAP Forum naming convention.

The WAP Forum technical realization of the MM1 interface (stage 3) is derived from 3GPP functional requirements (stage 2), defined in [3GPP-23.140]. Note that 3GPP functional requirements defined in release 99 for the MM1 interface are not defined in great detail. Consequently, the WAP Forum has had to develop, almost independently, the technical realization of the MM1 interface in WAP MMS 1.0. 3GPP functional requirements of the MM1 interface in release 4 are defined extensively and constitute the basis for the development of the MM1 interface in WAP MMS 1.1. In June 2002, the 3GPP finalized the definition of the release 5 MM1 interface functional requirements and this should constitute the basis for the development of the MM1 interface in the next version of the MMS specifications for the WAP environment (planned publication date: 2003). Regarding these relationships between 3GPP and WAP Forum specifications, in this section, the composition of responses and responses corresponding to operations that can be invoked over the MM1 interface is given for 3GPP release 4 and 5 (stage 2) and also for WAP MMS 1.0 and 1.1 (stage 3).

Note that MM1 operations related to network-based persistent storage (see Section 6.21) are not presented in this section.

Operations that can be invoked over the MM1 interface are shown in Table 6.17.

In the WAP MMS 1.0 technical realization, communications between the MMS user agent and the MMSC are performed over WSP from the MMS user agent to the WAP gateway and over HTTP from the WAP gateway to the MMSC (with WSP/HTTP transcoding at the WAP gateway). In the WAP MMS 1.1 technical realization, communications between the MMS user agent and the MMSC may alternatively be carried out directly over HTTP without the need for WSP transcoding. WSP is defined in the WAP recommendation [WAP-203], HTTP/1.1 has been published by the IETF in [RFC-2616] and the wireless profile for HTTP has been defined by the WAP Forum in [WAP-229].

At the application-level, the MMS Protocol Data Unit (PDU) consists of a PDU header and a PDU data section. Note that many MMS PDUs are composed of the PDU header only. The PDU may be a single-part [RFC-822][RFC-2822] or a multipart (MIME) message. To ensure transport efficiency, the MMS PDU is binary encoded according to [WAP-209] for WAP MMS 1.0 and [WAP-276] for WAP MMS 1.1. This binary encoding process is also known as *encapsulation* and is presented in Appendix E. The binary encoded MMS PDU is inserted in

Table 6.17 Operations over the MM1 interface

Operation	Description	3GPP Status		WAP Status	
		Rel-4	Rel-5	MMS 1.0	MMS 1.1
Message submission	Submission of a message from the MMS user agent to the MMSC	■	■	■	■
Message notification	Notification of the arrival of a new message. Notification sent from the MMSC to the MMS user agent	■	■	■	■
Message retrieval	Retrieval of a message from the MMSC by the MMS user agent	■	■	■	■
Message forwarding	Forwarding of a message. The request is sent from the MMS user agent to the MMSC	■	■		■
Delivery report	Delivery of a delivery report from the MMSC to the MMS user agent	■	■	■	■
Read reply report	Delivery of a ready-reply report from the MMSC to the MMS user agent[a] and from the MMS user agent to the MMSC	■	■	■	■

[a] The exchange of read-reply reports is possible in MMS 1.0 but in the form of normal multimedia message. In MMS 1.1, a read-reply report is transport with a dedicated MMS PDU.

the data section of a WSP or HTTP request/response. The content type of WSP/HTTP requests/responses containing an MMS PDU is always `application/vnd.wap.mms-message`. The process of including MMS PDUs in WSP/HTTP requests/responses is depicted in Figure 6.34.

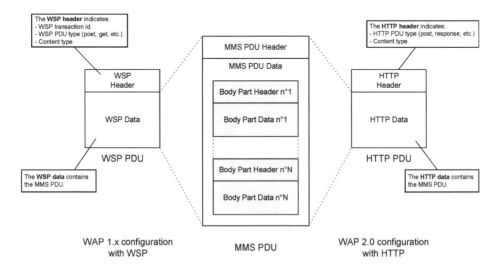

Figure 6.34 Encapsulation of the MMS PDU in HTTP and WSP PDUs

Table 6.18 MM1 stage 3 generic parameters

Stage 3 MIME parameter	Description
`X-MMS-Message-Type`	This parameter indicates the transaction type. The following values can be used:
	• `M-send-req` for message submission request
	• `M-send-conf` for message submission response
	• `M-notification-ind` for message notification indication
	• `M-notifyresp-ind` for message notification response
	• `M-retrieve-conf` for message retrieval response
	• `M-acknowledge-ind` for message retrieval acknowledgement
	• `M-delivery-ind` for delivery/retrieval report request
	• `M-read-rec.ind` for read report request (recipient MMSE)
	• `M-read-orig.ind` for read report request (originator MMSE)
	• `M-forward.req` for forward request
	• `M-forward.conf` for forward response
	Note that the only exception is for the delivery request in the WAP MMS technical realization. This request does not contain the `X-MMS-Message-Type` parameter
`X-MMS-Transaction-ID`	A unique identifier for the transaction. This transaction identification allows communicating entities to correlate a request with a corresponding response. Note that the following transactions do not contain this field, either for resource optimization or because transactions do not need to be confirmed upon processing:
	• delivery/retrieval request
	• read report indication (for recipient and originator MMSEs)
`X-MMS-MMS-Version`	The WAP MMS version number (e.g. 1.0 or 1.1). Transactions that do not specify the transaction identifier, do not specify the version of MMS which is supported

This section presents the information elements/parameters that can be contained in MMS PDUs. The generic parameters listed in Table 6.18 are always included in the header of most MMS PDUs (stage 3).

If the MMS PDU contains a multimedia message with a presentation part (scene description), then the message should be structured as a multipart message. For this purpose, the MMS PDU content type can appropriately be set up to `application/vnd.wap.multipart/related`. Theoretically, the order in which the different message parts appear in the MMS PDU is of no importance. In practice, initial MMS handsets that do not support scene descriptions in the form of MMS SMIL reconstruct a message presentation according to the order in which the different body parts appear in the multipart message. Consequently, for the generation of messages, it is recommended to include message elements in the order of appearance in the message presentation (first slide elements followed by second slide elements and so on). The message may contain more than one presentation part. In this case, the presentation part which is referred to by the Start parameter of the content type is used. If the Start parameter is not present, then the presentation part which appears first in the multipart message is used.

For the stage 2 definition of the MM1 interface, MMS PDUs are regarded as a set of information elements. In stage 3 technical realizations, to each stage 2 information element corresponds either an RFC 822 parameter, as defined by the IETF, or an MMS-specific parameter, as defined by the WAP Forum (names of MMS-specific parameters are prefixed with X-MMS).

6.30.1 Message Submission

The message submission refers to the submission of a multimedia message from an originator MMS user agent to the originator MMSC. The transaction flow in Figure 6.35 shows the interactions between the originator MMS user agent and the originator MMSC for the submission of a multimedia message.

In the WAP technical realizations, prior to the message submission, the originator MMS user agent establishes a data connection and a WSP session (unless one is already active). This may be a Circuit Switched Data (CSD) connection or a packet-switched connection over GPRS, for instance.

For the submission of the message, the request MM1_submit.REQ (known as the M-Send.req request in the WAP technical realizations) is composed of the MMS PDU information elements/parameters listed in Table 6.19.

If the date is not provided as part of the submission request, then the originator MMSC has the responsibility to specify the date of message submission assigned to the Date parameter. If a date is specified by the MMS user agent, then the MMSC can still overwrite the submission date. Note that if the MMSC assigns or overwrites the date, then there is no means for the originator MMS user agent to be informed of the assigned date. This can be seen as a misconception of MMS. In comparison, in the design of SMS, the date for a submitted

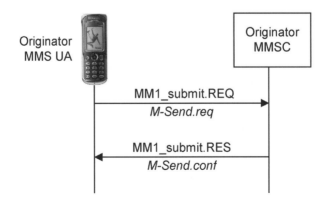

Figure 6.35   MM1 message submission

Table 6.19 MM1 message submission request

Stage 2 Information elements	Description	Stage 3 WAP implementation (MIME parameter)	3GPP status		WAP status	
			Rel-4	Rel-5	MMS 1.0	MMS 1.1
Recipient address	Address of the recipient(s)	`To, Cc, Bcc`[a] Type: string	●	●	●	●
Sender address	Address of the sender	`From`[b] Type: string	○	○	●	●
Date and time	Date and time of message submission	`Date` Type: integer	○	○	○	○
Time of expiry	Message time of expiry	`X-MMS-Expiry` Type: date	○	○	○	○
Earliest delivery time	Message earliest desired time of delivery	`X-MMS-Delivery-Time` Type: date Default: `immediate`	○	○	○	○
Reply charging	Request for reply charging	`X-MMS-Reply-Charging` Values: `Requested` `Requested text only`	○	○	–	○
Reply deadline	The latest time for submitting a reply message granted to the recipient(s)	`X-MMS-Reply-Charging-Deadline` Type: date	○	○	–	○
Reply charging size	The maximum size for a reply-message	`X-MMS-Reply-Charging-Size` Type: integer	○	○	–	○
Reply charging identification	The identification of the message to which a reply is associated	`X-MMS-Reply-Charging-ID` Type: string	○	○	–	○
Delivery report	Request for delivery report	`X-MMS-Delivery-Report` Values: `Yes` `No`	○	○	○	○

Table 6.19 (*continued*)

Stage 2 Information elements	Description	Stage 3 WAP implementation (MIME parameter)	3GPP status		WAP status	
			Rel-4	Rel-5	MMS 1.0	MMS 1.1
Read reply	Request for read-reply report	`X-MMS-Read-Reply` Values: `Yes` `No`	○	○	○	○
Message class	The message class (automatic or personal)	`X-MMS-Message-Class` Values: `Personal` (default) `Auto`	○	○	○	○
Priority	The message priority	`X-MMS-Priority` Values: `Low` `Normal` (default) `High`	○	○	○	○
Sender visibility	Request to show/hide the sender identity	`X-MMS-Sender-Visibility` Values: `Hide` `Show` (default)	○	○	○	○
Subject	Message subject	`Subject` Type: string	○	○	○	○
Content type	Message content type	`Content-type` Type: string	●	●	●	●
Content	Content of the message	Message body	○	○	○	○

[a] At least one of these address fields must be present (`To`, `Cc` or `Bcc`).
[b] This field is mandatory in the WAP MMS definition of the submission request but optional in functional requirements defined by the 3GPP.

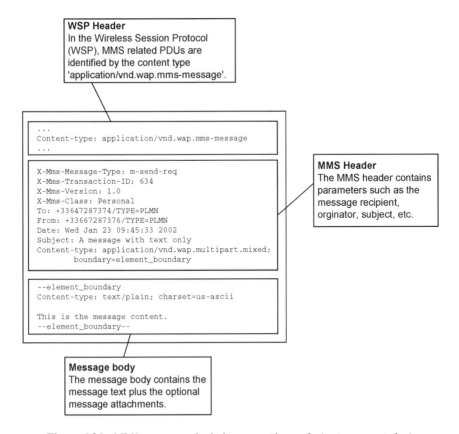

WSP Header
In the Wireless Session Protocol
(WSP), MMS related PDUs are
identified by the content type
'application/vnd.wap.mms-message'.

```
...
Content-type: application/vnd.wap.mms-message
...
```

```
X-Mms-Message-Type: m-send-req
X-Mms-Transaction-ID: 634
X-Mms-Version: 1.0
X-Mms-Class: Personal
To: +33647287374/TYPE=PLMN
From: +33667287376/TYPE=PLMN
Date: Wed Jan 23 09:45:33 2002
Subject: A message with text only
Content-type: application/vnd.wap.multipart.mixed;
        boundary=element_boundary
```

MMS Header
The MMS header contains
parameters such as the
message recipient,
orginator, subject, etc.

```
--element_boundary
Content-type: text/plain; charset=us-ascii

This is the message content.
--element_boundary--
```

Message body
The message body contains the
message text plus the optional
message attachments.

Figure 6.36   MM1 message submission request/example (text representation)

message is always provided by the SMSC as part of the submission acknowledgement (submission response).

The message class for a message is 'auto' if the message is automatically generated by the MMS user agent (e.g. automatic generation of a read-reply report). If the message has been composed by the subscriber, then the message class is 'personal'. If no class is specified for a message submission, then personal is considered as the default class for the message.

In the situation where the message originator requested reply-charging for a message submission and if reply charging is not supported by the originator MMSC, then the MMSC rejects the submission request by generating a negative response.

Figure 6.36 shows an example of MMS PDU corresponding to a message submission request before binary encoding. The corresponding MMS PDU in a binary format is represented in Figure 6.37.

The MMSC acknowledges the submission request with the MM1_submit.RES response (known as the M-Send.conf confirmation in the WAP technical realizations). This response is composed of the information elements/parameters listed in Table 6.20.

If the message submission request is accepted by the originator MMSC, then a message identification is provided back to the originator MMS user agent as part of the submission

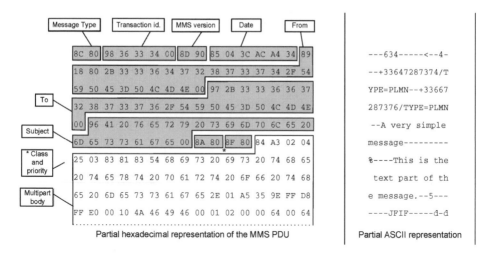

Figure 6.37 MM1 message submission request/example (binary representation)

response. This message identification ensures that associated reports or a message reply (reply charging) that may be received later can be correlated to the original message.

In the event of the message request not being accepted by the originator MMSC, the MMSC indicates the reason for request rejection. The error code provided by the MMSC may be of transient nature (e.g. temporary unavailability of the MMSC) or permanent nature (e.g. message request is badly formatted). Values that can be assigned to the request status parameter (X-MMS-Response-status) are listed below (note that some values only apply to the submission of a forward request):

- Ok: this status code indicates that the corresponding request has been accepted without errors. The binary representation of this status code is 0x80 (decimal 128).

Table 6.20 MM1 message submission response

Stage 2 Information elements	Description	3GPP status		Stage 3 WAP implementation (MIME parameter)	WAP status	
		Rel-4	Rel-5		MMS 1.0	MMS 1.1
Request status	Status of the submission request	●	●	X-MMS-Response-status Values: Ok Error codes in Tables 6.21 and 6.22	●	●
Request status text	Textual description of the status	O	O	X-MMS-Response-text Type: string	O	O
Message identification	Identification of the submitted message[a]	C	C	Message-ID Type: string	O	O

[a] *Condition:* present only if the MMSC accepts the submission request.

Table 6.21   MM1 message submission response/transient errors

Binary	Status code	Description
0xC0 (192)	`Error-transient-failure`	The request is valid but the MMSC is unable to process it, due to some temporary conditions
0xC1 (193)	`Error-transient-sending-` `address-unresolved`	Due to some temporary conditions, the MMSC is unable to resolve an address specified in the request
0xC2 (194)	`Error-transient-message-not-` `found`	Due to some temporary conditions, the MMSC is unable to retrieve the message (applicable to message forward only)
0xC3 (195)	`Error-transient-network-` `problem`	Due to some temporary conditions (capacity overload for instance), the MMSC is unable to process the request

The group of transient errors is provided in Table 6.21. The group of permanent errors is provided in Table 6.22. The following status codes are now obsolete:

- `Error-unspecified` (value 0x81, decimal 129)
- `Error-service-defined` (value 0x82, decimal 130)
- `Error-message-format-corrupt` (value 0x83, decimal 131)
- `Error-sending-address-unresolved` (value 0x84, decimal 132)
- `Error-message-not-found` (value 0x85, decimal 133)
- `Error-network-problem` (value 0x86, decimal 134)
- `Error-content-not-accepted` (value 0x87, decimal 135)
- `Error-unsupported-message` (value 0x88, decimal 136).

Note that values from 0xC4 to 0xDF (196 to 223, decimal) are reserved for future use for the representation of transient failures. Values from 0xEA to 0xFF (234 to 255, decimal) are reserved for future use for the representation of permanent failures.

6.30.2 Message Notification

Once a message has been received by the recipient MMSC, the multimedia message is stored and the MMSC builds a compact notification to be delivered to the recipient MMS user agent. With the notification, the recipient MMS user agent can retrieve the corresponding message immediately upon receipt of the notification (immediate retrieval) or later (deferred retrieval). The transaction flow in Figure 6.38 shows the interactions between the recipient MMSC and the recipient MMS user agent for the provision of a notification.

As shown in Figure 6.38, the notification is implicitly acknowledged with the retrieval request in the immediate retrieval case. In the deferred retrieval case, the notification response is explicitly provided by the MMS user agent to the recipient MMSC.

In the WAP technical realizations, the recipient MMSC delivers the message notification to the recipient MMS user agent as part of a push request.

Table 6.22   MM1 message submission response/permanent errors

Binary	Status code	Description
0xE0 (224)	`Error-permanent-failure`	An unspecified permanent error occurred during the processing of the request by the MMSC
0xE1 (225)	`Error-permanent-service-denied`	The request is rejected because of service authentication and authorization failure(s)
0xE2 (226)	`Error-permanent-message-format-corrupt`	The MMSC detected a problem in the request format
0xE3 (227)	`Error-permanent-sending-address-unresolved`	The MMSC is unable to resolve one of the addresses specified in the corresponding request
0xE4 (228)	`Error-permanent-message-not-found`	The MMSC is unable to retrieve the associated message (for message forward only)
0xE5 (229)	`Error-permanent-content-not-accepted`	The MMSC cannot process the request because of the message size, incompatible media types or copyright issues
0xE6 (230)	`Error-permanent-reply-charging-limitations-not-met`	The request does not meet the reply charging requirements
0xE7 (231)	`Error-permanent-reply-charging-request-not-accepted`	The MMSC supports reply charging but the request is rejected because of incompatibility with the service or user profile configuration
0xE8 (232)	`Error-permanent-reply-charging-forward-denied`	The forwarding request is for a message containing reply charging requirements
0xE9 (233)	`Error-permanent-reply-charging-not-supported`	The MMSC does not support reply charging

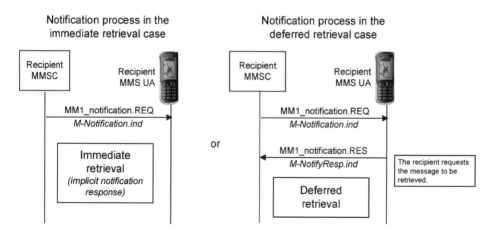

Figure 6.38   MM1 notification

After receiving the notification, the recipient MMS user agent can perform the following actions on the corresponding multimedia message:

- *Message rejection:* the message is rejected without being retrieved by the recipient MMS user agent.
- *Immediate message retrieval:* the MMS user agent retrieves immediately the message.
- *Deferred message retrieval:* the MMS user agent indicates to the MMSC that the multimedia message will be retrieved later by the recipient.

The message notification request informs on the corresponding message content (content type, element descriptors, message qualifiers, etc.). An important parameter of the notification is the message reference used by the recipient MMS user agent to identify a message to be retrieved or rejected.

The notification request `MM1_notification.REQ` (also known as the `M-Notification.ind` indication in the WAP technical realizations) is composed of the information elements/parameters listed in Table 6.23.

Note that the MMS PDU representing a message notification only contains a PDU header (no PDU body). The message reference (content location) is represented in the form of a Uniform Resource Identifier (URI). The URI format is defined in [RFC-2396]. An example of message reference assigned to the `X-MMS-Content-Location` parameter is

```
X-MMS-Content-Location: http://mmsc.operator.net/message-id-634
```

The recipient MMS user agent acknowledges the notification delivery with the `MM1_notification.RES` response (also known as the `M-NotifyResp.ind` indication in the WAP technical realizations). This response is composed of the information elements/parameters listed in Table 6.24.

The `X-MMS-Status` parameter indicates the status of the associated multimedia message. The values that can be assigned to this parameter are given in Table 6.25. Upon notification request, the recipient MMSC waits for a notification response or a message immediate retrieval. After a predefined period of time, if the recipient MMS user agent has not provided a notification response or has not retrieved the message, then the MMSC considers that the message notification has been lost. In this case, the MMSC re-attempts to deliver the message notification a number of times.

Figure 6.39 shows an example of a notification request along with the corresponding notification response (before binary encoding). It may occur that the recipient MMS user agent is temporarily unable to handle notifications (memory full for instance). The MMS user agent has two methods to cope with this situation:

- *Notification discard and retransmit:* this method consists of discarding the notification and allowing further notification re-attempts by the recipient MMSC. For this purpose, the recipient MMS user agent does not deliver the corresponding notification response (`MM1_notification.RES` response). Consequently, the recipient MMSC re-attempts to deliver the notification a number of times until the recipient MMS user agent provides the corresponding notification response.
- *Flow control on the SMS bearer:* the push request containing the notification request may be performed over the SMS bearer. It was shown in Chapter 3 that an SME can indicate to the SMSC that it is no longer able to handle SMS messages (no more storage space left). In

Table 6.23 MM1 notification request

Stage 2 Information elements	Description	3GPP status		Stage 3 WAP implementation (MIME parameter)	WAP status	
		Rel-4	Rel-5		MMS 1.0	MMS 1.1
Sender address	Address of the sender[a]	C	C	From Type: string	○	○
Message class	Message class (personal, advertisement, informational, etc.)	●	●	X-MMS-Message-Class Values: Personal (default) Advertisement Informational Auto	●	●
Time of expiry	Message time of expiry	●	●	X-MMS-Expiry Type: date	●	●
Delivery report	Request for delivery report[b]	○	○	X-MMS-Delivery-Report Values: Yes No	–	○
Reply charging	Request for reply charging	○	○	X-MMS-Reply-Charging Values: Requested Requested text only	–	○

Information element	Description	Header field				
Reply charging identification	Identification of the message to which a reply is associated	X-MMS-Reply-Charging-ID Type: string	○	○	—	○
Reply deadline	The latest time for submitting a reply message granted to the recipient(s)	X-MMS-Reply-Charging-Deadline Type: date	○	○	—	○
Reply charging size	Maximum size for a reply-message	X-MMS-Reply-Charging-Size Type: integer Unit: octet	○	○	—	○
Priority	Message priority	Not defined yet	—	○	—	—
Subject	Message subject	Subject Type: string	○	○	○	○
Message distribution indicator	Indication from the VAS provider about whether or not the message can be redistributed	Not defined yet	—	○	—	—
Element descriptor	Reference of an element of the message along with a description (e.g. name, size, type, format)	Not defined yet	—	○	—	—
Message size	Approximate size of the multimedia message	X-MMS-Message-Size Type: integer Unit: octet	●	●	●	●
Message reference	A reference for identifying uniquely the message for further actions (e.g. retrieval, forwarding)	X-MMS-Content-Location Type: string (URI)	●	●	●	●

a *Condition*: provided only if address hiding is not requested by the message originator.

b In MMS 1.0, the indication that the message originator wishes a delivery report to be generated is only present in the retrieved message.

Table 6.24 MM1 notification response

Stage 2 Information elements	Description	3GPP status		Stage 3 WAP implementation (MIME parameter)	WAP status	
		Rel-4	Rel-5		MMS 1.0	MMS 1.1
Message status	The status of the message retrieval	○	○	X-MMS-Status Values: Retrieved Rejected Deferred Forwarded Unrecognised	●	●
Delivery report allowed	Indicates whether or not the generation of a delivery report is allowed	○	○	X-MMS-Report-Allowed Values: Yes (default) No	○	○

this situation, the SME notifies the SMSC when it has retrieved the ability to handle messages.

Note that both methods have drawbacks. With the 'notification discard and retransmit' method, it is difficult for the recipient MMSC to determine when the notification request should be re-transmitted. The recipient MMSC only re-attempts to deliver notifications a certain number of times. There is therefore a risk of losing the notification and consequently the associated multimedia message. This method also leads to a waste of resources in trans-

Table 6.25 MM1 notification response/status codes

Binary	Status code	Description
0x81 (129)	Retrieved	This status code means that the associated multimedia message has already been retrieved by the recipient MMS user agent (immediate retrieval)
0x82 (130)	Rejected	This status code means that the MMS user agent is not willing to retrieve the message. In other words, the message has been rejected by the message recipient
0x81 (131)	Deferred	This status code means that the MMS user agent is not willing to retrieve immediately the message. The multimedia message will be retrieved later by the recipient MMS user agent (deferred retrieval)
0x81 (132)	Unrecognised	This status code is used for version management purpose only. For instance, this status code is used by MMS user agents upon reception of requests which are not recognized

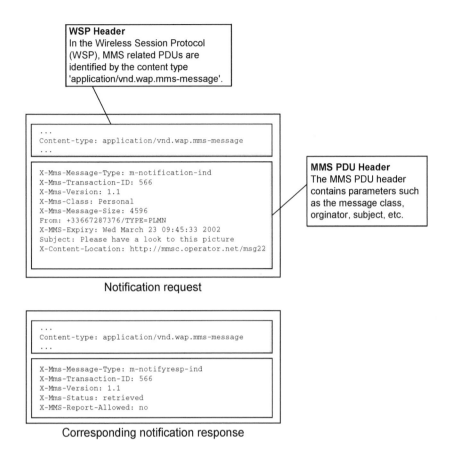

Figure 6.39 MM1 notification request and response/example

ferring all notifications that will be discarded by the recipient MMS user agent. On the other hand, the 'flow control on the SMS bearer' method means that all SMS messages are temporarily stored in the SMSC from the time the SME indicates that it is unable to handle messages. Consequently, even SMS/EMS messages to be delivered to other applications are blocked at the SMSC level.

Because of these drawbacks, more efficient solutions are being elaborated by standardization development teams. Consequently, a more robust solution should be defined in forth-coming releases and/or versions of 3GPP and WAP Forum specifications.

In the WAP technical realizations, the notification request is pushed to the mobile device. This can be performed over SMS (if no connection is available) or over WDP/UDP (if a connection is available). Theoretically, a device should always acknowledge a notification request to the MMSC either by retrieving the message (immediate retrieval) or by providing an explicit notification response (deferred retrieval). This enables the MMSC to detect noti-fication losses over the MM1 interface in order to re-transmit lost notifications. In practise, in the deferred retrieval case, MMSCs do not always expect a notification response from the mobile device if the corresponding request was pushed over SMS (already an acknowledged

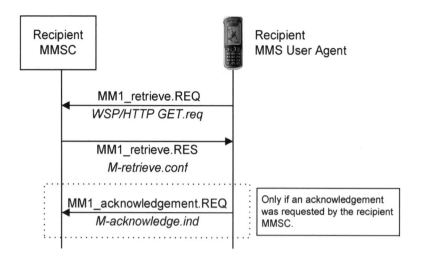

Figure 6.40 MM1 message retrieval

bearer). In this situation, the mobile device is not required to establish a connection to the network to acknowledge the notification request. This leads to a reduction of signalling traffic in the network.

6.30.3 Message Retrieval

Upon receipt of a message notification, the recipient MMS user agent has access to two methods for retrieving the corresponding message: immediate and deferred retrieval. Figure 6.40 shows the transaction flow for the generic retrieval process.

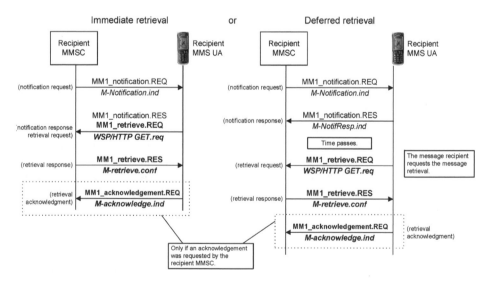

Figure 6.41 MM1 immediate and deferred message retrievals

Table 6.26 MM1 message retrieval request

Stage 2 Information elements	Description	3GPP status		Stage 3 WAP implementation (MIME parameter)	WAP status	
		Rel-4	Rel-5		MMS 1.0	MMS 1.1
Message reference	Reference of the message to be retrieved	●	●	The WSP/HTTP GET.req contains the URI identifying the message in the recipient MMSC message store	N/A	N/A

In the WAP technical realizations, the notification and retrieval transaction flows are tightly interrelated. Figure 6.41 shows the transactions flows for the two retrieval processes: immediate and deferred retrieval.

The final acknowledgement of retrieval (MM1_acknowledgement.REQ) is generated by the recipient MMS user agent, only if requested by the recipient MMSC (as part of the corresponding retrieval response).

In WAP technical realizations, prior to message retrieval, the recipient MMS user agent establishes a WAP connection. This may be a Circuit Switched Data (CSD) connection or a packet-switched connection over GPRS, for instance. The MM1_retrieve.REQ request does not correspond to an MMS-specific PDU in the WAP technical realizations. Indeed, the request is mapped to the existing WSP/HTTP GET.req request. This operation is composed of the WSP/HTTP PDU parameters listed in Table 6.26.

Upon receipt of the retrieval request, the recipient MMSC retrieves the message from its local store (according to the message reference contained in the message request). If for any reason, the message cannot be retrieved, then the MMSC provides a retrieval response with a negative status (represented with a MMS PDU without body). If the message can be successfully retrieved from the local store, then the message is included in the retrieval response MM1_retrieve.RES (also known as the M-retrieve.conf in the WAP technical realizations). In this case, the message content is encapsulated in the body of the corresponding retrieval response. In both cases, the retrieval response contains the information elements/parameters listed in Table 6.27.

If the message cannot be retrieved, then the recipient MMSC indicates to the recipient MMS user agent the nature of the problem with the value assigned to the retrieve status. Values that can be assigned to the retrieve status information element (X-MMS-Retrieve-status parameter) are listed below:

- Ok: this status code indicates that the corresponding request has been processed without errors. In this case, the multimedia message is included in the retrieval response. The binary representation of this status code is 0x80 (decimal 128).

The group of transient errors is provided in Table 6.28. The group of permanent errors is provided in Table 6.29. Note that values from 0xC3 to 0xDF (195 to 223, decimal) are reserved for future use for the representation of transient failures. Values from 0xE4 to

Table 6.27 MM1 message retrieval response

Stage 2 Information elements	Description	3GPP status		WAP implementation (MIME parameter)	WAP status	
		Rel-4	Rel-5		MMS 1.0	MMS 1.1
Message identification	Message identification.	●	●	Message-ID[a] Type: string	○	○
Recipient address	Address of the recipient(s)	○	○	To, Cc Types: string	○	○
Sender address	Address of the sender[b]	C	C	From Type: string	○	○
Message class	Message class (personal, advertisement, informational, etc.)	○	○	X-MMS-Message-Class Values: Personal (default) Advertisement Informational Auto	○	○
Date and time	Date and time of message submission or date and time of message forwarding	●	●	Date Value: date	●	●
Delivery report	Request for delivery report	○	○	X-MMS-Delivery-Report Values: Yes No (default)	○	○
Reply charging	Request for reply charging	○	○	X-MMS-Reply-Charging Values: Requested Requested text only	–	○
Reply deadline	Latest time for submitting a reply message granted to the recipient(s)	○	○	X-MMS-Reply-Charging-Deadline Type: date	–	○

Information element	Description			Field format		
Reply charging size	Maximum size for a reply-message	O	O	X-MMS-Reply-Charging-Size Type: string Unit: octet	—	O
Reply charging identification	Identification of the original message to which the delivered message is a reply to	O	O	X-MMS-Reply-Charging-ID Type: string	—	O
Priority	Message priority[c]	C	C	X-MMS-Priority Values: Low Normal (default) High	O	O
Read reply report	Request for read-reply report[c]	C	C	X-MMS-Read-Reply Values: Yes No (default)	O	O
Subject	Message subject[c]	C	C	Subject Type: string	O	O
Retrieve status	The status of the retrieve request	O	O	X-MMS-Retrieve-Status Values: Ok Error codes defined in Tables 6.28 and 6.29	—	O
Status text	Textual description of the request status	O	O	X-MMS-Retrieve-Text Type: string	—	O
Previously sent by	Address(es) of MMS user agent(s) that have handled (submitted or forwarded) the message prior to the manipulation by the MMS user agent, whose address is assigned to the sender information element	O	O	X-Mms-Previously-Sent-By Type: string with index e.g.: 1, armel@armorepro.com 2, gwenael@le-bodic.net	—	O

Table 6.27 (*continued*)

Stage 2 Information elements	Description	3GPP status		WAP implementation (MIME parameter)	WAP status	
		Rel-4	Rel-5		MMS 1.0	MMS 1.1
Previously sent date and time	Date and time when the message was submitted or forwarded	O	O	X-Mms-Previously-Sent-Date-and-Time Type: date with index e.g.: 1, Mon Jan 21 09:45:33 2002 2, Wed Jan 23 18:06:21 2002	–	O
Content type	Message content type	●	●	Content-type Type: string e.g. text/plain	●	●
Content	The message content[c]	C	C	Message body	O	O

[a] This parameter is always present if the message originator requested a read-reply report.
[b] *Condition:* available only if the originator did not request address hiding.
[c] *Condition:* available only if specified by the originator MMS user agent.

Table 6.28 MM1 message retrieval response/transient errors

Binary	Status code	Description
0xC0 (192)	`Error-transient-failure`	The request is valid but the MMSC is unable to process it due to some temporary conditions
0xC1 (193)	`Error-transient-message-not-found`	Due to some temporary conditions, the MMSC is unable to retrieve the message (for message forward only)
0xC2 (194)	`Error-transient-network-problem`	Due to some temporary conditions (capacity overload for instance), the MMSC is unable to process the request

0xFF (228 to 255, decimal) are reserved for future use for the representation of permanent failures.

Figure 6.42 shows an example of MMS PDU corresponding to a message retrieval response (before binary encoding). The recipient MMSC may request an acknowledgement for the message delivery. The recipient MMSC may require this acknowledgement to provide a delivery report back to the message originator or to delete the message from its local store after successful retrieval by the message recipient. The recipient MMSC requests a confirmation by including a transaction identification (X-Mms-Transaction-ID parameter) in the retrieval response. The retrieval acknowledgement MM1_acknowledgement.REQ (also known as M-acknowledge.ind in the WAP technical realizations) is composed of the information elements/parameters listed in Table 6.30.

Note that the transaction identification of the retrieval acknowledgement is the same as the one that was provided by the recipient MMSC for the corresponding retrieval response.

Table 6.29 MM1 message retrieval response/permanent errors

Binary	Status code	Description
0xE0 (224)	`Error-permanent-failure`	An unspecified permanent error occurred during the request processing by the MMSC
0xE1 (225)	`Error-permanent-service-denied`	The request is rejected because of service authentication and authorization failure(s)
0xE2 (226)	`Error-permanent-message-not-found`	The MMSC is unable to retrieve the associated message
0xE3 (227)	`Error-permanent-content-unsupported`	The MMSC indicates to the MMS user agent that the message cannot be retrieved because media types for one or message elements are not supported by the MMS user agent and content adaptation cannot be performed by the MMSC

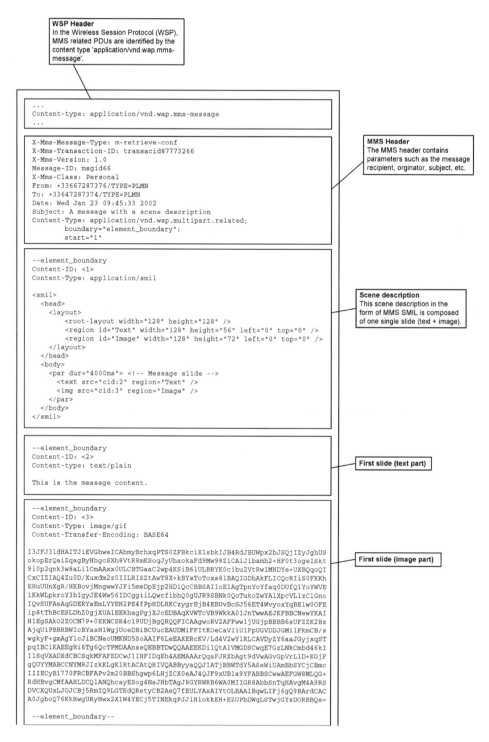

WSP Header
In the Wireless Session Protocol (WSP), MMS related PDUs are identified by the content type 'application/vnd.wap.mms-message'.

```
...
Content-type: application/vnd.wap.mms-message
...
```

MMS Header
The MMS header contains parameters such as the message recipient, orginator, subject, etc.

```
X-Mms-Message-Type: m-retrieve-conf
X-Mms-Transaction-ID: transacid87773266
X-Mms-Version: 1.0
Message-ID: msgid66
X-Mms-Class: Personal
From: +33667287376/TYPE=PLMN
To: +33647287374/TYPE=PLMN
Date: Wed Jan 23 09:45:33 2002
Subject: A message with a scene description
Content-Type: application/vnd.wap.multipart.related;
         boundary="element_boundary";
         start="1"
```

Scene description
This scene description in the form of MMS SMIL is composed of one single slide (text + image).

```
--element_boundary
Content-ID: <1>
Content-Type: application/smil

<smil>
  <head>
    <layout>
        <root-layout width="128" height="128" />
        <region id="Text" width="128" height="56" left="0" top="0" />
        <region id="Image" width="128" height="72" left="0" top="0" />
    </layout>
  </head>
  <body>
    <par dur="4000ms"> <!-- Message slide -->
      <text src="cid:2" region="Text" />
      <img src="cid:3" region="Image" />
    </par>
  </body>
</smil>
```

First slide (text part)

```
--element_boundary
Content-ID: <2>
Content-type: text/plain

This is the message content.
```

First slide (image part)

```
--element_boundary
Content-ID: <3>
Content-Type: image/gif
Content-Transfer-Encoding: BASE64
```

```
I3JFJ3ldHAITJiEVGhweICAhmyBrhxqPTS0ZFBkciX1sbkIJB4RdJBUWpx2bJSQjIZyJghUS
okopErQaiZqagByHhgcSXh8VtR8mKSoqJyUhaokaFd9Mw98ZiCAiJibamh2+HF0t3ogeISkt
9i0p2qnk3w8aLilCmAAxx0ULCBTGaaC2wp4KSiB6lULBRYK0cibu2VtRwlMHDYe+UXBQqoQI
CxCIZIAQ4Zu0D/Xuxdm2z0IILRISZtAwT8X+kBYsToToxs8lBAQIGDhAkFLICQcRIiS0FKKh
EHuUUnXgR/HKBovjMngwwYJFi5meDpEjp2HDiQcCBBSAIIoE1AgTpnYoYfaq0DUfQ1YoYWVE
1KkWLpkroY3b1gyJE4Ww56IDCggiiLQwcfibhQ0gUJR9SBNk0QoTukoZwYAlXpcVL1zClGno
IQvSUFAeAgGDERYaEmLYYEH2PE4fPpHDLRKCzygrEjB4EBUvBcSJ56ET4WvyoxYqBElw0OFE
ip8tThBcESLDhZ0gjXUAlEEkhagPgjXJcEDBAqXVWTcVB9WkkA0IJnTwwAEJKFBBCNewYKAI
H1EgSAkO2ZOCN7P+0EKNCSR4c19UDjBgQRQQFICAAgwoRV2AFFwwljUGjpBBBB6sUFZZK2Bz
AjqUiPBBRBWIcBYasH1WgjUoeDBiBCUucEAUDMiFFItKOeCaVIiUIFpUGVDDJGMilFkmCB/s
wgkyF+gmAgYlcJiBCNeoUMKND5SoAAIF6LeEAXERcKV/Ld4V2wY1RLCAVDyZY6aaJGyjxqST
pqIBCiKAEEgHi6Tg6QcTPMDAAnseQEBBTDwQQAAEEKDi1QtA1VMGDSCwqE7GzLNkCmbd46kI
I1SqVXADXdCBCSqKMFAFEDCwJ1IHFIDqEh4AEMAAArQqaFJRXbAgt9dVwAGvGpVrL1D+KGjF
gQUYYMABCCNYM8JIzkKLgK1RtACAtQHIVQABByyaQQJIATjBBWTdY5ASeWiUAmBbSYCjCBmc
IIIECyB1770FRCBFAPv2m20BBShgwp6LHjZCX0eAJ4QJF9xUB1a9YFABBSCwwAEFGW8MLQG+
RdHBvgCMfAAHLDCQ1ANQhcayESog4NsJHbTAgJ8GYRWRB6WA0MIIGR8AbbSnTqHAvgM4A9RS
DVCXQUxLJOJCBj5RmIQ9LGTKdQRetyCB2AeQ7fEULYAsAIYtOLBAA1BqwLIFj6gQ9BArdCAC
A0JgboQ76KkHwgURyHwx2X1W4YECj5TINEkqPJJiHiokkEH+EZUPbDWqLSTwjGYsDORBBQs=
```

```
--element_boundary--
```

Figure 6.42 MM1 message retrieval response/example

Table 6.30 MM1 message retrieval acknowledgement

Stage 2 Information elements	Description	3GPP status		Stage 3 WAP implementation (MIME parameter)	WAP status	
		Rel-4	Rel-5		MMS 1.0	MMS 1.1
Report allowed	This parameter indicates whether or not the message recipient allows a delivery report to be sent back to the message originator	○	○	X-MMS-Report-Allowed Values: Yes (default) No	○	○

Depending on MMSC implementations, this transaction identification may also be the one given to the corresponding notification request/response.

6.30.4 Message Forwarding

The recipient MMS user agent may have the possibility of forwarding a message without prior message retrieval. This feature is defined in 3GPP technical specifications from release 4 and is therefore only available from WAP MMS 1.1. The forward request can be submitted only if the corresponding message notification has first been received by the recipient MMS user agent. The forward request contains the message reference specified as part of the corresponding message notification. The transaction flow for the forward of a message is presented in Figure 6.43.

The forward request `MM1_forward.REQ` (also known as the `M-Forward.req` request

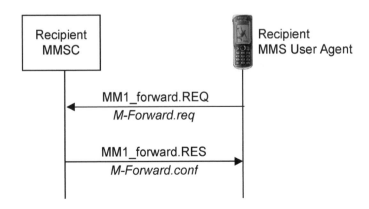

Figure 6.43 MM1 message forward

Table 6.31 MM1 message foward request

Stage 2 Information elements	Description	3GPP status		Stage 3 WAP implementation (MIME parameter)	WAP status	
		Rel-4	Rel-5		MMS 1.0	MMS 1.1
Recipient address	Address of the recipient(s) of the forwarded message	●	●	To, Cc, Bcc Types: string	–	○
Forwarding address	Address of the forwarding MMS user agent	○	○	From Type: string	–	●
Date and time	Date and time of the forwarding of the message	○	○	Date Value: date	–	○
Time of expiry	The message time of expiry	○	○	X-MMS-Expiry Type: date	–	○
Earliest delivery time	The message earliest desired time of delivery	○	○	X-MMS-Delivery-Time Type: date	–	○
Delivery report	Request for delivery report for the forwarded message	○	○	X-MMS-Delivery-Report Values: Yes (default) No	–	○
Read reply	Request for read-reply report for the forwarded message	○	○	X-MMS-Read-Reply Values: Yes No	–	○
Message reference	Reference of the message to be forwarded (the one specified in the message notification)	●	●	X-MMS-Content-Location Type: string (URI)	–	●

Table 6.32 MM1 message forward response

Stage 2 Information elements	Description	3GPP status		Stage 3 WAP implementation (MIME parameter)	WAP status	
		Rel-4	Rel-5		MMS 1.0	MMS 1.1
Request status	Status of the forward request	●	●	X-MMS-Response-status Values: Ok Error codes defined in Tables 6.21 and 6.22	–	●
Request status text	Textual description of the status	○	○	X-MMS-Response-text Type: string	–	○
Message identification	Identification of the message to be forwarded	●	●	Message-ID[a] Type: string	–	○

[a] In the WAP MMS 1.1 technical realization, the message identification is always provided if the forward request is accepted by the MMSC.

in the WAP technical realization) is associated with the information elements/parameters listed in Table 6.31.

Upon receipt of the forward request, the recipient MMSC responds to the recipient MMS user agent with the MM1_forward.RES response (also known as the M-Forward.conf confirmation in the WAP technical realization). The forward response is composed of the information elements/parameters listed in Table 6.32.

Values that can be assigned to the X-MMS-Response-status parameter are the same as the ones that can be assigned to the response of the submission request (Ok or error codes in Tables 6.21 and 6.22).

During the message forwarding process, the MMSC overwrites the value assigned to the From field with the address of the forwarding MMS user agent and adds a pair of 'previously sent by'/'previously sent date and time' information elements to the message. Values assigned to the set of 'previously sent' information elements relate to all MMS user agents which submitted or forwarded the message before the forward request (values that were assigned to the From and Date fields before update during the forwarding process).

6.30.5 Delivery Report

When submitting a message, an originator MMS user agent can request a delivery report to be generated upon message delivery or message deletion. The recipient MMS user agent can deny the generation of a delivery report back to the originator MMS user agent. The report is generated by the recipient MMSC and forwarded to the originator MMSC. The originator MMSC then delivers it to the originator MMS user agent. The delivery report indicates whether or not the multimedia message was successfully delivered to the recipient MMS

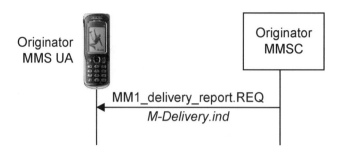

Figure 6.44   MM1 delivery report

user agent. The transaction flow in Figure 6.44 shows the interactions between the originator
MMS user agent and the originator MMSC for the delivery of the report over the MM1
interface.

The delivery report request `MM1_delivery_report.REQ` (also known as the
`M-Delivery.ind` indication in the WAP technical realizations) is composed of the information elements/parameters listed in Table 6.33.

Table 6.33   MM1 delivery report request

Stage 2 Information elements	Description	3GPP status		Stage 3 WAP implementation (MIME parameter)	WAP status	
		Rel-4	Rel-5		MMS 1.0	MMS 1.1
Message identification	Identification of the original message. This identification was provided by the originator MMSC to the originator MMS UA during message submission	●	●	Message-ID Type: string	●	●
Recipient address	Address of the original message recipient	●	●	To Type: string	●	●
Event date	Date and time when the message was handled (retrieved, expired, rejected, etc.)	●	●	Date Type: date	●	●
Message status	Status of the message	●	●	X-MMS-Status Values: Expired Retrieved Rejected Indeterminate Forwarded	●	●

Table 6.34 MM1 delivery report status codes

Status code	Status code	Description
0x80 (128)	Expired	This status code indicates that the multimedia message validity has expired before the message could be retrieved by the message recipient. This means that the multimedia message has been deleted and will not be delivered to the recipient MMS user agent
0x81 (129)	Retrieved	This status code means that the associated multimedia message has successfully been retrieved by the recipient MMS user agent
0x82 (130)	Rejected	This status code means that the MMS user agent is not willing to retrieve the message. In other words, the message has been rejected by the message recipient
0x85 (133)	Indeterminate	This status code indicates that it is unknown whether or not the associated multimedia message has been delivered to the recipient MMS user agent. This may happen if the associated multimedia message is transferred to a domain where the concept of delivery report is not supported (e.g. the Internet domain). This status code is supported from WAP MMS 1.1
0x86 (134)	Forwarded	This status code indicates that the associated message has been forwarded. This status code is supported from WAP MMS 1.1

The X-MMS-Status parameter of the delivery report request indicates what the status of the associated multimedia message is. The values listed in Table 6.34 can be assigned to this parameter.

6.30.6 Read-reply Report

When submitting a message, the originator MMS user agent can request to be notified when the message is deleted or read by the message recipient. In this situation, the recipient MMS user agent generates a report upon reading of the message by the message recipient. The recipient MMS user agent submits the read-reply report to the recipient MMSC over the MM1 interface (see (1) in the transaction flow shown in Figure 6.45). The recipient MMSC

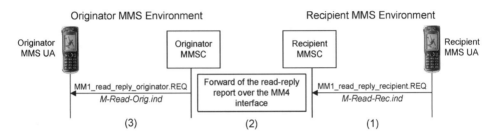

Figure 6.45 MM1 read-reply report

Table 6.35 MM1 read-reply report request (1)

Stage 2 Information elements	Description	3GPP status Rel-4	3GPP status Rel-5	Stage 3 WAP implementation (MIME parameter)	WAP status MMS 1.0	WAP status MMS 1.1
Recipient address	Address of the message recipient	●	●	`To` Type: string	–	●
Originator address	Address of the message originator	●	●	`From` Type: string	–	●
Message identification	Message identification	●	●	`Message-ID` Type: string	–	●
Date and time	Date and time when the message was handled by the recipient MMS user agent	○	○	`Date` Type: date	–	○
Status	Status of the message	●	●	`X-MMS-Read-Status` Values: `Read` `Deleted without being read`	–	●

transfers the report to the originator MMSC over the MM4 interface (see (2) in Figure 6.45). Finally, the originator MMSC delivers the read-reply report to the originator MMS user agent (see (3) in Figure 6.45).

In the WAP MMS 1.0 technical realization, the read-reply report is generated by the recipient MMS user agent in the form of an ordinary multimedia message. In the WAP MMS 1.1 technical realization, read-reply reports may be generated in the same way or using pairs of dedicated MMS PDUs over the MM1 interface (indications `M-Read-Rec.ind` and `M-Read-Orig.ind`) as shown in Figure 6.45.

The read-reply report request `MM1_read_reply_recipient.REQ` in the recipient MMS environment (also known as the `M-Read-Rec.ind` in the WAP technical realizations) is composed of the information elements/parameters listed in Table 6.35.

In the originator MMS environment, the originator MMSC delivers the read-reply report to the originator MMS user agent with the `MM1_read_reply_originator.REQ` request (also known as the `M-Read-Orig.ind` in the WAP MMS 1.1 technical realization). The request is composed of the information elements/parameters listed in Table 6.36.

If a date and time is not provided by the recipient MMS user agent, the recipient MMSC provides this information (timestamp). This explains why the date and time information element is optional in the read-reply report conveyed in the recipient MMSE (3) and mandatory when conveyed in the originator MMSE.

The `X-MMS-Read-Status` information element indicates whether or not the message

Table 6.36   MM1 read-reply report request (3)

Stage 2 Information elements	Description	3GPP status		Stage 3 WAP implementation (MIME parameter)	WAP status	
		Rel-4	Rel-5		MMS 1.0	MMS 1.1
Recipient address	Address of the message recipient.	●	●	To Type: string	–	●
Originator address	Address of the message originator.	●	●	From Type: string	–	●
Message identification	Message identification.	●	●	Message-ID Type: string	–	●
Date and time	Date and time when the message was handled by the recipient MMS user agent.	●	●	Date Type: date	–	●
Status	Status of the message.	●	●	X-MMS-Read-Status Values: Read Deleted without being read	–	●

Table 6.37   MM1 read-reply report status codes

Binary	Read status	Description
0x80 (128)	Read	This status indicates that the message recipient has read the multimedia message
0x80 (129)	Deleted without being read	This status indicates that the multimedia message has been deleted without being read by the message recipient

has been read by the recipient MMS user agent. The values listed in Table 6.37 can be assigned to this information element.

6.30.7 Availability Matrix of PDU Parameters

Table 6.38 indicates the set of parameters composing each request/response for operations that can be invoked over the MM1 interface.

6.31 MM2 Interface MMS Relay–MMS Server

The MM2 interface is an MMSC internal interface. The interface is required if MMS Relay and MMS Server are provided as two separate entities (e.g. developed by two different

Table 6.38 MM1 PDU parameters[a]

Parameter	A	B	C	D	E	F	G	H	I	J	K	L	New in WAP MMS 1.1
Bcc	○										○		
Cc	○										○		
Content-Type	●			○							○		
Content	○			●									
Date	○			○					○	●			
From	●			●		○		●	●	●	●		
Message-ID	○	○		○					●	●		○	
Subject	○			○		○			●	●			
To				○				●					
X-MMS-Content-Location	○			○		●		●	●	●	○		
X-MMS-Delivery-Report	○			○		○					●		
X-MMS-Delivery-Time	○					●					○		
X-MMS-Expiry	○			○		●					○		
X-MMS-Message-Class						●							
X-MMS-Message-Size						●							
X-MMS-Message-Type	●	●	●	●	●	●	●	●	●	●	●	●	

Field	A	B	C	D	E	F	G	H	I	J	K	L
X-MMS-Previously-Sent-By											○	Yes
X-MMS-Previously-Sent-Date											○	Yes
X-MMS-Priority	○	○		○							○	
X-MMS-Read-Reply	○	○		○							○	
X-MMS-Read-Status									●	●		Yes
X-MMS-Reply-Charging	○	○		○							○	Yes
X-MMS-Reply-Charging-Deadline	○	○		○							●	Yes
X-MMS-Reply-Charging-ID	○	○		○							○	Yes
X-MMS-Reply-Charging-Size	○	○		○							●	Yes
X-MMS-Report-Allowed				○				○		○		
X-MMS-Response-Status		●									●	○
X-MMS-Response-Text		●									●	
X-MMS-Retrieve-Status				○								
X-MMS-Retrieve-Text				○								
X-MMS-Sender-Visibility	○											
X-MMS-Status						●		●				
X-MMS-Transaction-ID	●	●		●	●	●	●				●	●
X-MMS-Version	●	●		●	●	●	●				●	●

[a] A, M-Send.req; B, M-Send.conf; C, WSP/HTTP GET.req; D, M-Retrieve.conf; E, M-Acknowledge.ind; F, M-notification.ind; G, M-NotifyResp.ind; H, M-Delivery.ind; I, M-Read-Rec.ind; J, M-Read-Orig.ind; K, M-Forward.req; L, M-Forward.conf. ●, Mandatory; ○, Optional; C, Conditional; otherwise, not applicable.

manufacturers). However, available commercial implementations combine MMS Relay and MMS Server into a single entity, called MMSC. In this situation, the MM2 interface is implemented in a proprietary manner. At the time of writing, no technical realization of the MM2 interface has been specified by standardization organizations.

6.32 MM3 Interface MMSC–External Servers

The MM3 interface allows the MMSC to exchange messages with external servers such as Email servers and SMSCs. This interface is typically based on existing IP-based transport protocols (e.g. HTTP or SMTP).

When sending a message to an external messaging system, the MMSC converts the multimedia message into an appropriate format supported by the external messaging system. For instance, the exchange of a message between an MMSC and an Internet Email server can be performed by converting the multimedia message from its MM1 binary multipart representation (WAP MMS 1.0 and 1.1) into a text-based MIME representation for transfer over SMTP. In order to receive a message from an external messaging system, the MMSC converts incoming messages to the format supported by receiving MMS user agents.

Several mechanisms can be used for discovering incoming messages from external messaging servers. These mechanisms include:

- Forwarding of the multimedia message from the external messaging server to the MMSC.
- The external messaging server notifies the MMSC that a message is waiting for retrieval. In this configuration, it is the responsibility of either the MMSC or the recipient MMS user agent to explicitly retrieve the message.
- The MMSC can periodically poll for incoming messages in the external messaging server.

6.33 MM4 Interface MMSC–MMSC

The MM4 interface allows interactions between two MMSCs. Such interactions are required in the situation where multimedia messages and associated reports are exchanged between

Table 6.39 Operations over the MM4 interface

Operation	Description	3GPP Status	
		Rel-4	Rel-5
Routing forward a message	The routing forward of a multimedia message from the originator MMSC to the recipient MMSC	■	■
Routing forward a delivery report	The routing forward of a delivery report from the recipient MMSC to the originator MMSC	■	■
Routing forward a read-reply report	The routing forward of a read-reply from the recipient MMSC to the originator MMSC	■	■

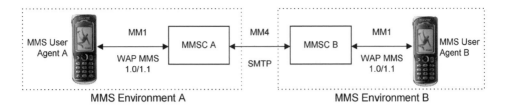

Figure 6.46   Interworking between two distinct MMS environments

subscribers belonging to distinct MMS environments. The operations listed in Table 6.39 can be invoked over the MM4 interface.

At the time of writing this book, the only standardized technical realization available for the MM4 interface is the one defined by the 3GPP in [3GPP-23.140] release 5. In this technical realization, operations specified for the MM4 interface are performed over the Simple Mail Transfer Protocol (SMTP) as shown in Figure 6.46.

As for the MM1 interface, each operation in the MM4 interface is usually composed of a request and a response. The 3GPP naming convention for naming requests and responses is used in this book.

6.33.1 Introduction to SMTP

Simple Mail Transfer Protocol is a basic protocol for exchanging messages between Mail User Agents (MUAs). Mail User Agent is the name given to applications in charge of managing Email messages exchanged over the Internet. In the Internet domain, SMTP has become the de facto Email transfer protocol for exchanging messages. MUAs have similar responsibilities as those of MMS user agents in an MMS environment. Specifications of SMTP have been published by the IETF in [RFC-821] and [RFC-821bis] (also known as STD 10).

The SMTP model is based on an interconnection of Message Transfer Agents (MTAs). In a typical SMTP transaction, a Mail Transfer Agent is the sender-SMTP if it originates the SMTP commands, or the receiver-SMTP if it handles the received SMTP commands. An MTA can usually play both roles, the sender-SMTP client and the receiver-SMTP server. SMTP defines how senders and receivers can initiate a transfer session, can transfer messages(s) over a session and can tear down an open session. Note that how the message is physically transferred from the sender to the receiver is not defined as part of the SMTP specifications. SMTP only defines the set of commands, and corresponding responses, for controlling the transfer of messages over sessions. SMTP is a stateful protocol, meaning that sender and receiver involved in operations over a session, maintain a current context for a session. Consequently, commands requested over SMTP have different results according to the session state.

In the context of MMS, SMTP is used to transfer multimedia messages, delivery reports and read-reply reports between MMSCs. For this purpose, originator and recipient MMSCs are MTAs playing the roles of sender-SMTP and receiver-SMTP, respectively.

An SMTP command is a four letter command such as HELO or DATA. The response to such a command consists of a three digit code followed by some optional human readable text. The status code is formatted according to the convention shown in Figure 6.47.

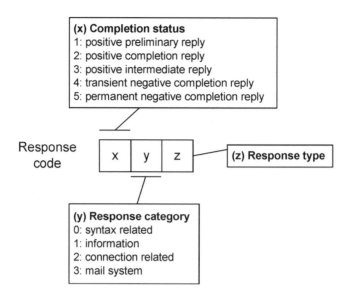

Figure 6.47   Structure of an SMTP reply

The minimum set of SMTP commands that should be supported by MMSCs is the following:

- HELO: this command (abbreviation for 'Hello') is used for initiating a session.
- QUIT: this command is used to tear down a session.
- MAIL: this command tells the SMTP-receiver that a message transfer is starting and that all state tables and buffers should be re-initialized. This command has one parameter named FROM which identifies the address of the message originator (if the message originator did not request address hiding).
- RCPT: this command (abbreviation for 'Recipient') has one parameter, named TO, which identifies the address for one of the message recipients. If the message is sent to several recipients, then this command may be executed several times for one single message transfer.
- DATA: this command is used for transferring the message itself.

MSISDN and RFC 822 addresses can be used for addressing recipients in an MMS environment. With SMTP, only RFC 822 addresses are supported for routing purpose. Consequently, MSISDN addresses specified as part of MAIL and RCPT parameters, need to be adapted from original and recipient MMS user agent addresses. For instance, the MSISDN address + 3367287376 for an originator MMS user agent is converted to a Fully Qualified Domain Name (FQDN) RFC 822 address in the form: (Figure 6.48).

For sending a message to an external MMS environment, the originator MMSC needs to resolve the recipient MMSC domain name to an IP address. If the recipient address is an RFC 822 address, then the originator MMSC can obtain the recipient MMSC IP address by interrogating a Domain Name Server (DNS) with the recipient address. If the recipient address is in the form of an MSISDN address, then two resolution methods have been identified by the 3GPP in [3GPP-23.140] release 5:

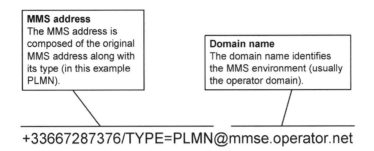

Figure 6.48   SMTP address conversion

- *DNS-ENUM-based method:* the IETF has defined a method for identifying a device associated with an MSISDN address with a Domain Name Server (DNS) [RFC-2916]. This method, known as DNS-ENUM, allows DNS records (including the associated RFC 822 address) to be retrieved with the recipient MSISDN address. With the RFC 822 address, the originator MMSC determines the recipient MMSC IP address with the method described above.
- *IMSI-based method:* this method for identifying the recipient MMSC IP address complies with the Mobile Number Portability (MNP) requirements. The MMSC interrogates the recipient HLR in order to get the IMSI associated with the recipient MSISDN (via the MM5 interface). From the IMSI, the originator MMSC extracts the Mobile Network Code (MNC) and the Mobile Country Code (MCC). With the MNC and MCC, the originator MMSC obtains the recipient MMSC Fully Qualified Domain Name (FQDN) by interrogating a local database (look-up table). With the MMSC FQDN, the originator MMSC retrieves the MMSC IP address by interrogating a DNS.

MMSCs can support the following SMTP extensions:

- SMTP service extension for message size declaration.
- SMTP service extension for 8-bit MIME transport.

The support of additional commands is also possible but their use has not yet been covered in 3GPP technical specifications. An example for transferring a message over SMTP between

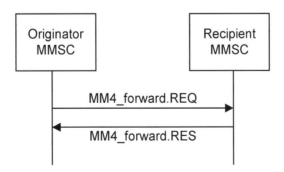

Figure 6.49   MM4 message forward

two distinct MMS environments is provided in Section 6.33.5. The following sections present transaction flows for message forward, delivery report forward and read-reply report forward.

6.33.2 Routing Forward a Message

The request for forwarding a message (`MM4_forward.REQ` request) over the MM4 interface allows a multimedia message to be transferred between two MMS environments. If requested by the originator MMSC, the recipient MMSC acknowledges the forward request with a forward response (`MM4_forward.RES` response). The transaction flow for the transfer of a message over the MM4 interface is shown in Figure 6.49.

The `MM4_forward.REQ` request is composed of the information elements/parameters listed in Table 6.40. If requested by the originator MMSC, the recipient MMSC acknowledges the forward request with a forward response (`MM4_forward.RES` response). The response is composed of the information elements/parameters listed in Table 6.41.

Unlike forward requests, the addressing of forward responses is related to neither the message originator nor the message recipient. Instead, the addressing of a forward response is related to special system addresses. The value to be assigned to the `To` parameter is the value assigned to the `X-MMS-Originator-System` parameter of the corresponding forward request (usually a special system address identifying the originator MMSC). The value to be assigned to the `Sender` parameter is a special address identifying the recipient MMSC. It is suggested that special system addresses should be formatted in the form:

system-user@mms-relay-host.operatorX.net

If the forward request has been processed without error by the recipient MMSC, the following value is assigned to the request status code information element (`X-Mms-Request-Status-Code` parameter).

- `Ok`: this status code indicates that the corresponding request has been processed without errors.

If errors occurred during the processing of the forward request, the codes listed in Table 6.42 can be assigned to the request status code information element.

6.33.3 Routing Forward a Delivery Report

The request for forwarding a delivery report (`MM4_delivery_report.REQ` request) over the MM4 interface allows the transfer of a delivery report between two MMS environments. If requested by the recipient MMSC, the originator MMSC acknowledges the forward request with a forward response (`MM4_delivery_report.RES` response). The transaction flow for the transfer of a delivery report over the MM4 interface is shown in Figure 6.50. The `MM4_delivery_report.REQ` request is composed of the information elements/ parameters listed in Table 6.43. The `MM4_delivery_report.RES` response is composed of the information elements/parameters listed in Table 6.44.

For the forward response, the value to be assigned to the `To` parameter is the value assigned to the `Sender` parameter of the corresponding forward request. The value to be assigned to the `Sender` parameter is the system address of the recipient MMSC.

Table 6.40 MM4 message forward request

Stage 2 IE	Description	3GPP status		Stage 3 STD11 Header
		Rel-4	Rel-5	
3GPP MMS version	MMS version of the MMSC	●	●	X-Mms-3GPP-MMS-Version Type: string Example: 5.2.0
Message type	Type of MM4 operation (request for the forward of a message)	●	●	X-Mms-Message-Type Type: string Value: MM4_forward.REQ
Transaction identification	Identification of the forward transaction	●	●	X-Mms-Transaction-ID Type: string
Message identification	Identification of the multimedia message being forwarded	●	●	X-Mms-Message-ID Type: string
Recipient address	Address(es) of the recipient(s) of the original message	●	●	To, Cc Types: string
Sender address	Address of the sender of the original message	●	●	From Type: string
Message subject	Message subject[a]	C	C	Subject Type: string
Message class	Class of the message[a]	C	C	X-Mss-Message-Class Type: string Values: Personal Advertisement Informational Auto
Message date and time	Date and time the original message was handled (retrieved, expired, rejected, etc.)	●	●	Date Type: date
Time of expiry	Time of expiry of the message[a]	C	C	X-Mms-Expiry Type: date or duration

Table 6.40 (*continued*)

Stage 2 IE	Description	3GPP status		Stage 3 STD11 Header
		Rel-4	Rel-5	
Delivery report	Whether or not a delivery report is requested[a]	C	C	X-Mms-Delivery-Report Values: Yes No
Read-reply report	Whether or not a read-reply report is requested[a]	C	C	X-Mms-Read-Reply Values: Yes No
Priority	Priority of the message being forwarded[a]	C	C	X-Mms-Priority Values: Low Normal High
Sender visibility	Whether or not the sender requested sender details to be hidden from recipients[a]	C	C	X-Mms-Sender-Visibility Values: Hide Show
Forward counter	Counter indicating how many times the message has been forwarded[a]	C	C	X-Mms-Forward-Counter Type: integer
Previously sent by	Address(es) of MMS user agents that have handled (submitted or forwarded) the message prior to the manipulation by the MMS user agent whose address is assigned to the sender information element	O	O	X-Mms-Previously-Sent-By Type: string with index e.g.: 1, armel@armorepro.com 2, gwenael@le-bodic.net
Previously sent date and time	Date and time when the message was handled	O	O	X-Mms-Previously-Sent-Date-and-Time

Information element	Description			Corresponding field
				Type: date with index e.g.: 1, Mon Jan 21 09:45:33 2002 2, Wed Jan 23 18:06:21 2002
Request for an acknowledgement	Whether or not an acknowledgement for the forward request is requested	○	○	X-Mms-Ack-Request Values: Yes No
No corresponding information element	Originator address as determined by the SMTP MAIL FROM command	●	●	Sender Type: string
No corresponding information element	System address to which the requested forward response should be sent[b]	C	C	X-Mms-Originator-System Type: string
No corresponding information element	Each SMTP request/response has a unique reference assigned to the Message-ID parameter	●	●	Message-ID Type: string
Content-type	Message content type	●	●	Content-Type Type: string
Content	Message content[a]	C	C	Message body

[a] *Condition*: available only if specified by the originator MMS user agent.
[b] *Condition*: required if a forward response is requested.

Table 6.41 MM4 message forward response

Stage 2 IE	Description	3GPP status		Stage 3 STD11 Header
		Rel-4	Rel-5	
3GPP MMS version	MMS version of the MMSC	●	●	X-Mms-3GPP-MMS-Version Type: string Example: 5.2.0
Message type	Type of MM4 operation (response for the forward of a message)	●	●	X-Mms-Message-Type Value: MM4_forward.RES
Transaction identification	Identification of the forward transaction	●	●	X-Mms-Transaction-ID Type: string
Multimedia message identification	Identification of the multimedia message being forwarded	●	●	X-Mms-Message-ID Type: string
Request status code	Status code of the request to forward the message	●	●	X-Mms-Request-Status-Code Values: Ok Error codes in Table 6.42
Status text	Optional status text	○	○	X-Mms-Status-Text Type: string
No corresponding information element	System address	●	●	Sender Type: string
No corresponding information element	Address of the recipient MMSC System address	●	●	To Type: string
No corresponding information element	Address of the originator MMSC	●	●	
No corresponding information element	Each SMTP request/response has a unique reference assigned to the Message-ID parameter	●	●	Message-ID Type: string
No corresponding information element	Date provided by the recipient MMSC	●	●	Date Type: date

Table 6.42 MM4 message forward response/error codes

Status code	Description
`Error-unspecified`	An unspecified error occurred during the processing of the request
`Error-service-denied`	The request was rejected due to a failure during service authentication or authorization
`Error-message-format-corrupt`	The content of the request is badly formatted
`Error-sending-address-unresolved`	The recipient MMS environment was unable to resolve one of the recipient addresses (values assigned to `From`, `To` and `Cc` parameters)
`Error-network-problem`	The recipient MMSC was unable to process the request due to capacity overload
`Error-content-not-accepted`	The request was not accepted because of issues with the associated message size or format, or because of copyright issues
`Error-unsupported-message`	The recipient MMSC does not support the request type
`Error-message-not-found`	This status code is now obsolete

6.33.4 Routing Forward a Read-reply Report

The request for forwarding a read-reply report (`MM4_read_reply_report.REQ` request) over the MM4 interface allows a read-reply report to be transferred between two distinct MMS environments. If requested by the recipient MMSC, the originator MMSC acknowledges the forward request with a forward response (`MM4_read_reply_report. RES` response). The transaction flow for the transfer of a read-reply report over the MM4 interface is shown in Figure 6.51. The `MM4_read_reply_report.REQ` request is composed of the information elements/parameters listed in Table 6.45. The `MM4_read_ reply_report.RES` response is composed of the information elements/parameters listed in Table 6.46.

For the response, the value to be assigned to the `To` parameter is the value assigned to the

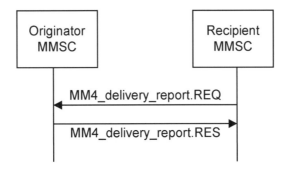

Figure 6.50 MM4 delivery report

Table 6.43 MM4 delivery report forward request

Stage 2 IE	Description	3GPP status		Stage 3 STD11 Header
		Rel-4	Rel-5	
3GPP MMS version	MMS version of the MMSC	●	●	`X-Mms-3GPP-MMS-Version` Type: string Example: `5.2.0`
Message type	Type of MM4 operation (request for the forward of a delivery report)	●	●	`X-Mms-Message-Type` Value: `MM4_delivery_report.REQ`
Transaction identification	Identification of the forward transaction	●	●	`X-Mms-Transaction-ID` Type: string
Multimedia message identification	Identification of the multimedia message to which the delivery report corresponds	●	●	`X-Mms-Message-ID` Type: string
Recipient address	Address of the recipient of the original message	●	●	`From` Type: string
Sender address	Address of the sender of the original message	●	●	`To` Type: string
Message date and time	Date and time the original message was handled (retrieved, expired, rejected, etc.)	●	●	`Date` Type: date
Request for an acknowledgement	Whether or not an acknowledgement of the forward request is requested	○	○	`X-Mms-Ack-Request` Values:

				Yes No
Message status code	Status of the corresponding multimedia message	●	●	X-Mms-MM-Status-Code Type: string Values: Expired Retrieved Rejected Deferred Indeterminate Forwarded Unrecognised
Message status text	Text corresponding to the status code	○	○	X-MM-Status-Text Type: string
No corresponding information element	System address Address to which the requested response should be sent	●	●	Sender Type: string
No corresponding information element	Each SMTP request/response has a unique reference assigned to the Message-ID parameter	●	●	Message-ID Type: string

Table 6.44 MM4 delivery report forward response

Stage 2 IE	Description	3GPP status Rel-4	3GPP status Rel-5	Stage 3 STD11 Header
3GPP MMS version	MMS version of the MMSC	●	●	X-Mms-3GPP-MMS-Version Type: string Example: 5.2.0
Message type	Type of MM4 operation (response for the forward of a delivery report)	●	●	X-Mms-Message-Type Type: string Values: MM4_delivery_report.RES
Transaction identification	Identification of the forward transaction	●	●	X-Mms-Transaction-ID Type: string
Multimedia message identification	Identification of the multimedia message being forwarded	●	●	X-Mms-Message-ID Type: string
Request status code	Status code of the request to forward the message	●	●	X-Mms-Request-Status-Code Values: Ok Error codes defined in Tables 6.42
Status text	Address of the sender of the original message	○	○	X-Mms-Status-Text Type: string
No corresponding information element	System address. Address of the recipient MMSC	●	●	Sender Type: string
No corresponding information element	System address. Address of the originator MMSC	●	●	To Type: string
No corresponding information element	Each SMTP request/response has a unique reference assigned to the Message-ID parameter	●	●	Message-ID Type: string
No corresponding information element	Date provided by the recipient MMSC	●	●	Date Type: string

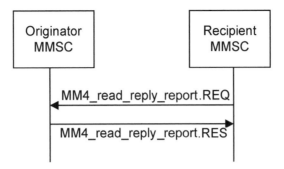

Figure 6.51   MM4 read-reply report

Sender parameter of the corresponding forward request. The value to be assigned to the Sender parameter is the system address of the recipient MMSC.

6.33.5 Example for Message Transfer with SMTP

Figure 6.52 shows the sequence of SMTP instructions required to open an SMTP session, transferring a message and tearing down the session. Note that values assigned to From, To and Cc fields are not used for routing purpose over SMTP. These values are transferred transparently over SMTP. Consequently, these values may be formatted as RFC 822 addresses or as MSISDN addresses. Values, used for routing purpose in SMTP, are those assigned to MAIL and RCPT fields.

6.33.6 Availability Matrix of PDU Parameters

Table 6.47 indicates the set of parameters composing each request/response for operations that can be invoked over the MM4 interface (3GPP releases 4 and 5).

6.34 MM5 Interface MMSC–HLR

The MM5 interface enables interactions between an MMSC and other network entities such as the HLR. Operations that can be invoked over the MM5 interface include:

- Interrogating the HLR to obtain routing information for the purpose of forwarding a message from one MMSC to another one over the MM4 interface.
- Determination of the recipient handset's location (e.g. if the subscriber is roaming).
- etc.

If the MM5 interface is present in the MMSE, then it is usually implemented on the basis of existing MAP operations. At the time of writing, no technical realization of the MM5 interface has been specified by standardization organizations.

Table 6.45 MM4 read-reply report forward request

Stage 2 Information elements	Description	3GPP status		Stage 3 STD11 Header
		Rel-4	Rel-5	
3GPP MMS version	MMS version of the MMSC	●	●	X-Mms-3GPP-MMS-Version Type: string Example: 5.2.0
Message type	Type of MM4 operation (request for the forward of a read-reply report)	●	●	X-Mms-Message-Type Type: string Value: M4_read_reply_report.REQ
Transaction identification	Identification of the forward transaction	●	●	X-Mms-Transaction-ID Type: string
Multimedia message identification	Identification of the multimedia message to which the read-reply report corresponds	●	●	X-Mms-Message-ID Type: string
Recipient address	Address of the recipient of the original message	●	●	From Type: string
Sender address	Address of the sender of the original message	●	●	To Type: string

Message date and time	Date and time the original message was handled (read or deleted)	●	●	Date Type: string
Request for an acknowledgement	Whether or not an acknowledgement of the forward request is requested	○	○	X-Mms-Ack-Request Values: Yes No
Read status	Status of the message	●	●	X-Mms-Read-Status-Code Type: string Values: Read Delete without being read
Status text	Text corresponding to the status code	○	○	X-MM-Status-Text Type: string
No corresponding information element	System address	●	●	Sender Type: string
No corresponding information element	Each SMTP request/response has a unique reference assigned to the Message-ID parameter	●	●	Message-ID Type: string

Table 6.46 MM4 read-reply report forward response

Stage 2 IE	Description	3GPP status Rel-4	3GPP status Rel-5	Stage 3 STD11 Header
3GPP MMS version	The MMS version of the MMSC	●	●	X-Mms-3GPP-MMS-Version Type: string Example: 5.2.0
Message type	The type of MM4 operation	●	●	X-Mms-Message-Type Type: string Value: M4_read_reply_report.RES
Transaction identification	The identification of the forward transaction	●	●	X-Mms-Transaction-ID Type: string
Request status code	The status code of the request to forward the message	●	●	X-Mms-Request-Status-Code Values Ok Error codes in Table 6.42
Status text	Address of the sender of the original message	○	○	X-Mms-Status-Text Type: string
No corresponding information element	System address	●	●	Sender
No corresponding information element	Address of the recipient MMSC	●	●	To Type: string
No corresponding information element	System address	●	●	Type: string
No corresponding information element	Address of the recipient MMSC	●	●	Message-ID Type: string
No corresponding information element	Each SMTP request/response has a unique reference assigned to the Message-ID parameter	●	●	Date Type: string
No corresponding information element	No corresponding information element. Value provided by the recipient MMSC	●	●	

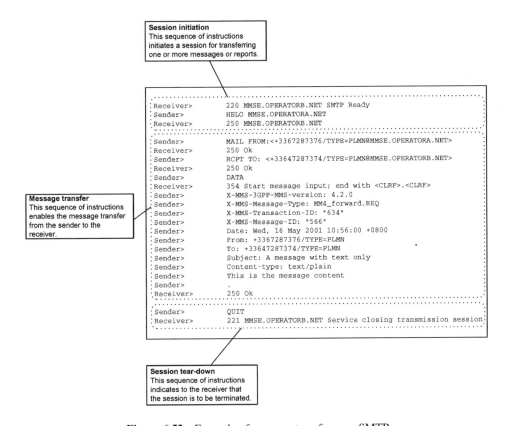

Figure 6.52   Example of message transfer over SMTP

6.35 MM6 Interface MMSC–User Databases

The MM6 interface allows interactions between the MMSC and user databases. Unfortunately, the MM6 interface has yet to be standardized. Consequently, this interface is not covered in this book.

6.36 MM7 Interface MMSC–VAS Applications

The MM7 interface enables interactions between Value Added Service (VAS) applications and an MMSC. Table 6.48 lists the operations that are available on the MM7 interface. At the time of writing, the only standardized technical realization for the MM7 interface is the one published by the 3GPP in [3GPP-23.140] release 5 (stages 2 and 3). This technical realization is based on the Simple Object Access Protocol (SOAP) with HTTP at the transport layer. Figure 6.53 shows a typical network configuration allowing a VAS application to interact with several MMS user agents. In this configuration, the VAS application and the MMSC play dual roles of sender and receiver of SOAP messages.

Table 6.47 MM4 PDU parameters[a]

Parameter	MM4_Forward. REQ	MM4_Forward. RES	MM4_Delivery_ Report.REQ	MM4_Delivery_ Report.RES	MM4_Read_Reply_ Report.REQ	MM4_Read_Reply_ Report.RES
Cc	O					
Content	C					
Content-type	●					
Date	●	●	●	●	●	●
From	●		●		●	
Message-ID	●	●	●	●	●	●
Sender	●	●	●	●	●	●
Subject	C					
To	O	●	●	●	●	●
X-MMS-3GPP-Version	●	●	●	●	●	●
X-MMS-Ack-Request	O		○		○	
X-MMS-Delivery-Report	C					
X-MMS-Expiry	C					
X-MMS-Forward-Counter	C					
X-MMS-Message-Class	C					
X-MMS-Message-ID	●	●	●	●	●	●
X-MMS-Message-Type	●	●	●	●	●	●
X-MMS-MM-Status-Code			●			
X-MMS-Originator-System	●					
X-MMS-Previously-Sent-By	O					
X-MMS-Previously-Sent-Date-and-Time	O					
X-MMS-Priority	C					
X-MMS-Read-Reply	C					
X-MMS-Read-Status					●	
X-MMS-Request-Status-Code		●		●		●
X-MMS-Sender-Visibility	C					
X-MMS-Status-Text		○	○	○	○	○
X-MMS-Transaction-ID	●	●	●	●	●	●

[a] ●, Mandatory; O, Optional; C, Conditional; otherwise, not applicable.

Table 6.48 Operations over the MM7 interface

Operation	Description	3GPP status Rel-5
Message submission	Submission of a message from a VAS application to the MMSC	■
Message delivery	Delivery of a message from an MMSC to a VAS application	■
Cancellation	Cancel the delivery of a previously submitted message	■
Replacement	Replace a previously submitted message with a new message	■
Delivery report	Delivery reporting from the MMSC to a VAS application	■
Read reply report	Delivery of a read-reply report from the MMSC to the VAS application	■
MMSC error	Error indication from the MMSC to the VAS application	■
VASP error	Error indication from the VAS application to the MMSC	■

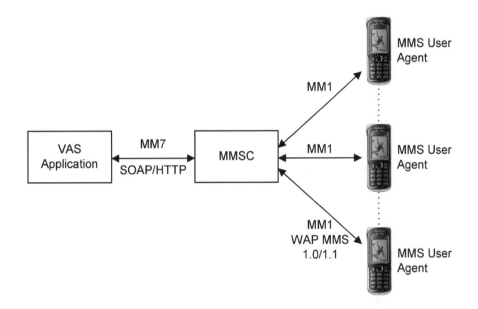

Figure 6.53 Network architecture with a VAS application

As for other interfaces, each operation invoked over the MM7 interface is composed of a request and a corresponding response. HTTP-level mechanisms[5] can be used in order to authenticate parties communicating over the MM7 interface. Additionally, messages over the MM7 interface can be transported over TLS to ensure confidentiality between communicating parties.

---

[5] For instance, the access authentication mechanism described in [RFC-2617] can be used for authenticating parties communicating over the MM7 interface.

The support of the MM7 interface is optional for the MMSC. However, if such an interface is supported, then message submission, message delivery, operations related to the provision of delivery reports and the management of errors are mandatory operations to be supported by the MMSC. The support of other operations such as message cancellation, replacement and operations related to the management of read-reply reports is optional for the MMSC.

The two addressing modes, MSISDN [ITU-E.164] and Email address [RFC-2822], can be used to identify entities communicating over the MM7 interface. Communicating entities include MMS user agents, VAS applications and the MMSC.

SOAP messages are structured according to SOAP technical specifications published by the W3C in [W3C-SOAP] and [W3C-SOAP-ATT]. In addition, the XML schema for formatting MMS-specific SOAP messages is published by the 3GPP at the following location: http://www.3gpp.org/ftp/Specs/archive/23_series/23.140/schema/REL-5-MM7-1-0.

6.36.1 Introduction to SOAP

SOAP is a lightweight protocol for the exchange of information in distributed environments such as the MMS environment. All SOAP messages are represented using XML. SOAP specifications consist of three distinct parts:

- *Envelope:* this part defines a framework for describing the content of a SOAP message and how to process it.
- *Set of encoding rules:* encoding rules are used for expressing instances of application-defined data types.
- *Convention for representing remote procedure calls:* this convention helps entities in a distributed environment to request services from each other in an interoperable manner.

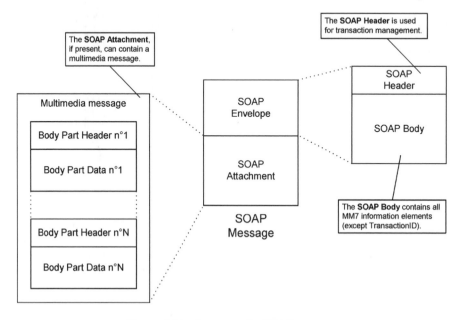

Figure 6.54 Structure of a SOAP message

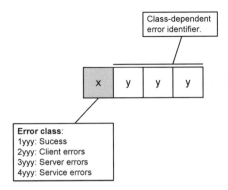

Figure 6.55   Classification of MM7 error codes

SOAP may be used over a variety of transport protocols. In the MMS environment, for the realization of the MM7 interface, SOAP is used over the HTTP transport protocol. With this configuration, MM7 request messages are transferred in HTTP POST requests whereas corresponding MM7 response messages are transferred as part of HTTP Response messages.

A SOAP message, represented using XML, consists of a SOAP envelope, a SOAP header, a SOAP body and an optional SOAP attachment as shown in Figure 6.54. For messages containing a SOAP envelope only, then the media type `text/xml` is used. If the SOAP message also contains an attachment, then the media type `multipart/related` is used and the SOAP envelope is identified with the `Start` field of the content type. Each part of the SOAP message has, at least, the two fields `Content-Type` and `Content-ID`.

The SOAP envelope is the first element to appear in HTTP POST requests and corresponding responses. The `SOAPAction` parameter is set to the 'Null string'. The MMSC or the VAS server is identified uniquely with a URI placed in the host header field of the HTTP POST method.

A request status is provided as part of the corresponding response. The status can be of three types:

- *Success or partial success:* this status indicates the successful or partial treatment of a request. This status is composed of three information elements identified by parameters `StatusCode` (numerical code), `StatusText` (human readable textual description) and `Details` (optional human readable detailed textual description). The classification of four-digit error identifiers to be assigned to the `StatusCode` is provided in Figure 6.55.
- *Processing error:* this status indicates that a fault occurred while parsing SOAP elements. Possible errors include `faultcode`, `faultstring` and `detail` elements as defined in [W3C-SOAP].
- *Network error:* this status indicates that an error occurred at the HTTP level.

6.36.2 Message Submission

In the context of value added services, message submission refers to the submission of a message from an originator VAS application to an MMSC. The message is addressed to a single recipient, to multiple recipients or to a distribution list managed by the MMSC. If the

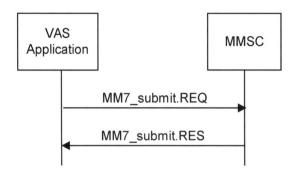

Figure 6.56   MM7 message submission

MMSC accepts the submission request, then the MMSC sends back a positive response. This indicates that the submission request is accepted but does not indicate that the message has been successfully delivered to the message recipients. The transaction flow in Figure 6.56 shows interactions between the VAS application and the MMSC for the submission of a multimedia message over the MM7 interface.

In tables describing the content of requests and responses for the MM7 interface, the column named 'location' indicates whether the corresponding information element is placed in the SOAP header ('H'), SOAP body ('B') or SOAP attachment ('A').

The submission request `MM7_submit.REQ` is composed of the information elements/ parameters listed in Table 6.49. The MMSC acknowledges the submission request with the `MM7_submit.RES` response. This response is composed of the information elements/parameters listed in Table 6.50.

Figure 6.57 shows an example of submission request embedded in an HTTP POST request, whereas Figure 6.58 shows the corresponding response.

6.36.3 Message Delivery

In the context of value added services, message delivery refers to the delivery of a multimedia message from the MMSC to a VAS application. The MMSC may deliver the message to the VAS application along with a linked identification. This identification can be conveyed as part of a subsequent message submission from the VAS application to indicate that the submitted message is related to a previously delivered message. The transaction flow in Figure 6.59 shows interactions between the MMSC and the VAS application for the delivery of a multimedia message over the MM7 interface.

The submission request `MM7_deliver.REQ` is composed of the information elements/ parameters listed in Table 6.51.

The VAS application acknowledges the delivery request with the `MM7_deliver.RES` response. This response is composed of the information elements/parameters listed in Table 6.52.

6.36.4 Message Cancellation

In the context of value added services, message cancellation refers to the possibility for a

Table 6.49 MM7 message submission request

Stage 2 IE	Status	Location	Description	Stage 3 Name in the SOAP message
Transaction identification	●	H	Identification of the submission transaction	TransactionID
Message type	●	B	Type of MM7 operation (message submission request)	MessageType Value: SubmitReq Root element of the SOAP body
MM7 version	●	B	Version of the MM7 interface supported by the VAS application	MM7Version e.g. 5.3.0
VASP identification	○	B	Identification of the VAS provider	VASPID
VAS identification	○	B	Identification of the values added service	VASID
Sender address	○	B	Address of the message originator	SenderAddress
Recipient address	●	B	Address of the message recipient. Multiple recipient addresses can be specified	Recipients
Service code	○	B	Service code for charging purpose	ServiceCode
Linked identification	○	B	Relation with another message	LinkedID
Message class	○	B	Message class	MessageClass Values: Informational Advertisement Auto
Date and time	○	B	Date and time of message submission	TimeStamp
Time of expiry	○	B	Time of expiry for the message	ExpiryDate
Earliest delivery of time	○	B	Earliest time of delivery for the message	EarliestDeliveryTime
Delivery report	○	B	Request for a delivery report	DeliveryReport Values: True False
Read reply	○	B	Request for a read-reply report	ReadReply Values: True False

Table 6.49 (*continued*)

Stage 2 IE	Status	Location	Description	Stage 3 Name in the SOAP message
Reply charging	O	B	Request for reply-charging	ReplyCharging No value. Presence implies that reply charging is enabled
Reply deadline	O	B	Reply charging deadline	ReplyDeadline Absolute or relative date format
Reply charging size	O	B	Reply charging message size	ReplyChargingSize
Priority	O	B	Message priority	Priority Values: High Normal Low
Subject	O	B	Message subject	Subject
Adaptations	O	B	Whether or not content adaptation is allowed (default value is True)	allowAdaptations Values: True False
Message distribution indicator	O	B	Whether or not the message can be redistributed freely	DistributionIndicator Values: True False
Charged party	O	B	Indication of the party(ies) that should be charged for the cost of handling the message	ChargedParty Values: Sender Recipient Both Neither
Content type	●	A	Message content type	Content-type
Content	O	B	Message content	Content

Table 6.50   MM7 message submission response

Stage 2 IE	Status	Location	Description	Stage 3 Name in the SOAP message
Transaction identification	●	H	Identification of the submission transaction	`TransactionID`
Message type	●	B	Type of MM7 operation (message submission response)	`MessageType` Value: `SubmitRsp` Root element of the SOAP body
MM7 version	●	B	Version of the MM7 interface supported by the MMSC	`MM7Version` e.g. `5.3.0`
Message identification	C	B	Message identification generated by the MMSC[a]	`MessageID`
Request status	●	B	Status of the request completion.	`StatusCode`
Request status text	○	B	Textual description of the status of the request completion.	`StatusText`
Request status details	○	B	Human readable detailed textual description of the corresponding request status	`Details`

[a] *Condition:* available only if the MMSC accepts the corresponding forward request.

VAS application to cancel a previously submitted multimedia message prior to its delivery. Upon receipt of such a cancellation request, the MMSC cancels the delivery of the associated message to all message recipients to which the associated notification has not yet been sent out. The transaction flow in Figure 6.60 shows interactions between the MMSC and the VAS application for the cancellation of a previously submitted multimedia message over the MM7 interface.

The cancellation request `MM7_cancel.REQ` is composed of the information elements/ parameters shown in Table 6.53. The MMSC acknowledges the cancellation request with the `MM7_cancel.RES` response. This response is composed of the information elements/para-meters listed in Table 6.54.

6.36.5 Message Replacement

In the context of value added services, message replacement refers to the possibility for a VAS application to replace a previously submitted multimedia message prior to its delivery. Upon receipt of such a replacement request, the MMSC replaces the previously submitted message with the new message specified as part of the replacement request. Only messages that have not yet been retrieved or forwarded can be replaced. The replacement request contains a number of information elements that overwrite the ones associated with the previously submitted message. If an information element was associated with the previously submitted message but is not provided as part of the replacement request, then this informa-tion element is retained for the new message. The transaction flow in Figure 6.61 shows

SOAP Envelope

```
POST /mmsc/mm7 HTTP/1.1
Host: mms.operator.com

Content-Type: Multipart/Related; boundary=PrimaryBoundary; type=text/xml;
     start="<id_start>"
Content-Length: nnnn
SOAPAction: ""

--PrimaryBoundary
Content-Type: text/xml; charset="utf-8"
Content-ID: <id_start>

<?xml version='1.0' ?>
<env:Envelope xmlns:env="http://schemas.xmlsoap.org/soap/envelope/">

        <env:Header>
                <mm7:TransactionID xmlns="//www.3gpp.org/MMS/Rel5/MM7" env:mustUnderstand="1">
                 env:mustUnderstand="1" />
                            vasp0324-sub
                </mm7:TransactionID>
        </env:Header>

        <env:Body>
                <mm7:SubmitReq xmlns:mm7="//www.3gpp.org/MMS/Rel5/MM7">
                        <MM7Version>5.3.0</MM7Version>
                        <SenderIdentification>
                                <VASPID>w-up</VASPID>
                                <VASID>Weather</VASID>
                        </SenderIdentification>
                        <Recipients>
                                <To>
                                        <Number>0147287355</Number>
                                        <RFC2822Address>0147287366@operator.com</RFC2822Address>
                                </To>
                                <Cc>
                                        <Number>0147243477</Number>
                                </Cc>
                                <Bcc>
                                        <Number>0147243488</Number>
                                </Bcc>
                        </Recipients>
                        <ServiceCode>weather-france</ServiceCode>
                        <LinkID>mms00000006</LinkID>
                        <MessageClass>Informational</MessageClass>
                        <TimeStamp>2002-05-05T09:30:45-01:00</TimeStamp>
                        <EarliestDeliveryTime>2002-05-05T09:30:45-01:00</EarliestDeliveryTime>
                        <ExpiryDate>P5D</ExpiryDate>
                        <DeliveryReport>True</DeliveryReport>
                        <Priority>Normal</Priority>
                        <Subject>Weather for today</Subject>
                        <Content href="cid:weatherpic@w-up.com" ; allowAdptations="true" />
                </mm7:SubmitReq>
        </env:Body>
</env:Envelope>
```

SOAP Attachment

```
--PrimaryBoundary
Content-Type: multipart/mixed; boundary=SecondaryBoundary
Content-ID: <weatherpic@w-up.com>

--SecondaryBoundary
Content-Type: text/plain; charset="us-ascii"

The weather picture.

--SecondaryBoundary
Content-Type: image/gif
Content-ID: <weather.gif>

R0IGODfg...

--SecondaryBoundary--
--PrimaryBoundary--
```

Figure 6.57   Example of message submission request over the MM7 interface

```
HTTP/1.1 200 OK
Content-Type: text/xml; charset="utf8"
Content-Length: nnnn

<?xml version='1.0' ?>
<env:Envelope xmlns:env="http://schemas.xmlsoap.org/soap/soap-envelope">
        <env:Header>
                <mm7:TransactionID xmlns:mm7="//www.3gpp.org/MMS/Rel5/MM7" env:mustUnderstand="1">
                        vasp0324-sub
                </mm7:TransactionID>
        </env:Header>
        <env:Body>
                <mm7:SubmitRsp xmlns:mm7="//www.3gpp.org/MMS/Rel5/MM7">
                        <MM7Version>5.3.0</MM7Version>
                        <Status>
                                <StatusCode>1000</StatusCode>
                                <StatusText>Message sent</StatusText>
                                <MessageID>123456</MessageID>
                        </Status>
                </mm7:SubmitRsp>
        </env:Body>
</env:Envelope>
```

Figure 6.58   Example of message submission response over the MM7 interface

interactions between the MMSC and the VAS application for the replacement of a previously submitted multimedia message over the MM7 interface.

The replacement request `MM7_replace.REQ` is composed of the information elements/parameters listed in Table 6.55. The MMSC acknowledges the replacement request with the `MM7_replace.RES` response. This response is composed of the information elements/parameters listed in Table 6.56.

6.36.6 Delivery Report

In the context of value added services, a VAS application has the ability to request, as part of a message submission request, the generation of a delivery report. If allowed by the message recipient, the MMSC generates a delivery report upon message retrieval, forwarding, deletion, rejection, etc. Note that if the message was submitted to multiple recipients, then several delivery reports may be received by the originator VAS application. The transaction flow in

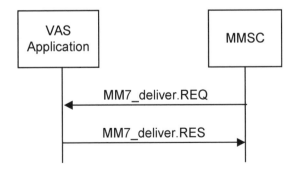

Figure 6.59   MM7 message delivery

Table 6.51 MM7 message delivery request

Stage 2 IE	Status	Location	Description	Stage 3 Name in the SOAP message
Transaction identification	●	H	Identification of the delivery transaction	TransactionID
Message type	●	B	Type of MM7 operation (message delivery request)	MessageType Value: DeliverReq Root element of the SOAP body
MM7 version	●	B	Version of the MM7 interface supported by the MMSC	MM7Version e.g. 5.3.0
MMSC identification	○	B	Identification of the MMSC	MMSRelayServerID
Sender address	●	B	Address of the message originator	SenderAddress
Recipient address	○	B	Address of the message recipient. Multiple recipient addresses can be specified	Recipients
Linked identification	○	B	Relation with another message	LinkedID
Date and time	○	B	Date and time of message submission	TimeStamp
Reply charging identification	○	B	Identification of the original message to which this reply is in response	ReplyChargingID
Priority	○	B	Message priority	Priority Values: High Normal Low
Subject	○	B	Message subject	Subject
Content type	●	A	Message content type	Content-type
Content	○	B	Message content	Content

Figure 6.62 shows interactions between the MMSC and the VAS application for the provision of a delivery report over the MM7 interface.

The request for providing a delivery report MM7_delivery_report.REQ is composed of the information elements/parameters listed in Table 6.57. The MMSC acknowledges the request with the MM7_delivery_report.RES response. This response is composed of the information elements/parameters listed in Table 6.58.

6.36.7 Read-reply Report

In the context of value added services, a VAS application has the ability to request, as part of

Table 6.52 MM7 message delivery response

Stage 2 IE	Status	Location	Description	Stage 3 Name in the SOAP message
Transaction identification	●	H	Identification of the delivery transaction	`TransactionID`
Message type	●	B	Type of MM7 operation (message delivery response)	`MessageType` Value: `DeliverRsp` Root element of the SOAP body
MM7 version	●	B	Version of the MM7 interface supported by the VAS application	`MM7Version` e.g. `5.3.0`
Service code	○	B	Service code for charging purpose	`ServiceCode`
Request status	●	B	Status of the request completion	`StatusCode`
Request status text	○	B	Textual description of the status of the request completion	`StatusText`
Request status details	○	B	Human readable detailed textual description of the corresponding request status	`Details`

a message submission request, the generation of a read-reply report. If allowed by the message recipient, the recipient MMS user agent generates a read-reply report upon message reading, deletion, etc. Note that, if the message was submitted to multiple recipients, then several read-reply reports may be received by the originator VAS application. The transaction flow in Figure 6.63 shows interactions between the MMSC and the VAS application for the provision of a read-reply report over the MM7 interface.

The request for providing a delivery report `MM7_read_reply_report.REQ` is composed of the information elements/parameters listed in Table 6.59. The MMSC acknowledges the request with the `MM7_read_reply_report.RES` response. This response is composed of the information elements/parameters listed in Table 6.60.

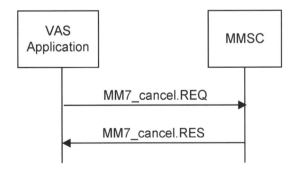

Figure 6.60 MM7 message cancellation

Table 6.53 MM7 message cancellation request

Stage 2 IE	Status	Location	Description	Stage 3 Name in the SOAP message
Transaction identification	●	H	Identification of the cancellation transaction	`TransactionID`
Message type	●	B	Type of MM7 operation (message cancellation request)	`MessageType` Value: `CancelReq` Root element of the SOAP body
MM7 version	●	B	Version of the MM7 interface supported by the VAS application	`MM7Version` e.g. `5.3.0`
VASP identification	○	B	Identification of the VAS provider	`VASPID`
VAS identification	○	B	Identification of the value added service	`VASID`
Sender address	○	B	Address of the message originator	`SenderAddress`
Message identification	●	B	Identification of the message for which the delivery is to be cancelled. This identification was provided by the MMSC in the response of the associated submission request	`MessageID`

Table 6.54 MM7 message cancellation response

Stage 2 IE	Status	Location	Description	Stage 3 Name in the SOAP message
Transaction identification	●	H	Identification of the cancellation transaction	`TransactionID`
Message type	●	B	Type of MM7 operation (message cancellation response)	`MessageType` Value: `CancelRsp` Root element of the SOAP body
MM7 version	●	B	Version of the MM7 interface supported by the MMSC	`MM7Version` e.g. `5.3.0`
Request status	●	B	Status of the request completion	`StatusCode`
Request status text	○	B	Textual description of the status of the request completion	`StatusText`
Request status details	○	B	Human readable detailed textual description of the corresponding request status	`Details`

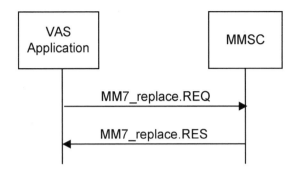

Figure 6.61   MM7 message replacement

Table 6.55   MM7 message replacement request

Stage 2 IE	Status	Location	Description	Stage 3 Name in the SOAP message
Transaction identification	●	H	Identification of the replacement transaction	`TransactionID`
Message type	●	B	Type of MM7 operation (message replacement request)	`MessageType` Value: `ReplaceReq` Root element of the SOAP body
MM7 version	●	B	Version of the MM7 interface supported by the VAS application	`MM7Version` e.g. `5.3.0`
VASP identification	○	B	Identification of the VAS provider	`VASPID`
VAS identification	○	B	Identification of the value added service	`VASID`
Message identification	●	B	Identification of the message which is to be replaced. This identification was provided by the MMSC in response of the submission request	`MessageID`
Service code	○	B	Service code for charging purpose	`ServiceCode`
Date and time	○	B	Date and time of message replacement	`TimeStamp`
Earliest delivery of time	○	B	Earliest time of delivery for the message	`EarliestDeliveryTime`
Read reply	○	B	Request for a status report	`ReadReply` Values: `True` `False`

Table 6.55 (*continued*)

Stage 2 IE	Status	Location	Description	Stage 3 Name in the SOAP message
Adaptations	O	B	Whether or not content adaptation is allowed (default value is True)	allowAdaptations Values: True False
Content type	C	A	Message content type	Content-type
Content	O	B	Message content	Content

Table 6.56   MM7 message replacement response

Stage 2 IE	Status	Location	Description	Stage 3 Name in the SOAP message
Transaction identification	●	H	Identification of the replacement transaction	TransactionID
Message type	●	B	Type of MM7 operation (message replacement response)	MessageType Value: ReplaceRsp Root element of the SOAP body
MM7 version	●	B	Version of the MM7 interface supported by the MMSC	MM7Version e.g. 5.3.0
Request status	●	B	Status of the request completion	StatusCode
Request status text	O	B	Textual description of the status of the request completion	StatusText
Request status details	O	B	Human readable detailed textual description of the corresponding request status	Details

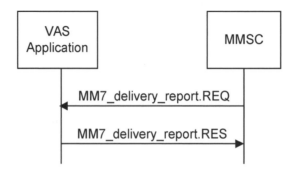

Figure 6.62   MM7 delivery report

Table 6.57 MM7 delivery report request

Stage 2 IE	Status	Location	Description	Stage 3 Name in the SOAP message
Transaction identification	●	H	Identification of the delivery report transaction	`TransactionID`
Message Type	●	B	Type of MM7 operation (delivery report request)	`MessageType` Value: `DeliveryReportReq` Root element of the SOAP body
MM7 version	●	B	Version of the MM7 interface supported by the MMSC	`MM7Version` e.g. `5.3.0`
MMSC identification	○	B	Identification of the MMSC	`MMSRelayServerID`
Message identification	●	B	Identification of the message to which the delivery report relates	`MessageID`
Recipient address	●	B	Address of the message recipient	`Recipient`
Sender address	●	B	Address of the VAS application which previously submitted the message	`SenderAddress`
Date and time	○	B	Date and time when the associated message was handled	`TimeStamp`

Delivery status	●	B	Status of the associated message	`MMStatus`
				Values:
				`Expired`
				`Retrieved`
				`Rejected`
				`Indeterminate`
				`Forwarded`

Status text	○	B	Textual description of the status of the associated message	`StatusText`

6.36.8 Generic Error Handling

In the situation where the MMSC or the VAS application receives a request which it cannot process, then a generic error notification can be used. The generic error notification always contains the identification of the corresponding request (value assigned to the `TransactionID` parameter) to which it relates. Two generic error notifications can be used as shown in the transaction flows in Figure 6.64.

The MMSC notifies a VAS application of a generic error with the `MM7_RS_error.RES` response. This response is composed of the information elements/parameters listed in Table 6.61. The VAS application notifies the MMSC of a generic error with the `MM7_VASP_`

Table 6.58   MM7 delivery report response

Stage 2 IE	Status	Location	Description	Stage 3 Name in the SOAP message
Transaction identification	●	H	Identification of the delivery report transaction	`TransactionID`
Message type	●	B	Type of MM7 operation (delivery report response)	`MessageType` Value: `DeliveryReportRsp` Root element of the SOAP body
MM7 version	●	B	Version of the MM7 interface supported by the VAS application	`MM7Version` e.g. `5.3.0`
Request status	●	B	Status of the request completion	`StatusCode`
Request status text	○	B	Textual description of the status of the request completion	`StatusText`
Request status details	○	B	Human readable detailed textual description of the corresponding request status	`Details`

`error.RES` response. This response is composed of the information elements/parameters listed in Table 6.62.

6.36.9 Availability Matrix of PDU Parameters

Table 6.63 indicates the set of parameters composing each request/response for operations that can be invoked over the MM7 interface.

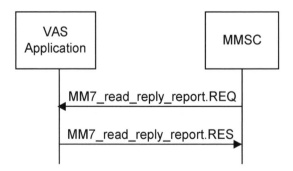

Figure 6.63   MM7 read-reply report

Table 6.59   MM7 read-reply report request

Information elements	Status	Location	Description	Stage 3 Name in the SOAP message
Transaction identification	●	H	Identification of the read-reply report transaction	`TransactionID`
Message type	●	B	Type of MM7 operation (read-reply report request)	`MessageType` Value: `ReadReplyReq` Root element of the SOAP body
MM7 version	●	B	Version of the MM7 interface supported by the VAS application	`MM7Version` e.g. `5.3.0`
MMSC identification	○	B	Identification of the MMSC	`MMSRelayServerID`
Message identification	●	B	Identification of the message to which the delivery report relates	`MessageID`
Recipient address	●	B	Address of the message recipient	`Recipient`
Sender address	●	B	Address of the VAS application that previously submitted the message	`SenderAddress`
Date and time	○	B	Date and time when the associated message was handled	`TimeStamp`
Read status	●	B	Status of the associated message	`MMStatus` Values: `Indeterminate` `Read` `Deleted`
Status text	○	B	Textual description of the status of the associated message	`StatusText`

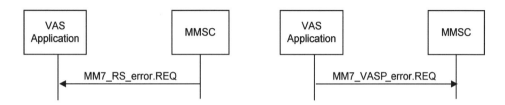

Figure 6.64   MM7 generic errors

Table 6.60   MM7 read-reply report response

Stage 2 IE	Status	Location	Description	Stage 3 Name in the SOAP message
Transaction identification	●	H	Identification of the read-reply report transaction	`TransactionID`
Message type	●	B	Type of MM7 operation (read-reply report response)	`MessageType` Value: `ReadReplyRsp` Root element of the SOAP body
MM7 version	●	B	Version of the MM7 interface supported by the MMSC	`MM7Version` e.g. `5.3.0`
Request status	●	B	Status of the request completion	`StatusCode`
Request status text	○	B	Textual description of the status of the request completion	`StatusText`
Request status details	○	B	Human readable detailed textual description of the corresponding request status	`Details`

Table 6.61   MM7 MMSC error request (from MMSC to VAS application)

Stage 2 IE	Status	Location	Description	Stage 3 Name in the SOAP message
Transaction identification	●	H	Identification of the corresponding request	`TransactionID`
Message type	●	B	Type of MM7 operation (generic error from MMSC)	`MessageType` Value: `RSErrorRsp` Root element of the SOAP body
MM7 version	●	B	Version of the MM7 interface supported by the MMSC	`MM7Version` e.g. `5.3.0`
Error status	●	B	Status of the request completion	`StatusCode`
Error status text	○	B	Textual description of the error	`StatusText`
Request status details	○	B	Human readable detailed textual description of the corresponding request status	`Details`

Table 6.62   MM7 VAS application error request (from VAS application to MMSC)

Stage 2 IE	Status	Location	Description	Stage 3 Name in the SOAP message
Transaction identification	●	H	Identification of the corresponding request	`TransactionID`
Message type	●	B	Type of MM7 operation (generic error from VAS application)	`MessageType` Value: `VASPErrorRsp` Root element of the SOAP body
MM7 version	●	B	Version of the MM7 interface supported by the VAS application	`MM7Version` e.g. `5.3.0`
Error status	●	B	Status of the request completion	`StatusCode`
Error status text	○	B	Textual description of the error	`StatusText`
Request status details	○	B	Human readable detailed textual description of the corresponding request status	`Details`

6.37 Content Authoring and Testing Tools

Several manufacturers have designed content authoring tools to facilitate the introduction of MMS in the market. This section outlines the features offered by the most relevant tools.

6.37.1 Sony-Ericsson Tools

Sony-Ericsson offers several EMS and MMS development tools from the Ericsson developer on-line resources, called Ericsson Mobility World, at http://www.ericsson.com/mobility-world. After registration to access these resources, the following development tools can be downloaded:

- *MMS composer:* the composer is a PC tool allowing the creation of multimedia messages containing a scene description (MMS SMIL). After creation with the tool, messages can later be retrieved from a Sony-Ericsson handset by connecting the PC to the handset via a serial cable.
- *Conversion tool for AMR and WAV:* this PC tool allows the conversion of an AMR file to a WAV file, and vice versa. This tool has a graphical interface but can also be executed without the graphical interface, in a DOS prompt mode.

6.37.1 Nokia Tools

Nokia provides a set of development guidelines and tools from the Nokia developer forum, at http://www.forum.nokia.com. After registration with the Nokia developer Forum, the following MMS development tools can be downloaded:

Table 6.63   MM7 PDU parameters[a]

	A	B	C	D	E	F	G	H	I	J	K	L	M	N
allowAdaptations	O						O							
ChargedParty	O													
Content	O		O				O							
Content-Type	●		●				C							
Details		O		O		O		O		O		O	O	O
EarliestDeliveryTime	O						O							
ExpiryDate	O													
DeliveryReport	O													
DistributionIndicator	O													
LinkedId	O		O											
MessageClass	O													
MessageID		●			●		●		●		●			
MessageType	●	●	●	●	●	●	●	●	●	●	●	●	●	●
MM7Version	●	●	●	●	●	●	●	●	●	●	●	●	●	●
MMSRelayServerID			O						O		O			
MMStatus									●		●			
Priority	O		O											
ReadReply	O						O							
Recipient	●		O						●		●			
ReplyCharging	O													
ReplyChargingID			O											
ReplyChargingSize	O													
ReplyDeadline	O													
SenderAddress	O		●		O				●		●			
ServiceCode	O			O			O							
StatusCode		●		●		●		●		●		●	●	●
StatusText		O		O		O		O	O	O	O	O	O	O
Subject	O		O											
TimeStamp	O		O				O		●		●			
TransactionID	●	●	●	●	●	●	●	●	●	●	●	●	●	●
VASID	O			O			O							
VASPID	O			O			O							

[a] A, MM7_submit.REQ; B, MM7_submit.RES; C, MM7_deliver.REQ; D, MM7_deliver.RES; E, MM7_cancel.REQ; F, MM7_cancel.RES; G, MM7_replace.REQ; H, MM7_replace.RES; I, MM7_delivery_report.REQ; J, MM7_delivery_report.RES; K, MM7_read_reply.REQ; L, MM7_read_reply_report.RES; M, MM7_RS_error.RES; N, MM7_VASP_error.RES. ●, Mandatory; O, Optional; C, Conditional; otherwise, not applicable.

- *Emulator for Nokia MMSC interface:* this tool emulates the interface of Nokia's MMSC for VAS applications. As described in this chapter, this interface is known as the MM7 interface. Note that this tool does not yet support the MM7 interface as described in this book. At the time of writing, the interface between VAS applications and the Nokia MMSC is based on a set of Nokia proprietary commands conveyed over HTTP.

- *MMS Java library:* the Java library can be used for developing MMS-based applications in Java. This library includes the following features:

  - message creation and encoding
  - message decoding
  - message sending to Nokia MMSC or to Nokia emulator MMSC.

- *Developer's suite for MMS:* this is a software add-on for Adobe GoLive 6.0. Adobe Golive, with this add-on, becomes a powerful authoring tool facilitating the generation of multimedia messages with rich multimedia presentations.
- *Series 60 SDK Symbian OS and MMS extension:* this software development kit allows the development of applications for devices based on the Series 60 platform. This SDK includes a Series 60 emulator enabling the development and testing of applications with a PC.

Box 6.10 Web Resources for MMS Authoring

Nokia developer Forum at http://www.forum.nokia.com
Ericsson Mobility World at http://www.ericsson.com/mobilityworld

---

6.38 MMS Features Supported by Commercial Devices

At the time of writing, four MMS-capable handsets are available on the market. The characteristics of these devices are given in Table 6.64.

Other products, not yet available on the market, are planned for commercial release by the end of 2002: the A820 from Motorola and the P800 from Ericsson. Furthermore, other major manufacturers, such as Alcatel and Siemens, are likely to release MMS-capable devices on the market by the end of 2002 or during 2003.

6.39 Pros and Cons of MMS

MMS represents a huge evolutionary step from EMS in the messaging roadmap. In comparison with SMS and EMS, MMS has the following advantages:

- Multimedia messages can contain a wide range of content formats such as colour images/ animations, video, etc. Additionally, messages can be organized as compact multimedia slideshows.
- Compared with other messaging technologies, interoperability between MMS and the Email service has been significantly improved.
- MMS is well integrated with innovative devices/services such as camera accessories, remote photo albums.
- MMS is a future-proof technology and will be the subject of many evolutions in the near future. Several of these evolutions are introduced in the next section.

On the other hand, the following drawbacks have been identified for MMS:

Table 6.64 MMS features supported by available handsets

	Sony-Ericsson T68i	Nokia 3510	Nokia 7210	Nokia 7650
Message sending	●	●	●	●
Message retrieval	●		●	●
Screen size	101 × 80	96 × 65	128 × 128	174 × 142
Camera	Accessory	Accessory	Accessory	Built-in
Supported formats				
Images	JPEG, GIF and WBMP	JPEG, GIF, WBMP, Windows BMP and OTA BMP	JPEG, GIF, PNG, WBMP and Windows BMP	JPEG, GIF, PNG and WBMP
Audio	iMelody and AMR	SP-MIDI	SP-MIDI	AMR and WAV
Scene description	MMS SMIL partial support for incoming and outgoing messages (no support for layout information)	No	MMS SMIL support for outgoing messages only	MMS SMIL support for incoming and outgoing messages
Miscellaneous	vCard, vCalendar and vNote	No	Java application descriptor, WML content	vCard, vCalendar, Java application descriptor, WML content, Symbian installation package

- MMS is a very recent service and MMS-capable devices (handsets and MMSCs) are not yet widely available. This prevents global introduction of MMS for all market segments.
- Unlike EMS, MMS requires several network infrastructure extensions in order to operate properly (e.g. MMSC, additional WAP gateway). Nevertheless, at the time of writing, mobile operators seem to be prepared for significant investment in the service.

Overall, MMS is a compelling sophisticated service. However, this service still has to be accepted by the mass market.

6.40 The Future of MMS

As shown in this chapter, two technical realizations (of the MM1 interface) of MMS are available for the WAP environment. A third is under development in the scope of the WAP Forum standardization process. This third technical realization, codenamed MMS Cubed (likely to become WAP MMS 2.0), will fulfil the functional requirements defined by the 3GPP in release 5. Technical specifications for this realization should be available by 2003 and devices compliant with this realization should appear on the market by the end of 2003 or early in 2004. In the meantime, the 3GPP has already initiated work on the definition of the functional requirements for MMS release 6. MMS release 6 technical specifications should be published by mid-2003 and devices compliant with MMS release 6 should appear on the market by 2005. In the scope of the 3GPP standardization process, work on MMS release 6 will focus on the following aspects:

- *Enhancements of the MM1 interface* (MMS user agent/MMSC) including:
  - optimized flow control of notifications
  - management of requests for message replacement or cancellation by MMS user agents (currently supported over the MM7 interface only)
  - use of 3GPP push instead of WAP push
  - enhanced MMbox management (selective retrieval of message, etc.)
  - enhancement of USIM/USAT aspects of MMS
  - investigation of technical realizations in non-WAP environments (e.g. IMAP and HTTP technical realizations of the MM1 interface).

- *Investigate the need for the MM2 interface* (MMS Relay/MMS Server).
- *Definition of the MM3 interface for interworking with other messaging systems* including:
  - Email system
  - voice-mail systems/voice messaging
  - SMS and EMS.

- *Enhancements of the MM4 interface* (MMSC/MMSC) including discovery of MMSC capabilities.
- *Definition of the MM6 interface* (MMSC/User databases).
- *Enhancement of the MM7 interface* (MMSC/VAS application) including:
  - management of mailing lists
  - improvement of authentication/authorization procedures
  - convergence of MM7 interface with generic access defined in OSA.

- *Definition of the MM8 interface* (MMSC/billing system).
- *End-to-end service issues* including:
  - digital rights management
  - addressing enhancements
  - support for end-to-end security and privacy
  - filtering mechanisms for spam control
  - interface to systems managing presence-related data (relationships with Instant Messaging)
  - enhancement of streaming delivery of messages
  - improvement of media types/formats supported in MMS
  - over-the-air provisioning of MMS settings to mobile devices.

Further Reading

P. Loshin, 1999, Essential Email Standards, Wiley.

H. Schulzrinne and J. Rosenberg, The IETF Internet telephony architecture and protocols, IEEE Network, May/June, 1999.

D.C.A. Bulterman (Ed. Peiya Liu), SMIL 2.0, part I: overview, concepts and structure, IEEE Multimedia Magazine, October-December, 2001.

D.C.A. Bulterman (Ed. Peiya Liu), SMIL 2.0, part II: examples and comparisons, IEEE Multimedia Magazine, January-March, 2002.

7

Other Mobile Messaging Services

This chapter outlines several other messaging services relevant to the world of mobile communications. First, services and technologies being developed for the support of Immediate Messaging and Presence Services (IMPS) are introduced. IMPS offer subscribers the ability to manage interactive discussion rooms (also known as 'chat' rooms) and to determine other subscribers' status (e.g. 'buddies list'). The Wireless Village is one of the most active organizations working on the definition of IMPS for mobile telecommunications. This chapter also describes how the Email service available for Internet users can be adapted for mobile users. The Blackberry service, designed by the Canadian company Research in Motion (RIM) and presented in this chapter, is an interesting adaptation of the Email service to the constraints of mobile devices and networks. This service has become popular for itinerant professional users in North America. Finally, this chapter also introduces the work recently initiated by the 3GPP for the definition of future messaging services in the scope of the IP Multimedia Subsystem (IMS).

7.1 Immediate Messaging and Presence Services

Immediate Messaging and Presence Services (IMPS) are very popular in the Internet world. IMPS offers subscribers the ability to manage interactive discussion rooms (also known as 'chat' rooms) and subscriber presence states (e.g. 'buddies list'). Three major Internet Instant Messaging (IM) services are currently available: AIM operated by AOL, MSN Messenger operated by Microsoft and Yahoo Messenger operated by Yahoo. Unfortunately, these services are closed systems using proprietary client-server protocols and consequently do not operate well.

Several standardization development organizations and forums have considered the support of IMPS in the mobile environment. The most notable are the 3GPP, the IETF and the Wireless Village Initiative. The 3GPP initiated standardization work with the definition of the concept of presence (stage 1, defined in [3GPP-22.141]/stage 2, defined in [3GPP-23.841]). However, this work is currently at an early stage. In parallel, the following IETF working groups have been actively involved in defining IM protocols for the IP domain:

- Instant Messaging and Presence Protocol (IMPP)
- Application Exchange (APEX)

- Presence and Instant Messaging Protocol (PRIM)
- SIP for Instant Messaging and Presence Leveraging Extensions (SIMPLE)

SIMPLE is an IMPS protocol based on the Session Initiation Protocol (SIP) [RFC-3261] and is particularly well suited for UMTS phase 2. SIMPLE defines a mechanism for subscribing to a service which manages subscriber status changes over a SIP network.

The Wireless Village Initiative was recently set up by several manufacturers to consider the support of IMPS in the mobile environment. The prime objective of this initiative is to develop an industry standard to guarantee interoperability between various IMPS services that could be deployed in the near future. The IMPS model, proposed by the Wireless Village, includes four primary features:

- *Presence:* this feature manages information such as device availability (e.g. mobile device has been switched on/off, voice call under progress), user status (e.g. available, not available, having a meeting), user location, mobile device capabilities, personal status/moods (e.g. happy, angry), etc. To ensure confidentiality, presence information is provided according to user instructions only.
- *Immediate Messaging:* this feature allows the exchange of messages delivered instantaneously to the recipient(s). This means that, upon sending by the originator, the message is perceived as being immediately delivered to the message recipient(s). Compared with traditional messaging services, instant messaging allows the establishment of interactive messaging sessions between users. These sessions are usually displayed using a threaded conversational interface, also known as *chat*.
- *Groups:* this feature allows users to create and manage their own groups. The manager of a group can invite other users to chat with the group members. Network operators can create general interest groups.
- *Shared Content:* this feature provides users with storage zones where pictures, audio and other multimedia contents can be posted and retrieved. These contents can be shared among several users during group messaging sessions.

The general architecture of the Wireless Village initiative solutions is depicted in Figure 7.1. In the architecture shown in the figure, the *Wireless Village server* is a central element. This server manages the following services: presence, instant messaging, groups and content sharing. Two types of Wireless Village clients can communicate with the server: the Wireless Village embedded client and the Wireless Village Command Line Interface (CLI) client. The *Wireless Village embedded client* is a dedicated application embedded in a mobile device. This client communicates with the server over the Client Server Protocol (CSP). On the other hand, the *Wireless Village CLI client* uses text messages (e.g. SMS) to communicate with the server using the Command Line Protocol (CLP). A CLI client is usually a legacy mobile device. Two Wireless Village servers, located in the same provider domain or in two distinct provider domains, can communicate with the Server-Server Protocol (SSP). A Wireless Village server may also be connected to a proprietary IMPS system via a proprietary gateway. In this configuration, the gateway transcodes SSP instructions into instructions supported by the proprietary system, and vice versa. The Server-Mobile Core Network Protocol (SMCNP) allows the server to obtain presence information and service capabilities from the mobile core network. In addition, the SMCNP can also be used for authentication and authorization of users, clients and servers.

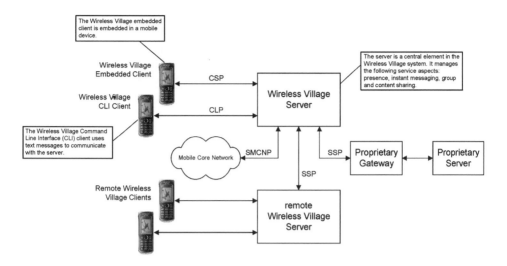

Figure 7.1   General Wireless Village architecture

Wireless Village technical specifications can be downloaded from http://www.wireless-village.org.

7.2 Mobile Email

Email is the de facto messaging service on the Internet. However, due to the bandwidth limitations of mobile systems and the fact that mobile devices are seldom permanently connected to the network, Email is not widely used in the mobile telecommunications domain. Nevertheless, several manufacturers have designed Email clients for mobile devices using standard Internet messaging protocols such as the Post Office Protocol-3 (POP3), defined in [RFC-1939] and the Interactive Mail Access Protocol (IMAP), defined in [RFC-1730]. These solutions have the advantage of allowing mobile subscribers to communicate seamlessly with remote Internet users (using the same message formats and server access protocols). However, these solutions have proven to be very impractical to use without a minimum adaptation to the constraints of mobile devices and networks. The major barriers to the success of these solutions are the 'pull' model for retrieving messages which requires frequent accesses to the Email server and the fact that server access protocols are not resource efficient.

In order to offer an Email service adapted to the requirements of mobile subscribers, the company Research in Motion (RIM) designed a set of extensions for the existing Email service. This extended Email service, offered to subscribers under the denomination 'Blackberry service', bypasses Email inadequacies to the mobile domain by enabling:

- a 'push' model for message retrieval
- a compression of messages
- an encryption of messages.

Two main configurations are available for the Blackberry service. The first configuration

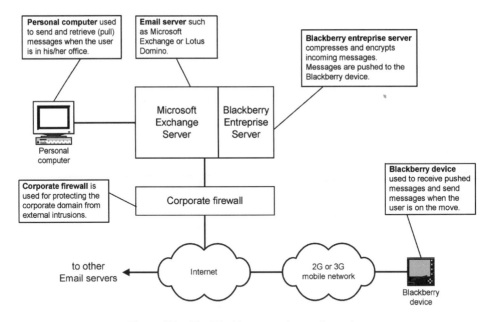

Figure 7.2   The Blackberry service configuration

limits the impact on existing Email architectures by integrating a 'desktop' Blackberry application (the Blackberry desktop redirector) in the user's personal computer used for accessing Email messages. When the user is on the move, the desktop application intercepts incoming messages, compresses them, encrypts them and pushes them to the Blackberry device via a mobile network. The other way round, the user can compose a new message with the Blackberry device. The message is compressed and encrypted by the device and sent via the mobile network to the desktop application. The desktop application receives the message (by polling the Email server), decompresses and decrypts it and sends it normally to the message recipients as if the message had been sent out directly by the user from his/her personal computer. A more sophisticated configuration of the Blackberry service consists of installing an extension to the Email server itself (the Blackberry enterprise server). Basically, in the second configuration, the user's personal computer does not have to be left running when the user is on the move. With this configuration, messaging functions performed by the desktop application in the first configuration are performed here by the server extension. In addition, this configuration also allows the synchronization of calendaring and scheduling data between shared corporate databases and remote Blackberry devices. The enterprise configuration of the Blackberry service is depicted in Figure 7.2.

The Blackberry service is already available in North America and is currently being deployed in other countries in Europe (United Kingdom and France). The service fulfils particularly well the needs of itinerant professional users, who avoid using laptop computers while on the move (because of long dial-up time for accessing Email servers, etc.). The Multimedia Messaging Service (MMS), described in the previous chapter, targets the mass market by supporting a messaging service, similar to the Email service, with small handsets. On the other hand, the Blackberry service targets the professional market with devices designed as Personal Digital Assistants. More information can be obtained on the Blackberry service from Research in Motion (RIM) at http://www.rim.com and Blackberry service at http://www.blackberry.net.

7.3 IMS Messaging

Chapter 1 introduced the two phases for UMTS. The second phase of UMTS is built on the IP-based Multimedia Service (IMS) based on the Session Initiation Protocol (SIP) for session/ call management. The 3GPP recently initiated work on messaging services based on a combination of IMS service capabilities (e.g. presence) and already defined messaging services (SMS, EMS and MMS). The scope of this work initially consists of identifying (in the release 6 timeframe) the requirements for messaging services in the IMS environment. These requirements will be detailed in the technical report [3GPP-22.940].

Appendices

A TP-PID Values for Telematic Interworking

For enabling SMS interworking with various telematic devices, the set of protocol identifiers (`TP-Protocol-Identifier`) listed in Table 1 can be used.

Table 1 Protocol identifiers for telematic interworking

TP-PID value (hex)	Description
0x20	Type of telematic device is defined by the message destination or originator address.
0x21	Telex (or teletex reduced to telex format)
0x22	Group 3 telefax
0x23	Group 4 telefax
0x24	Voice telephone (i.e. conversion to speech)
0x25	European Radio Messaging System (ERMES)
0x26	National paging system (type known to the service centre)
0x27	Videotext such as T.100 or T101
0x28	Teletex, carrier unspecified
0x29	Teletex, in PSPDN
0x2A	Teletex, in CSPDN
0x2B	Teletex, in analogue PSTN
0x2C	Teletex, in digital ISDN
0x2D	Universal Computer Interface (UCI)
0x2E...0x2F	*Reserved* (2 values)
0x30	Message handling facility (type known to the service centre)
0x31	Public X.400-based message handling system
0x32	Internet electronic mail
0x33...0x37	*Reserved* (5 values)
0x38...0x3E	SC specific use (7 values)
0x3F	GSM or UMTS mobile station. The SMSC converts the short message into a coding scheme which is understandable by the GSM/UMTS mobile station

B Numeric and Alphanumeric Representations/SMS

Various numeric values can be assigned to the parameters of an SMS TPDU. In this context, numeric values can be represented in three different ways:

- integer representation
- octet representation
- semi-octet representation.

B.1 Integer Representation

With the integer representation, a numeric value is represented with one or more octets (complete or in fractions). For such a representation, the following rules apply:

- *1st rule:* octets with the lowest octet indexes contain the most significant bits.
- *2nd rule:* bits with the highest bit indexes are the most significant bits.

The example in Figure 1 shows how the decimal number 987 351 is represented.

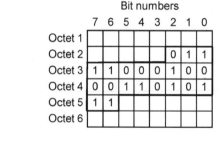

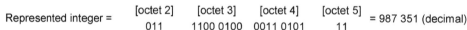

Figure 1   Integer representation/example

B.1.2 Octet Representation

With the octet representation, a numeric value is represented with one or more complete octets where each octet represents one decimal digit. The only rule to apply is that octets with the lowest octet indexes contain the most significant decimal digits. Each octet can take the values listed in Table 2. All other octet values are reserved. The example in Figure 2 shows how the decimal value 43 is represented.

Table 2 Octet representation

Octet value	Decimal digit
0000 0000	0
0000 0001	1
0000 0010	2
0000 0011	3
0000 0100	4
0000 0101	5
0000 0110	6
0000 0111	7
0000 1000	8
0000 1001	9

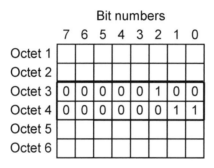

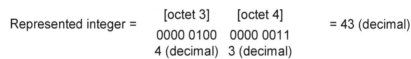

Figure 2 Octet representation/example

B.3 Semi-octet Representation

With the semi-octet representation, a numeric value is represented with one or more half-octets (4 bits). For such a representation, the following rules apply:

- *1st rule:* octets with the lowest octet indexes contain the most significant decimal digits.
- *2nd rule:* within one octet, the half-octet with bits numbered 0–3 represents the most significant digit.

Each half-octet can take the values listed in Table 3. The example in Figure 3 shows how the decimal value 431 is represented with four semi-octets.

Table 3 Semi-octet representation

Half-octet value	Decimal digit
0000	0
0001	1
0010	2
0011	3
0100	4
0101	5
0110	6
0111	7
1000	8
1001	9
1010	*
1011	#
1100	a
1101	b
1110	c
1111	Used as fill bits

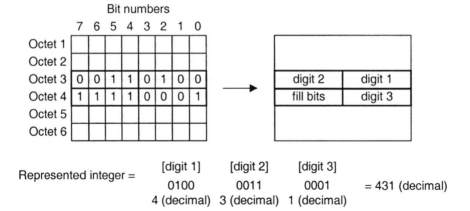

Figure 3 Semi-octet representation/example

C Character Sets and Transformation Formats

C.1 GSM 7-bit Default Alphabet

Table 4 presents all the characters in the GSM 7-bit default alphabet. Each character is represented with a septet (7 bits) for which the most significant bit is $b7$ and the least significant bit is $b1$.

Table 4 GSM 7 bit alphabet (first table)[a]

b4	b3	b2	b1	b7 b6 b5	0 0 0 **0**	0 0 1 **1**	0 1 0 **2**	0 1 1 **3**	1 0 0 **4**	1 0 1 **5**	1 1 0 **6**	1 1 1 **7**
0	0	0	0	**0**	@	Δ	SP	0	¡	P	¿	p
0	0	0	1	**1**	£	_	!	1	A	Q	a	q
0	0	1	0	**2**	$	Φ	"	2	B	R	b	r
0	0	1	1	**3**	¥	Γ	#	3	C	S	c	s
0	1	0	0	**4**	è	Λ	¤	4	D	T	d	t
0	1	0	1	**5**	é	Ω	%	5	E	U	e	u
0	1	1	0	**6**	ù	Π	&	6	F	V	f	v
0	1	1	1	**7**	ì	Ψ	'	7	G	W	g	w
1	0	0	0	**8**	ò	Σ	(	8	H	X	h	x
1	0	0	1	**9**	Ç	Θ	)	9	I	Y	i	y
1	0	1	0	**10**	LF	Ξ	*	:	J	Z	j	z
1	0	1	1	**11**	Ø	Esc[a]	+	;	K	Ä	k	ä
1	1	0	0	**12**	ø	Æ	,	<	L	Ö	l	ö
1	1	0	1	**13**	CR	æ	-	=	M	Ñ	m	ñ
1	1	1	0	**14**	Å	ß	.	>	N	Ü	n	ü
1	1	1	1	**15**	å	É	/	?	O	§	o	à

[a] Esc, the escape character indicates that the following character corresponds to an entry in the GSM 7 bits default alphabet extension table as defined in Table 5

Table 5 GSM 7 bit alphabet (extension table)[a]

b4	b3	b2	b1	b7 b6 b5	0 0 0 **0**	0 0 1 **1**	0 1 0 **2**	0 1 1 **3**	1 0 0 **4**	1 0 1 **5**	1 1 0 **6**	1 1 1 **7**
0	0	0	0	**0**					|			
0	0	0	1	**1**								
0	0	1	0	**2**								
0	0	1	1	**3**								
0	1	0	0	**4**		^						
0	1	0	1	**5**							€	
0	1	1	0	**6**								
0	1	1	1	**7**								
1	0	0	0	**8**				{				
1	0	0	1	**9**				}				
1	0	1	0	**10**	Page break							
1	0	1	1	**11**			Esc[a]					
1	1	0	0	**12**					[			
1	1	0	1	**13**					~			
1	1	1	0	**14**					]			
1	1	1	1	**15**			\					

[a] Esc, the escape character indicates that the following character corresponds to an entry in an additional GSM 7 bits default alphabet extension table. At the time of writing, such a table had not been defined.

C.2 US-ASCII

The US-ASCII character set [US-ASCII] is widely used for representing text in the Internet domain. This character set can be used for representing the text part of multimedia messages in MMS. The 128 characters composing the US-ASCII character set are listed in Table 6.

Table 6   US-ASCII character set

Dec.	Hex.	Char.	Dec.	Hex.	Char.	Dec.	Hex.	Char.
0	0x00	NUL	43	0x2b	+	86	0x56	V
1	0x01	SOH	44	0x2c	,	87	0x57	W
2	0x02	STX	45	0x2d	−	88	0x58	X
3	0x03	ETX	46	0x2e	.	89	0x59	Y
4	0x04	EOT	47	0x2f	/	90	0x5a	Z
5	0x05	ENQ	48	0x30	0	91	0x5b	[
6	0x06	ACK	49	0x31	1	92	0x5c	\
7	0x07	BEL	50	0x32	2	93	0x5d	]
8	0x08	BS	51	0x33	3	94	0x5e	^
9	0x09	HT	52	0x34	4	95	0x5f	_
10	0x0a	LF	53	0x35	5	96	0x60	`
11	0x0b	VT	54	0x36	6	97	0x61	a
12	0x0c	FF	55	0x37	7	98	0x62	b
13	0x0d	CR	56	0x38	8	99	0x63	c
14	0x0e	SO	57	0x39	9	100	0x64	d
15	0x0f	SI	58	0x3a	:	101	0x65	e
16	0x10	DLE	59	0x3b	;	102	0x66	f
17	0x11	DC1	60	0x3c	<	103	0x67	g
18	0x12	DC2	61	0x3d	=	104	0x68	h
19	0x13	DC3	62	0x3e	>	105	0x69	i
20	0x14	DC4	63	0x3f	?	106	0x6a	j
21	0x15	NAK	64	0x40	@	107	0x6b	k
22	0x16	SYN	65	0x41	A	108	0x6c	l
23	0x17	ETB	66	0x42	B	109	0x6d	m
24	0x18	CAN	67	0x43	C	110	0x6e	n
25	0x19	EM	68	0x44	D	111	0x6f	o
26	0x1a	SUB	69	0x45	E	112	0x70	p
27	0x1b	ESC	70	0x46	F	113	0x71	q
28	0x1c	FS	71	0x47	G	114	0x72	r
29	0x1d	GS	72	0x48	H	115	0x73	s
30	0x1e	RS	73	0x49	I	116	0x74	t
31	0x1f	US	74	0x4a	J	117	0x75	u
32	0x20	space	75	0x4b	K	118	0x76	v
33	0x21	!	76	0x4c	L	119	0x77	w
34	0x22	"	77	0x4d	M	120	0x78	x
35	0x23	#	78	0x4e	N	121	0x79	y
36	0x24	$	79	0x4f	O	122	0x7a	z
37	0x25	%	80	0x50	P	123	0x7b	{
38	0x26	&	81	0x51	Q	124	0x7c	|
39	0x27	'	82	0x52	R	125	0x7d	}
40	0x28	(	83	0x53	S	126	0x7e	~
41	0x29	)	84	0x54	T	127	0x7f	delete
42	0x2a	*	85	0x55	U			

C.3 Universal Character Set

The ISO has defined in [ISO-10646] the Universal Character Set (UCS), a multi-octet character set for representing most of the world's writing symbols. Two encoding methods are available for UCS:

- UCS2: a two-octet per symbol encoding.
- UCS4: a four-octet per symbol encoding.

UCS2 and UCS4 are difficult to use in systems on 7-bit and 8-bit transport. To cope with these difficulties, UCS transformation formats have been developed. These formats are defined in the following sections.

C.4 UCS Transformation Formats

The most commonly used UCS Transformation Formats (UTF) are:

- UTF8 [RFC-2279]: this transformation format is 8-bit aligned and has the key characteristic of preserving the US-ASCII value range. UTF8 encodes UCS2 and UCS4 with one or more octets per symbol (1 to 6 octets).
- UTF16: this transformation format is 16-bit aligned. UTF16 transforms UCS4 symbols into pairs of UCS2 values.

Symbols represented by one octet in UTF8 are US-ASCII characters. In this configuration, the most significant bit of the octet is set to 0. For other symbols represented with n octets, then the n most significant bits of the first octet representing the symbol are set to 1. For remaining octets, the most significant bit is set to 1 and the second most significant bit is set to 0. Table 7 summarizes the relationships between UTF8 and UCS4:

Table 7   Relationships between UTF8 and UCS4

UTF8 octet sequence (binary)	UCS4 range (hexadecimal)	Description
0xxxxxxx	0000 0000 to 0000 007F	Used to encode US-ASCII symbols over 1 octet
110xxxxx 10xxxxxx	0000 0080 to 0000 07FF	Used to encode symbols over 2 octets
1110xxxx 10xxxxxx 10xxxxxx	0000 0800 to 0000 FFFF	Used to encode symbols over 3 octets
11110xxx 10xxxxxx 10xxxxxx 10xxxxxx	0001 0000 to 001F FFFF	Used to encode symbols over 4 octets
111110xx 10xxxxxx 10xxxxxx 10xxxxxx 10xxxxxx	0020 0000 to 03FF FFFF	Used to encode symbols over 5 octets
1111110x 10xxxxxx 10xxxxxx 10xxxxxx 10xxxxxx 10xxxxxx	0400 0000 to 7FFF FFFF	Used to encode symbols over 6 octets

D iMelody Grammar

The iMelody format is used in EMS and MMS messaging systems. The BNF grammar of the iMelody format (version 1.2) is given below:

```
<imelody-object> =        "BEGIN:IMELODY"<cr><line-feed>
                          "VERSION:"<version><cr><line-feed>
                          "FORMAT:"<format><cr><line-feed>
                          ["NAME:"<characters-not-lf><cr><line-
                          feed>]
                          ["COMPOSER:"<characters-not-lf><cr><line-
                          feed>]
                          ["BEAT:"<beat><cr><line-feed>]
                          ["STYLE:"<style><cr><line-feed>]
                          ["VOLUME:"<volume><cr><line-feed>]
                          "MELODY:"<melody><cr><line-feed>
                          "END:IMELODY"<cr><line-feed>
<version> =               "1.2"
<format> =                "CLASS1.0" | "CLASS2.0"
<beat> =                  "25" | "26" | "27" |...| "899" | "900"
<style> =                 "S0" | "S1" | "S2"
<volume-modifier> =       "V+" | "V-"
<volume> =                "V0" | "V1" |...| "V15" | <volume-modifier>
<basic-note> =            "c" | "d" | "e" | "f" | "g" | "a" | "b"
<flat-note> =             "&d" | "&e" | "&g" | "&a" | "&b"
<sharp-note> =            "#c" | "#d" | "#f" | "#g" | "#a"
< note> =                 <basic-note> | <flat-note> | <sharp-note>
<octave-prefix> =         "*0" | "*1" | ... | "*8"
<duration> =              "0" | "1" | "2" | "3" | "4" | "5"
<duration-specifier> =    "." | ":" | ";"
<rest> =                  "r"
<led> =                   "ledoff" | "ledon"
<vibe> =                  "vibeon" | "vibeoff"
<backlight> =             "backon" | "backoff"
<full-note> =             [<octave-prefix>] <note> <duration>
                          [<duration-specifier>]
<silence> =               <rest> <duration> [<duration-specifier]
<repeat> =                "(" {<silence> | <full-note> | <led> | <vib> |
                          <volume> | <backlight> } + "@" <repeat-count>
                          [<volume-modifier>] ")"
<repeat-count> =          "0" | "1" | "2" | ...
<melody> =                { <silence> | <full-note> | <led> | <vib> |
                          <repeat> |<volume> | <backlight> } +
<characters-no-lf> =      Any character in the US-ASCII character-set
                          except <line-feed>
```

E MMS Binary Encoding for MMS PDUs

The process of representing the parameters of an MMS PDU in a binary form for transfer over the MM1 interface is called *encapsulation* and is defined in [WAP-209] for MMS 1.0 and [WAP-276] for MMS 1.1. Table 8 gives a summary of binary representations for all MMS PDU parameters.

Table 8   Encapsulation of MMS PDU parameters

Parameter			Values	
Name	Assigned number	Binary encoding	Possible values	Binary encoding
Bcc	0x01	0x81		Encoded string
Cc	0x02	0x82		Encoded string
X-Mms-Content-location	0x03	0x83		Text string
Content-type	0x04	0x84		Multipart as defined in [WAP-203].
Date	0x05	0x85		long integer
X-Mms-Delivery report	0x06	0x86	Yes	0x80 (decimal 128)
			No	0x81 (decimal 129)
X-Mms-Delivery time	0x07	0x87		Absolute or relative date
X-Mms-Expiry	0x08	0x88		Absolute or relative date
From	0x09	0x89		Address or 'insert' token
X-Mms-Message-class	0x0A	0x8A	Personal	0x80 (decimal 128)
			Advertisement	0x81 (decimal 129)
			Informational	0x82 (decimal 130)
			Auto	0x83 (decimal 131)
Message-ID	0x0B	0x8B		Text string
X-Mms-Message-Size	0x0E	0x8E		Long integer
X-Mms-Message-Type	0x0C	0x8C	m-send-req	0x80 (decimal 128)
			m-send-conf	0x81 (decimal 129)
			m-notification-ind	0x82 (decimal 130)
			m-notifyresp-ind	0x83 (decimal 131)
			m-retrieve-conf	0x84 (decimal 132)
			m-acknowledge-ind	0x85 (decimal 133)
			m-delivery-ind	0x86 (decimal 134)
			m-read-rec-ind	0x87 (decimal 135)
			m-read-orig-ind	0x88 (decimal 136)
			m-forward-req	0x89 (decimal 137)
			m-forward-conf	0x8A (decimal 138)
X-Mms-MMS-version	0x0D	0x8D		Short integer
X-Mms-Previously-Sent-By	0x20	0xA0		Indexed encoded string
X-Mms-Previously-Sent-Date	0x21	0xA1		Indexed date
X-Mms-Priority	0x0F	0x8F	Low	0x80 (decimal 128)
			Normal	0x81 (decimal 129)
			High	0x82 (decimal 130)

Table 8 (*continued*)

Parameter			Values	
Name	Assigned number	Binary encoding	Possible values	Binary encoding
X-Mms-Read-Reply	0x10	0x90	Yes	0x80 (decimal 128)
			No	0x81 (decimal 129)
X-Mms-Read-Status	0x1B	0x9B	Read	0x80 (decimal 128)
			Deleted without being read	0x81 (decimal 129)
X-Mms-Reply-Charging	0x1C	0x9C	Requested	0x80 (decimal 128)
			Request text only	0x81 (decimal 129)
			Accepted	0x82 (decimal 130)
			Accepted text only	0x83 (decimal 131)
X-Mms-Reply-Charging-Deadline	0x1D	0x9D		Absolute or relative date
X-Mms-Reply-Charging-ID	0x1E	0x9E		Text string
X-Mms-Reply-Charging-Size	0x1F	0x9F		Long integer
X-Mms-Report-Allowed	0x11	0x91	Yes	0x80 (decimal 128)
			No	0x81 (decimal 129)
X-Mms-Response-Status	0x12	0x92	Ok or error codes in Tables 6.21 and 6.22	
X-Mms-Response-Text	0x13	0x93		Encoded string
X-Mms-Retrieve-Status	0x19	0x99	Ok or error codes in Tables 6.28 and 6.29	
X-Mms-Retrieve-Text	0x1A	0xAA		Encoded string
X-Mms-Sender-Visibility	0x14	0x94	Hide	0x80 (decimal 128)
			Show	0x81 (decimal 129)
X-Mms-Status	0x15	0x95	Expired	0x80 (decimal 128)
			Retrieved	0x81 (decimal 129)
			Rejected	0x82 (decimal 130)
			Deferred	0x83 (decimal 131)
			Unrecognised	0x84 (decimal 132)
			Indeterminate	0x85 (decimal 133)
			Forwarded	0x86 (decimal 134)
Subject	0x16	0x96		Encoded string
To	0x17	0x97		Encoded string
X-Mms-Transaction-ID	0x18	0x98		Text string

References

Internet links/organizations involved in the development of standards

European Telecommunications Standard Institute (ETSI)	http://www.etsi.org
International Telecommunication Union (ITU)	http://www.itu.org
Internet Engineering Task Force	http://www.ietf.org
MIDI Manufacturers Association (MMA)	http://www.midi.org
Third Generation Partnership Project (3GPP)	http://www.3gpp.org
UMTS Forum	http://www.umts-forum.org
WAP Forum	http://www.wapforum.org
Wireless Village	http://www.wireless-village.org
World Wide Web Consoritum (W3C)	http://www.w3c.org
Open Mobile Alliance (OMA)	http://www.openmobilealliance.org

Documents

3GPP Documents

[3GPP-11.11]	3GPP TR 11.11: Specification of the SIM-ME interface.
[3GPP-21.101]	3GPP TS 21.101: 3rd generation mobile system release 1999 specifications.
[3GPP-21.102]	3GPP TS 21.102: 3rd generation mobile system release 4 specifications.
[3GPP-21.103]	3GPP TS 21.103: 3rd generation mobile system release 5 specifications.
[3GPP-21.900]	3GPP TR 21.900: 3GPP working methods.
[3GPP-21.905]	3GPP TR 21.905: Vocabulary for 3GPP specifications.
[3GPP-21.910]	3GPP TS 21.910: Multi-mode UE issues: categories, principles and procedures.
[3GPP-22.060]	3GPP TS 22.060: General Packet Radio Service (GPRS), stage 1.
[3GPP-22.105]	3GPP TS 22.105: Services and service capabilities.
[3GPP-22.121]	3GPP TS 22.121: The Virtual Home Environment (VHE), stage 1.
[3GPP-22.127]	3GPP TS 22.127: Open Service Access (OSA), stage 1.
[3GPP-22.140]	3GPP TS 22.140: Multimedia Messaging Service (MMS), stage 1.
[3GPP-22.141]	3GPP TS 22.141: Presence service, stage 1.
[3GPP-22.228]	3GPP TS 22.228: Service requirements for the IP multimedia core network subsystem, stage 1.
[3GPP-22.940]	3GPP TS 22.940: IMS messaging, stage 1.
[3GPP-23.002]	3GPP TS 23.002: Network architecture.
[3GPP-23.011]	3GPP TS 23.011: Technical realization of supplementary services.
[3GPP-23.038]	3GPP TS 23.038: Alphabets and language-specific information.
[3GPP-23.039]	3GPP TR 23.039: Interface protocols for the connection of SMSCs to SMEs.

[3GPP-23.040]	3GPP TS 23.040: Technical realization of the Short Message Service (SMS).
[3GPP-23.042]	3GPP TS 23.042: Compression algorithm for text messaging services.
[3GPP-23.060]	3GPP TS 23.060: General Packet Radio Service (GPRS), stage 2.
[3GPP-23.101]	3GPP TS 23.101: General UMTS architecture.
[3GPP-23.127]	3GPP TS 23.127: Virtual Home Environment/Open Service Access, stage 2.
[3GPP-23.140]	3GPP TS 23.140: Multimedia Messaging Service (MMS), stage 2.
[3GPP-23.228]	3GPP TS 23.228: IP multimedia subsystem, stage 2.
[3GPP-23.841]	3GPP TS 23.841: Presence service, stage 2.
[3GPP-24.011]	3GPP TS 24.011: Point-to-point SMS support on mobile radio interface.
[3GPP-26.140]	3GPP TS 26.140: Multimedia Messaging Service, media formats and codecs.
[3GPP-26.233]	3GPP TS 26.233: Transparent end-to-end packet-switched streaming service.
[3GPP-26.234]	3GPP TS 26.234: End-to-end transparent streaming service, protocols and codecs.
[3GPP-27.005]	3GPP TS 27.005: Use of DTE-DCE interface for CBS.
[3GPP-31.102]	3GPP TS 31.102: Characteristics of the USIM application.
[3GPP-31.111]	3GPP TS 31.111: USIM Application Toolkit (USAT).
[3GPP-32.200]	3GPP TS 32.200: Charging management, charging principles.
[3GPP-32.235]	3GPP TS 32.235: Charging management, charging data description for application services.
[3GPP-41.102]	3GPP TS 41.102: GSM release 4 specifications.
[3GPP-41.103]	3GPP TS 41.103: GSM release 5 specifications.
[3GPP-43.041]	3GPP TS 43.041: Example protocol stacks for interconnecting SMSCs and MSCs.
[3GPP-51.11]	3GPP TS 51.11: Specification of the SIM-ME Interface.

International Telecommunication Union (ITU) documents

[ITU-E.164]	ITU-T E.164: The international public telecommunication numbering plan, ITU, May 1997.
[ITU-E.212]	ITU-T E.212: The international identification plan for mobile terminals and mobile users, ITU, November 1998.
[ITU-H.263]	ITU-T H.263: Video coding for low bit rate communication, ITU, February 1998.
[ITU-I.130]	ITU-T I.130: Method for the characterization of telecommunication services supported by an ISDN and network capabilities of an ISDN, ITU, November 1988.

IETF documents

[RFC-821]	Simple Mail Transfer Protocol (SMTP), August 1982.
[RFC-822]	Standard for the format of ARPA Internet text messages, August 1982. Note that this RFC is replaced by [RFC-2822].
[RFC-1730]	Internet Message Access Protocol (IMAP), version 4, December 1994.
[RFC-1766]	Tags for the identification of languages, March 1995.
[RFC-1806]	Communicating presentation information in Internet messages: The Content-Disposition header, June 1995.
[RFC-1889]	RTP: A Transport Protocol for Real-Time Applications, January 1996.
[RFC-1939]	(STD 0053) Post Office Protocol (POP), version 3, May 1996.
[RFC-2026]	The Internet Standards Process – Revision 3, October 1996
[RFC-2045]	MIME Part 1: Format of Internet message bodies, November 1996.
[RFC-2046]	MIME Part 2: Media types, November 1996.
[RFC-2047]	MIME Part 3: Message header extensions for non-ASCII text, November 1996.
[RFC-2048]	MIME Part 4: Registration procedures, November 1996.
[RFC-2049]	MIME Part 5: Conformance criteria and examples, November 1996.
[RFC-2279]	UTF-8, a transformation format of ISO 10646, January 1998.

[RFC-2326]	Real Time Streaming Protocol (RTSP), April 1998.
[RFC-2327]	Session Description Protocol (SDP), April 1998.
[RFC-2387]	The MIME Multipart/Related Content-type, August 1998.
[RFC-2396]	Uniform Resource Identifiers (URI): Generic Syntax, August 1998.
[RFC-2616]	Hypertext Transfer Protocol – HTTP/1.1, June 1999.
[RFC-2633]	S/MIME message specification, version 3, June 1999.
[RFC-2822]	Internet message format, April 2001.
[RFC-2916]	E.164 number and DNS, September 2000.
[RFC-3261]	Session initiation protocol, June 2002.
[RFC-2617]	HTTP authentication: basic and digest access authentication, June 1999.
[RFC-2557]	MIME encapsulation of aggregate documents, March 1999.

World Wide Web Consortium (W3C) documents

[W3C-PNG]	(W3C recommendation) Portable Network Graphics (PNG), October 1996. http://www.w3.org/TR/REC-png.html
[W3C-SMIL]	(W3C recommendation) Synchronized Multimedia Integration Language (SMIL) 2.0, August 2001. http://www.w3.org/TR/smil20
[W3C-SOAP]	(W3C note) Simple Object Access Protocol (SOAP) 1.1, May 2000. http://www.w3.org/TR/SOAP/
[W3C-SOAP-ATT]	(W3C note) SOAP messages with attachments, December 2000. http://www.w3.org/TR/SOAP-attachments
[W3C-SVG]	(W3C recommendation) Scalable Vector Graphics (SVG) 1.0 Specification, September 2001. http://www.w3.org/TR/SVG/

WAP Forum documents

[WAP-174]	WAP-174-UAProf-19991110-a: WAP user agent profile, WAP Forum, November 1999.
[WAP-181]	WAP-181-TAWP-200001213-a: WAP work processes, WAP Forum, December 2000.
[WAP-190]	WAP-190-WAESpec-20000329-a: Wireless application environment specification, March 2000.
[WAP-203]	WAP-203-WSP-20000504-a: Wireless session protocol specification, May 2000.
[WAP-205]	WAP-205-MMSArchOverview-20010425-a: WAP MMS architecture overview, version 1.0, WAP Forum, April 2001.
[WAP-206]	WAP-206-MMSCTR-20010612-a: WAP MMS client transactions, version 1.0, WAP Forum, June 2001.
[WAP-209]	WAP-209-MMSEncapsulation-20010601-a: WAP MMS encapsulation protocol, version 1.0, WAP Forum, June 2001.
[WAP-229]	WAP-229-HTTP-20010329-a: Wireless-Profiled HTTP (WP-HTTP), WAP Forum, March 2001.
[WAP-250]	WAP-250-PushArchOverview-20010703-a: WAP push architectural overview, WAP Forum, July 2001.
[WAP-274]	WAP-274-MMSArchOverview-20020409-d: (draft) WAP MMS architecture overview, version 1.1, WAP Forum, April 2002 (also available from OMA under OMA-WAP-MMS-ARCH-v1_1-20020409-p).
[WAP-275]	WAP-275-MMSCTR-20020410-d: (draft) WAP MMS client transactions, version 1.1, WAP Forum, April 2002 (also available from OMA under OMA-WAP-MMS-CTR-v1_1-20020823-p).

[WAP-276]   WAP-276-MMSEncapsulation-20020409-d: (draft) WAP MMS encapsulation
                    protocol, version 1.1, WAP Forum, April 2001 (also available from OMA under
                    OMA-WAP-MMS-ENC-v1_1-20020823-p).

UMTS Forum documents

[UF-Rep9]   UMTS Forum Report no. 9: The UMTS third generation market – Phase I: struc-
                    turing the service revenues opportunities, UMTS Forum, October 2000.
[UF-Rep13]   UMTS Forum Report no. 13: The UMTS third generation market – Phase II:
                    structuring the service revenue opportunities, UMTS Forum, April 2001.

Other documents

[US-ASCII]   Coded Character Set–7-bit American Standard Code for Information Inter-
                    change, ANSI X3.4-1986.
[IRDA-iMelody]   iMelody, specifications for Ir Mobile Communications, version 1.2: Infrared
                    Association, Octobre 2000.
[ISO-10646]   ISO/IEC 10646-1: Universal multipleOctet Coded Character Set (UCS), 1993.
[MMA-SP-MIDI]   Scalable Polyphony MIDI specification and device profiles, version 1.0a, MIDI
                    Manufacturers Association, May 2002.
[MMA-MIDI-1]   The complete MIDI 1.0 detailed specification, version 96.1, MIDI Manufac-
                    turers Association, 1996.
[MMS-Conformance]   MMS conformance document, version 2.0, MMS Interoperability Group,
                    February 2002. Available from Nokia and Ericsson developer support sites at
                    respectively http://www.forum.nokia.com *and* http://www.ericsson.com/mobi-
                    lityworld
[VERSIT-vCalendar] vCalendar, version 2.1, The Internet Mail Consortium (IMC), September 1996.
[VERSIT-vCard]   vCard, version 1.0, The Internet Mail Consortium (IMC), September 1996.

Acronyms and Abbreviations

3GPP	Third Generation Partnership Project
AMR	Adaptive Multirate
APEX	Application Exchange
APN	Access Point Name
ARIB	Association of Radio Industries and Businesses
ARPANET	ARPA wide-area networking
AT	ATtention
BIFS	Binary Format for Scenes
BNF	Backus-Naur Form
bpm	beats per minute
BSC	Base Station Controller
BSS	Base Station Subsystem
BTS	Base Transceiver Station
CC/PP	Composite Capability/Preference Profiles
CDR	Charging Data Records
CEPT	Conférence Européenne des Postes et Télécommunications
CLI	Command Line Interface
CLP	Command Line Protocol
CN	Core Network
CO	Cache Operation
CPI	Capability and Preference Information
CR	Change Request
CSD	Circuit Switched Data
CSP	Client Server Protocol
CSPDN	Circuit Switched Public Data Network
CWTS	China Wireless Telecommunication Standard
DF	Dedicated File
DID	Document IDentifier
DNS	Domain Name Server
DRM	Digital Right Management
EDGE	Enhanced Data Rate for Global Evolution
EEPROM	Electrically Erasable Programmable Read Only Memory
EF	Elementary File
EFI	External Functionality Interface
EFR	Enhanced Full Rate
EMS	Enhanced Messaging Service
ESME	External SME
ETSI	European Telecommunications Standard Institute
FDD	Frequency Division Duplex
FQDN	Fully Qualified Domain Name

GGSN	Gateway GPRS Support Node
GM	General MIDI
GPRS	General Packet Radio Service
GPS	Global Positioning System
GSA	Global Mobile Suppliers Association
GSM	Global System for Mobile
HLR	Home Location Register
HSCSD	High Speed CSD
HTML	Hypertext Markup Language
HTTP	HyperText Transfer Protocol
IAB	Internet Architecture Board
IANA	Internet Assigned Numbers Authority
IE	Information Element
IED	Information Element Data
IEDL	Information Element Data Length
IEI	Information Element Identifier
IESG	Internet Engineering Steering Group
IETF	Internet Engineering Task Force
IM	Instant Messaging
IMAP	Interactive Mail Access Protocol
IMEI	International Mobile Equipment Identity
IMPP	Instant Messaging and Presence Protocol
IMPS	Immediate Messaging and Presence Services
IMS	IP Multimedia Subsystem
IMSI	International Mobile Subscriber Identity
IMT-2000	International Mobile Telecommunications 2000
ISDN	Integrated Services Digital Network
ISOC	Internet Society
ITU	International Telecommunication Union
LAN	Local Area Networks
LSB	Least Significant Bit
LZSS	Lempel-Ziv-Storer-Szymanski
MAP	Mobile Application Part
MCC	Mobile Country Code
MCEF	Mobile-station-memory-Capacity-Exceeded-Flag
MDI	Message Distribution Indicator
ME	Mobile Equipment
MExE	Mobile Execution Environment
MF	Master File
MIDI	Musical Instrument Digital Interface
MIME	Multipurpose Internet Mail Extensions
MIP	Maximum Instantaneous Polyphony
MMA	MIDI Manufacturers Association
MMoIP	Multimedia over IP
MMS	Multimedia Messaging Service
MMS UA	MMS User Agent
MMSC	MMS Centre
MMSE	MMS Environment
MNC	Mobile Network Code
MNP	Mobile Number Portability

MNRF	Mobile-station-Not-Reachable-Flag
MNRG	Mobile-station-Not-Reachable-for-GPRS
MNRR	Mobile-station-Not-Reachable-Reason
MPEG	Moving Pictures Expert Group
MS	Mobile Station
MSB	Most Significant Bit
MSC	Mobile Switching Centre
MSISDN	Mobile Station ISDN Number
MWI	Message Waiting Indication
MWIF	Mobile Wireless Internet Forum
ODI	Object Distribution Indicator
OMA	Open Mobile Alliance
OSA	Open Service Architecture
OSS	Operation subsystem
OTA	Over-The-Air
PAP	Push Access Protocol
PCG	Project Co-ordination Group
PDA	Personal Digital Assistant
PDC	Personal Data Cellular
PDP	Packet Data Protocol
PDU	Protocol Data Unit
PI	Push Initiator
PIM	Personal Information Manager
PKI	Public Key Infrastructure
PNG	Portable Network Graphic
POP3	Post Office Protocol-3
PPG	Push Proxy Gateway
PRIM	Presence and Instant Messaging Protocol
PSPDN	Packet Switched Public Data Network
PSS	Packet-switched Streaming Service
PSTN	Public Switched Telephone Network
RAM	Random Access Memory
RDF	Resource Description Framework
RFC	Request For Comments
RIM	Research in Motion
RNC	Radio Network Controllers
ROM	Read Only Memory
RTP	Real-Time Transport Protocol
RTSP	Real-Time Streaming Protocol
SAT	SIM Application Toolkit
SC	SMS Centre
SCD	Specification Change Document
SDO	Standard Development Organization
SDP	Session Description Protocol
SGSN	Serving GPRS Support Node
SI	Service Indication
SIM	Subscriber Identity Module
SIMPLE	SIP for Instant Messaging and Presence Leveraging Extensions
SIN	Specification Implementation Note
SIP	Session Initiation Protocol

SL	Service Loading
SMAP	Short Message Application Protocol
SMCNP	Server-Mobile Core Network Protocol
SME	Short Message Entities
SMF	Standard MIDI Files
SMIL	Synchronised Multimedia Integration Language
SM-MO	Short Message-Mobile Originated
SM-MT	Short Message-Mobile Terminated
SMPP	Short Message Peer to Peer
SMS	Short Message Service
SMSC	SMS Centre
SMS-GMSC	SMS gateway MSC
SMS-IWMSC	SMS InterWorking MSC
SP-MIDI	Scalable-Polyphony MIDI
SR	Status Report
SS7	Signalling System Number 7
SSL	Secure Socket Layer
SSP	Server-Server Protocol
SVG	Scalable Vector Graphics
sysex	system exclusive message
TDD	Time Division Duplex
TDMA	Time Division Multiple Access
TE	Terminal Equipment
TLS	Transport Layer Security
TP-CD	TP-Command-Data
TP-CDL	TP-Command-Data-Length
TP-CT	TP-Command-Type
TP-DA	TP-Destination-Address
TP-DCS	TP-Data-Coding-Scheme
TP-DT	TP-Discharge-Time
TPDU	Transfer Protocol Data Unit
TP-FCS	TP-Failure-Cause
TP-MMS	TP-More-Message-to-Send
TP-MN	TP-Message-Number
TP-MR	TP-Message-Reference
TP-MTI	TP-Message-Type-Indicator
TP-OA	TP-Originator-Address
TP-PI	TP-Parameter-Indicator
TP-PID	TP-Protocol-Identifier
TP-RA	TP-Recipient-Address
TP-RD	TP-Reject-Duplicates
TP-RP	TP-Reply-Path
TP-SCTS	TP-Service-Centre-Time-Stamp
TP-SRI	TP-Status-Report-Indication
TP-SRQ	TP-Status-Report-Qualifier
TP-SRR	TP-Status-Report-Request
TP-ST	TP-Status
TP-UD	TP-User-Data
TP-UDHI	TP-User-Data-Header-Indicator
TP-UDL	TP-User-Data-Length

TP-VP	TP-Validity-Period
TP-VPF	TP-Validity-Period-Format
TR	Technical Report
TS	Technical Specification
TSG	Technical Specifications Groups
TTA	Telecommunications Technology Association
TTC	Telecommunications Technology Committee
UAProf	User Agent Profile
UDH	User-Data-Header
UDHL	User-Data-Header-Length
UE	User Equipment
UICC	UMTS IC card
UMTS	Universal Mobile Telecommunications System
UPI	User Prompt Indicator
URI	Uniform Resource Identifier
URL	Uniform Resource Locator
UCS	Universal Character Set
USIM	UMTS Subscriber Identity Module
UTC	Co-ordinated Universal Time
UTF	UCS Transformation Format
UTRAN	Universal Terrestrial Radio Access Network
UWCC	Universal Wireless Communications Consortium
VAS	Value Added Service
VASP	Value Added Service Provider
VLR	Visitor Location Register
VoIP	Voice over IP
VPN	Virtual Private Networks
W3C	World Wide Web Consortium
WAE	Wireless Application Environment
WAP	Wireless Application Protocol
WBXML	WAP binary XML
WCDMA	Wideband CDMA
WDP	Wireless Data Protocol
WG	Working Groups
WIM	Wireless Identity Module
WML	Wireless Markup Language
WP-HTTP	Wireless Profiled HTTP
WP-TCP	Wireless Profiled TCP
WSP	Wireless Session Protocol
WTA	Wireless Telephony Application
WTLS	Wireless Transport Layer Security
WTP	Wireless Transaction Protocol
WVG	Wireless Vector Graphics
XHTML	Extensible HTML
XHTML MP	XHTML Mobile Profile
XML	Extensible Markup Language

A list of abbreviations and corresponding definitions used in 3GPP specifications is provided in [3GPP-21.905].

Index